MW00595451

Climate Analysis

Sensational images and stories about variations in Earth's climate and their impacts on society are pervasive in the media. The scientific basis for these stories is often not understood by the general public, nor even by those with a scientific background in fields other than climate science. This comprehensive book will enable the reader to understand and appreciate the significance of the flood of climate information. It is an excellent non-mathematical resource for learning the fundamentals of climate analysis, as well as a reference for non-climate experts who need to use climate information and data. The focus is on the basics of the climate system, how climate is observed, and how the observations are transformed into datasets useful for monitoring the climate. Each chapter contains Discussion Questions. This is an invaluable textbook on climate analysis for advanced students, and a reference textbook for researchers and practitioners.

Chester F. Ropelewski has 50 years of experience in climate analysis, including weather forecasting, development of climate datasets and real-time monitoring systems, research in atmospheric turbulence, tropical boundary layers, hurricanes, the North American monsoon, sea ice, snow cover, climate variability, El Niño/Southern Oscillation (ENSO), Quasi-Biennial Oscillation, hydrologic cycle, and droughts. He was awarded the Norbert Gerbier-Mumm International Award from the World Meteorological Organization in 1990, and was elected a fellow of the American Meteorological Society in 2001. He is the author of over 100 research papers, book chapters, and reports. He was the chief of the Analysis Branch of the Climate Prediction Center of the National Oceanic and Atmospheric Administration (NOAA), the director of the Climate Monitoring section of the International Research Institute for Climate and Society at Columbia University, and a senior policy analyst and advisor of The Earth Institute, Columbia University.

Phillip A. Arkin has over 40 years of experience as an innovative research scientist and leader at the National Oceanic and Atmospheric Administration (NOAA), Columbia University, and the University of Maryland, where he currently works. He has played a leading role in shaping the use of satellite data to estimate precipitation for climate studies and created the first system to use weather model analyses for climate monitoring and research. He has initiated and led several international research projects and has mentored a diverse collection of young scientists who have gone on to fruitful careers in climate science. He has published nearly 100 papers in the weather and climate literature, most of them focused on the climate datasets and their applications that are the subject of this book. He is a fellow of the American Meteorological Society (2000), recipient of the Hugh Robert Mill Award from the Royal Meteorology Society (2004), and a distinguished alumnus of the University of Maryland.

"Everyone – and I mean everyone – is increasingly asked to make important decisions about the future of our planet. These are decisions that affect our children and our children's children. This book is a one-of-a-kind presentation of the science underpinning climate variability and change that is accessible to all, ensuring that these critical decisions are supported by the best available science."

—Ben Kirtman, Rosenstiel School of Marine and Atmospheric Science, University of Miami

"A comprehensive, non-mathematical survey of essential concepts of climatic data analysis, filling the gap between introductory textbooks and detailed technical memos. Ropelewski and Arkin expertly place up-to-date analysis concepts into historical context, describing the dramatic evolution of this field over the past half-century – a history in which the authors have played integral roles."

—David S. Gutzler, The University of New Mexico

"A much needed and valuable resource for students and professionals alike. Observations and their analysis are key to understanding the climate system, and those interested in gaining that understanding will find their Rosetta stone here. Approachable by non-specialists across a broad range of disciplines, this work deserves to be widely read and studied."

—Richard D. Rosen, National Oceanic and Atmospheric Administration (retired)

"Climate science is data-driven. Models may be the loudest voice in the climate change discussion at the moment, but it is observations that cut through to the truth. They are the facts, but they come encumbered by noise, by gaps, by error. Ropelewski and Arkin – proven masters at extracting knowledge from what was merely data – tell you where the problems are and how to deal with them. This book is not theory and it is not statistical recipes. It the practical wisdom acquired in two lifetimes in climate science."

—Mark Cane, Lamont-Doherty Earth Observatory, Columbia University

"How to detect climate signals from data? This masterful and brilliant book tells you all. I heartily recommend *Climate Analysis* for each modern climatology course as a text, and for every climate scientist as a reference manual."

—Samuel Shen, Distinguished Professor, San Diego State University

Climate Analysis

CHESTER F. ROPELEWSKI

Retired (formerly The Earth Institute, Columbia University)

PHILLIP A. ARKIN

University of Maryland, College Park

Cambridge
UNIVERSITY PRESS

University Printing House, Cambridge CB2 8BS, United Kingdom

One Liberty Plaza, 20th Floor, New York, NY 10006, USA

477 Williamstown Road, Port Melbourne, VIC 3207, Australia

314–321, 3rd Floor, Plot 3, Splendor Forum, Jasola District Centre, New Delhi – 110025, India

79 Anson Road, #06–04/06, Singapore 079906

Cambridge University Press is part of the University of Cambridge.

It furthers the University's mission by disseminating knowledge in the pursuit of education, learning, and research at the highest international levels of excellence.

www.cambridge.org
Information on this title: www.cambridge.org/9780521896160
DOI: 10.1017/9781139034746

First published 2019

Printed in the United Kingdom by TJ International Ltd. Padstow Cornwall

A catalogue record for this publication is available from the British Library.

Library of Congress Cataloging-in-Publication Data
Names: Ropelewski, C. F., author. | Arkin, Phillip A., author.
Title: Climate analysis / Chester F. Ropelewski, Phillip A. Arkin.
Description: Cambridge, United Kingdom ; New York, NY, USA : Cambridge University Press, 2019. | Includes bibliographical references and index.
Identifiers: LCCN 2018029217 | ISBN 9780521896160 (hardback)
Subjects: LCSH: Climatology–Methodology. | Climatology–Data processing.
Classification: LCC QC981 .R66 2019 | DDC 551.6072/7–dc23
LC record available at https://lccn.loc.gov/2018029217

ISBN 978-0-521-89616-0 Hardback

To Eugene Rasmusson, who inspired us to learn the science, art, and practice of climate analysis and whose memory motivated us to begin this journey.

Contents

Foreword

During the course of my career, I always tried to confront the results of tropical ocean and climate models with real-world observations. As a graduate student in the 1970s, my major professor, Jim O'Brien, began to analyze thousands of shipboard wind observations from the tropical Pacific Ocean. This proved to be one of the first climate datasets of the ocean surface wind field suitable for numerical ocean modeling. Access to this climate dataset allowed me to hindcast and analyze the El Niño phenomenon during the 1960s and 1970s for my PhD thesis. Little did I appreciate at the time the future importance climate analysis would hold for climate monitoring and prediction efforts. Therefore, I was honored when Chet Ropelewski and Phil Arkin asked me to write this foreword for their book on climate analysis.

The authors have done a truly masterful job and service to the scientific community in capturing the essence of climate analysis. As the authors state, "Climate analysis is the science and practice of creating and examining complete datasets to describe, understand, and predict the current state and evolution of the climate system." This *tour de force* nicely fills a serious gap in present climate literature. While there are books on climate dynamics, climate physics, and climate change, there has been no dedicated text to date on the fundamental analysis of the data that underpins all climate science.

During the late twentieth century, as weather observations reached decades to a century in length, such data enabled analysis of the coupled climate system encompassing the atmosphere, oceans, land surface, and cryosphere. Who better to write the definitive book on how analysis of day-to-day weather was transformed into climate analysis than Arkin and Ropelewski. These two atmospheric scientists spent the better part of their distinguished careers with the National Ocean and Atmospheric Administration. To say they were among the pioneers in climate analysis is not an exaggeration, as they were both key contributors to the early development of climate diagnostics and climate analysis. It is not often that a scientist is able to participate in the genesis of a new avenue of science. Arkin and Ropelewski, to the benefit of the reader, have been able to draw on their perspectives in this regard in *Climate Analysis*. I have had the good fortune of working with both as colleagues over the years, and just like their analyses of climate datasets, this book is clear, concise, and to the point.

This book should be of broad interest to a wide range of readers. Its greatest use will be as a textbook for upper-level undergraduate students in the atmospheric and related Earth system sciences or entry-level graduate students. The appendices, glossary, and discussion questions prove a valuable complement to the main text. However, this book will also prove to be a worthwhile read to others interested in learning more about climate observations, and how datasets from these observations are used to analyze climate variability, climate change, and for use with climate models. My only regret is that I did not have this resource available to me when teaching an introductory course in Earth system science!

Antonio J. Busalacchi, Jr.
President, University Corporation for Atmospheric Research (UCAR)

Preface

Our title is *Climate Analysis*, but, unlike many recent works, we have written primarily about data and how the data are analyzed. It is the data, and their analysis, that underpin what we know about the climate system. We explain what climate is, what methods are used to understand it, and how observations of its components are turned into climate datasets that are used to learn more about the climate. In general, we concentrate on data from networks, like weather stations, that have a long period of record. This includes most meteorological climate variables, oceanic sea surface temperatures taken by ships and buoys, and satellite-derived data mostly dating from no earlier than the late 1970s. We also include descriptions of model-derived estimates of climate based on input from observations that may extend into the nineteenth century.

Our aim is for this text to be useful to a diverse community. Our intended audience includes students as well as those whose professions require familiarity with climate datasets. We believe that this book is suitable for an introductory course in climate science for nonspecialists, either at the advanced undergraduate or beginning graduate level, or as a supplementary text for more specialized curricula, for instance in architecture, agriculture, engineering, policy, public health, or urban planning. It might also be an accessible introduction to observational climate science for interested nonacademic readers.

Both of us are trained as atmospheric scientists and have spent most of our careers working on the development and use of climate datasets. We take a practical rather than theoretical approach to this material, but this is not a handbook – we do not provide a collection of recipes and algorithms. Nor does this text focus on the physics and dynamics of the climate system. Our goal is to provide general readers with a foundation that enables them to understand how climate datasets are created and to help scientists, engineers, and decision makers from other fields to understand and utilize climate datasets in their work.

We describe the basic steps necessary for the use of data in climate analysis and present examples of analyses used in routine climate monitoring as well as in research. Many of the examples rely on statistics but this is not a statistics textbook. Rather, we identify what statistics are commonly used in climate analysis, describe what these statistics do, identify pitfalls in their interpretation, and provide references to the standard statistical texts used in weather and climate analysis.

The book is organized as follows. Our first three chapters provide an overview of the climate system, describe how climate analyses are, or should be, performed, and discuss many of the observations and datasets that form the basis for climate analysis. The fourth chapter describes and illustrates commonly recurring climate patterns identified through analysis of climate datasets in the atmosphere and upper ocean. In Chapter 5, we discuss the drivers of climate change and note that the notion of a stationary background climate against which all observations, analyses, datasets, predictions, and applications occur is never fully accurate since climate on all time scales is continually changing. In Chapter 6 we discuss the topics that must be considered in building climate datasets, using atmospheric temperature over land as an example. In Chapter 7, using precipitation as an example, we describe the challenges that must be solved to use a variety of information in constructing climate datasets. Temperature and precipitation are singled out for special attention because of their profound impact on so many human activities at all time scales.

The first seven chapters concentrate on analysis of the atmospheric component of the climate system, reflecting our professional experience and expertise. In the next three chapters we introduce the ocean, cryosphere, and land components of the climate system. In Chapter 11 we discuss the use of climate models as sources of more complete historical climate data and as analysis tools. The final chapter describes how ongoing analyses of the most recent data are utilized for the routine monitoring of the current state of the climate system as well as to support climate prediction.

We include four short appendices that provide supplemental material on certain topics. Appendix A provides a description of some of the statistical techniques often used in climate analysis. The mathematical definitions of some of the quantities used to monitor the atmospheric circulation are presented in Appendix B for those interested in these details. Some of the practical issues that arise when dealing with gaps in the data are introduced in Appendix C. The mathematical underpinnings for analyzing the components of the atmospheric and terrestrial water budgets are outlined in Appendix D.

A glossary is provided to define terms or phrases that might not be obvious. Footnotes provide short additional material about terms and concepts introduced in the body of the text. We use boxes in most chapters to provide lengthier additional information on selected topics. A number of technical references in the scientific literature are provided for those interested in delving deeper into a topic. In general, we tend to reference older articles and summary papers on many of the topics since these are more likely to be accessible in the free open literature and since the most recent publications are more likely to be focused on details that would require the reader to return to the early literature to interpret. Books, articles, and websites that may be of general interest to our readers are listed as suggested further reading at the end of each chapter.

The purpose of the discussion questions given at the end of each chapter is to stimulate interest in the topics discussed; these questions do not necessarily have universally agreed upon answers. We encourage instructors using this book as a class text, as well as students and readers, to contact us with comments, questions, and suggestions concerning the discussion questions.

The analysis of climate is an application of applied physics and chemistry to the real world around us. We introduce an encyclopedic list of topics and point to references for further information as well as to many of the datasets discussed. This text should help to illustrate how much the climate system varies and will help readers to decide how they might include climate considerations in their professional endeavors. We also hope that readers will come away with an increasing understanding of and appreciation for the unceasing and fascinating variations in climate.

Acknowledgments

When we started this book we thought that we were well aware of the magnitude of the task that loomed ahead. As it turns out, we had greatly underestimated the effort and the time it would take. It is only due to the consistent encouragement that we received from friends and family as well as to the advice and material help that we received from colleagues and co-workers that we were able to complete the project. Sincere thanks go to Dr. Matt Lloyd at Cambridge University Press for continued faith that we would complete this book, even in the years when the focus of our efforts was elsewhere, and to Zoë Pruce for guiding us through the mechanics of producing a book.

We had the privilege of being introduced to climate analysis early in our careers by Gene Rasmusson, mentor, colleague, and later friend at NOAA's newly formed Climate Analysis Center (CAC) at a time when there was an explosion of interest and support for climate analysis. Gene and others at the CAC instilled in us the basics of climate analysis that have served as the inspiration for this book.

The book cuts a wide swath through much of climate science and while we are familiar with the many topics covered we are certainly not expert in all of them. It is only because of the freely given advice and comments from many experts in all aspects of climate analysis that we have been able to complete this book. In particular, we acknowledge the advice and chapter review comments from Matt Barlow, Tony Busalacchi, Ben DeVries, Wesley Ebisuzaki, Dave Gutzler, Mike Halpert, George Huffman, John Janowiak, Chris Justice, Monika Kopacz, Michelle L'Heureux, Brad Lyon, Rol Madden, Fernando Miralles-Wilhelm, David Parker, Tom Peterson, Allen Pope, Jerry Potter, David Robinson, Rick Rosen, Pingping Xie, and Fred Zimnoch. Their kind words and frank comments contributed greatly to improving the content of this book.

We greatly appreciate the contributions of Dr. Muthuvel Chelliah at NOAA's Climate Prediction Center and Michael Bell at the International Research Institute for Climate and Society, Columbia University, who produced a number of the figures for this book, particularly in Chapters 4 and 12. We also gratefully acknowledge figures and/or advice provided by Gerry Bell, Jane Beitler, Ray Bradley, Charlotte Demott, Mike Ek, Thomas Estilow, Carlyn Iverson, Ellen Keohane, George Kiladis, Jim Kinter, Jay Lawrimore, Rebecca Lindsey, Dai Mcclurg, Mike McPhaden, Walt Meier, Frank Niepold, Dick Reynolds, Megan Scandenberg, Claudia Schmid, Mark Serreze, Marshall Shepherd, Tom Smith, and Chidong Zhang.

Most particularly, we owe enormous gratitude to our wives, Marie Ropelewski and Margie Arkin, for their support and encouragement throughout our careers and especially during the decade it took us to produce this book. We could not have done it without them.

Obviously we are responsible for and regret any errors, omissions, or shortcomings that remain.

Abbreviations and Acronyms

4-D VAR	Four-Dimensional Variational Assimilation
AAO	Antarctic Oscillation
AMIP	Atmosphere Model Intercomparison Project
AMO	Atlantic Multi-Decadal Oscillation
AMOC	Atlantic Meridional Overturning Circulation
AMS	American Meteorological Society
AMSR-E	Advanced Microwave Scanning Radiometer – Earth Observing System
AOML	Atlantic Oceanographic and Meteorological Laboratory
AR	Arctic Oscillation
ARF	ARM Mobile Facilities
ARM	Atmospheric Radiation Program
ARs	IPCC Assessment Reports
ASOS	Automated Surface Observing System
ASR	Arctic System Reanalysis
AVHRR	Advanced Very High Resolution Radiometer
AWOS	Automated Weather Observation Stations (U.S. Federal Aviation Administration)
BCC	Beijing Climate Center
BHO	Blue Hill Observatory
BoM	Bureau of Meteorology (Australia)
BOREAS	Boreal Ecosystem-Atmosphere Study
CAC	Climate Analysis Center
CADB	Climate Anomaly Data Base
CAMS	Climate Anomaly Monitoring System
CDB	Climate Diagnostics Bulletin
CDDB	Climate Diagnostics Data Base
CDR	Climate Data Record
CEOS	Committee on Earth Observations
CESM	Community Earth System Model
CFC	Chlorofluorocarbon
CFSR	Climate Forecast System Reanalysis
CGMS	Coordination Group for Meteorological Satellites
CIIFIN	International Center of the Investigation of El Niño (South America)
CLIVAR	Climate and Ocean: Variability, Predictability, and Change Program

CMAP	Climate Analysis Center Merged Analysis of Precipitation
CMI	Crop Moisture Index
CMORPH	CPC MORPHing Analysis
CO_2	Carbon dioxide
COADS	Comprehensive Ocean Atmosphere Data Set
COF	Climate Outlook Forum
COLA	Center for Ocean-Land-Atmosphere Studies
CONUS	Continental United States
COOP	Cooperative Observer Program (NWS)
CPC	Climate Prediction Center
CPTEC	Center for Weather Foresting and Climate Studies (Brazil)
CREST	Core Research for Evolutional Science and Technology
CRU	Climatic Research Unit
CSP	Climate Services Partnership
DAAC	Distributed Active Archive Center
DAO	Data Assimilation Office (NASA)
DEM	Digital Elevation Model
DLR	German Aerospace Center
DMSP	Defense Meteorological Satellite Program
DPR	Duel-frequency Precipitation Radar
EC	European Community
ECMWF	European Centre for Medium-Range Weather Forecasts
ENSO	El Niño/Southern Oscillation
EOF	Empirical Orthogonal Function
EOS	Earth Observing System
EOSDIS	Earth Observing System Data and Information System
ERA-40	ECMWF 40-year Reanalysis
ESA	European Space Agency
ESGF	Earth System Grid Federation
ESMR	Electrically Scanning Microwave Radiometer
ESRL	Earth System Research Laboratory (NOAA)
EUMETSAT	European Organization for the Exploitation of Meteorological Satellites
FAR	Fourth Assessment Report (IPCC)
FIFE	First ISLSCP Field Experiment
FIRE	First ISCCP Regional Experiment
GCM	General Circulation Model
GCRP	Global Change Research Program
GDAR	Global Argo Data Repository
GDP	Global Drifter Program
GEO	Geostationary Orbit
GEO	Group on Earth Observations
GEOSS	Global Earth Observing System of Systems
GEWEX	Global Energy and Water Exchanges

GFDL	Geophysical Fluid Dynamics Laboratory
GHCN	Global Historical Climatological Network
GHG	Greenhouse Gas
GISS	Goddard Institute for Space Studies
GLACE	Global Land-Atmosphere Coupling Experiment
GLCF	Global Land Cover Facility
GMAO	Global Modeling and Assimilation Office
GMT	Greenwich Mean Time
GODAS	Global Ocean Data Assimilation System
GOES	Geostationary Operational Environmental Satellite
GPCC	Global Precipitation Climatology Center
GPCP	Global Precipitation Climatology Project
GPI	GOES Precipitation Index
GPM	Global Precipitation Measurement (NASA)
GPS	Global Positioning System
GRACE	Gravity Recovery and Climate Experiment
GrADS	Grid Analysis and Display System
GSFC	Goddard Space Flight Center
GSMaP	Global Satellite Mapping of Precipitation
GTS	Global Telecommunication System
HAPEX	Hydrological-Atmospheric Pilot Experiment
ICAO	International Civil Aviation Organization
ICDC	Integrated Climate Data Center
ICOADS	International Comprehensive Ocean Atmosphere Data Set
IGBP	International Geosphere-Biosphere Program
IIC	International Ice Center
IMD	India Meteorological Department
IMERG	Integrated Multi-SatellitE Retrievals for GPM
INMET	Brazilian Weather Service
INPE	National Institute for Space Research (Brazil)
IOCZ	Indian Ocean Convergence Zone
IPCC	Intergovernmental Panel on Climate Change
IR	Infrared
IRI	International Research Institute for Climate and Society
ISA	International Standard Atmosphere
ISCCP	International Satellite Cloud Climatology Project
ISLSCP	International Satellite Land Surface Climatology Project
ISMN	International Soil Moisture Network
ISPD	International Surface Pressure Databank
ITCZ	Inter-Tropical Convergence Zone
JAMSTEC	Japan Agency for Marine-Earth Science and Technology
JAXA	Japan Aerospace Exploration Agency
JCDAS	JMA Climate Data Assimilation System

JMA	Japan Meteorological Agency
JRA-25	JMA 25-year reanalysis
JRA-55	JMA 55-year reanalysis
JST	Japan Science and Technology Agency
LAI	Leaf Area Index
LDAS	Land Data Assimilation System
LEO	Low Earth Orbit
LIDAR	Light Detection and Ranging
LRPG	Long Range Prediction Group (NOAA)
LSM	Land Surface Model
LTDR	Long-Term Data Record
MERRA	Modern-Era Retrospective Analysis for Research and Applications (NASA)
METEOSAT	Geostationary meteorological satellites operated by EUMETSAT
METOP	Polar orbiting meteorological satellites operated by EUMETSAT
MJO	Madden-Julian Oscillation
MKS	Meter, Kilogram, Second
MME	Multi-Model Ensemble
MODIS	Moderate Resolution Imagine Spectroradiometer
MSG	METEOSAT Second Generation
MSLP	Mean Sea Level Pressure
MTPE	Mission to Planet Earth
NAM	Northern Annular Mode
NAO	North Atlantic Oscillation
NARR	North American Regional Reanalysis
NASA	National Aeronautics and Space Administration
NASMD	North American Soil Moisture Database
NCAR	National Center for Atmospheric Research
NCDC	National Climatic Data Center
NCEI	National Centers for Environmental Information
NCEP	National Centers for Environmental Prediction
NCP	National Climate Program
NDMC	National Drought Mitigation Center
NDVI	Normalized Difference Vegetation Index
NESDIS	National Environmental Satellite, Data, and Information Service
NGDC	National Geophysical Data Center
NIC	National Ice Center
NICL	National Ice Core Laboratory
NLCD	National Land Cover Database (USGS)
NMME	National Multi Model Ensemble (NOAA)
NOAA	National Oceanic and Atmospheric Administration
NODC	National Oceanographic Data Center
NRCS	National Resources Conservation Center

NSF	National Science Foundation
NSIDC	National Snow and Ice Data Center
NWP	Numerical Weather Prediction
NWS	National Weather Service (NOAA)
OLR	Outgoing Longwave Radiation
ONI	Oceanic Niño Index
OOI	Ocean Observatories Initiative
ORAS4	Ocean Reanalysis System 4
PCA	Principal Component Analysis
PCMDI	Program for Climate Model Diagnosis & Intercomparison
PDO	Pacific Decadal Oscillation
PDSI	Palmer Drought Severity Index
PERSIANN	Precipitation Estimation from Remotely Sensed Information using Artificial Neural Networks
PIP	Precipitation Intercomparison Project
PIRATA	Prediction and Research Moored Array in the Tropical Atlantic
PMEL	Pacific Marine Environmental Laboratory
PMM	Precipitation Measurement Missions
PNA	Pacific North American Pattern
PRISM	Parameter-Elevation Regressions on Independent Slopes Model
PSD	Physical Science Division, NOAA Environmental Research Laboratory
QBO	Quasi-Biennial Oscillation
RADAR	Radio Detection and Ranging
RAMA	Research Moored Array for African, Asian, Australian Monsoon Analysis and Prediction
RCOF	Regional Climate Outlook Forum
RF	Radio frequency
SACZ	South Atlantic Convergence Zone
SAF	Satellite Application Facility
SAM	Southern Annular Mode
SAR	Second Assessment Report (IPCC)
SAR	Synthetic Aperture Radar
SCAN	Soil Conservation Analysis Network
SLP	Sea Level Pressure
SMOPS	Soil Moisture Operational Products System
SNOTEL	Snow Telemetry observational network
SO	Southern Oscillation
SOD	Summary of the day
SPCZ	South Pacific Convergence Zone
SPI	Standardized Precipitation Index
SSM/I	Special Sensor Microwave Imager
SSMR	Scanning Multi-channel Microwave Radiomete
SST	Sea Surface Temperature

STAR	Center for Satellite Applications and Research
SWE	Snow-Water Equivalent
TAO	Tropical Atmosphere Ocean
TAR	Third Assessment Report (IPCC)
TMPA	TRMM Multi-Sensor Precipitation Analysis
TOGA	Tropical Ocean – Global Atmosphere program
TPR	TRMM Precipitation Radar
TRITON	Triangle Trans-Ocean Buoy Network
TRMM	Tropical Rainfall Measuring Mission
TSI	Total Solar Irradiance
UKMO	United Kingdom Meteorological Office
URL	Uniform Resource Locator
USCRN	U.S. Climate Reference Network
USDA	U.S. Department of Agriculture
USGS	U.S. Geological Survey
UTC	Coordinated Universal Time
VIC	Variable Infiltration Capacity model
WASP	Weighted Anomaly of Standardized Precipitation index
WCRP	World Climate Research Programme
WMO	World Meteorological Organization
WOCE	World Ocean Circulation Experiment

1 Earth's Climate System

1.1 Introduction

The climate system of Earth includes the surface of the globe, the oceans, and the atmosphere, together with the interactions among these across a range of time scales. Understanding and predicting the behavior of the climate system requires both observations of past behavior and objective models that project into the future. In this book, we attempt to explain and illustrate the methods used to create and use climate datasets based on the wide variety of relevant observations.

We use climate to denote the mean state and variations of the physical environment of Earth's surface and nearby atmosphere and ocean. "Mean state and variations" is not adequately well defined without further specification, but here we begin to see the nature of the problem. Hardly anyone would disagree with a statement that the monthly averages of the daily high and low temperatures at some location are climatic parameters. When we refer to a recent month as being relatively warm or cold, we are discussing recent climate events in the context of some longer time period, such as the average of the same parameters over the past 30 years. In the same sense, few who are familiar with the field would claim that yesterday's high and low temperatures are climate parameters – there is some minimum time span that implicitly separates climate from weather. The difficulty is made clearer when we try to unambiguously define a specific time that separates the two.

Similarly, we must clarify which elements of Earth's surface are part of the climate, as well as what we mean by "nearby" atmosphere and ocean. For example, most climate scientists would agree that vegetation growing on the land surface is part of the climate system and must be observed and simulated if we are to understand and predict the climate. Similarly, most would include as climate components water on the surface, in rivers and lakes, ground water within a meter or two of the surface, and snow and ice resting on the surface. The atmospheric part of the climate system certainly includes the lowest 50 or so kilometers, and the oceanic part includes at least the upper several hundred meters.

The climate system also responds to forcing from external agents, and we will discuss the most important ones. Radiation from the sun is a major influence, although we typically exclude radiation that enters the atmosphere from other sources, such as stars other than the sun. Most variations of the solid earth, such as earthquakes, are excluded as well, but large volcanic eruptions produce both particulates and gases in sufficient quantity to affect the climate and are generally thought of as forcing climate variations. We choose to include human activities as another forcing, as discussed in Section 1.4.

Analysis is the decomposition of a complex substance or topic into its component parts in order to improve understanding. While that definition certainly applies to our goals here, there is another quite specialized and idiosyncratic usage that is relevant as well. In the early days of weather forecasting, observations of conditions at a number of points were used to construct maps of, for example, the atmospheric pressure contours over a broad region. Such maps were used to locate significant weather features like storms and fronts. The process of locating and drawing lines connecting points with equal pressure values, or isobars, was referred to as analysis. The term came to be used more generally for the task of creating a complete field on a regular array of points from a sparse and irregularly distributed set of observations. This usage remains common in atmospheric and oceanic science and applications such as weather forecasting. The need for analysis in this sense is particularly critical in climate studies because individual observations are rarely complete in their original form.

So, climate analysis, for our purposes here, is the science and practice of creating and examining complete and comprehensive datasets to describe, understand, and predict the state and evolution of the climate system. The first two chapters provide an overview of the climate system as a whole and a discussion of how climate analyses are performed. Chapter 3 describes the observations and instruments used to form atmospheric climate datasets. Chapters 4 and 5 summarize the nature of climate variability and change. Chapters 6 through 10 discuss in some detail the data that describe the atmosphere, the oceans, the cryosphere and the Earth's surface. Chapter 11 illustrates how these datasets are used together with mathematical models of the climate system to improve our understanding of its behavior and to predict its future. Finally, Chapter 12 describes how the current state of the climate is monitored.

Chapter 1 continues with a very high-level overview of the climate system. We describe the components and their interactions, the external agents that drive changes in the climate, and the systematic behaviors, or cycles, that characterize the system. Since our focus is on climate datasets and the observations and techniques used to produce them, we will not spend much time on the theoretical understanding required to create predictive models – a number of excellent texts are available for that purpose.

1.2 Components of the Climate System

We defined the components of the climate system as the land surface, the upper parts of the ocean and the solid earth, including in particular water in both solid and liquid form, and most of the atmosphere (Figure 1.1).

As with any interactive system, determining where to begin a description is both arbitrary and significant: The order in which elements are discussed will inevitably influence how the discussion is perceived. Our arbitrary choice is to order the discussion

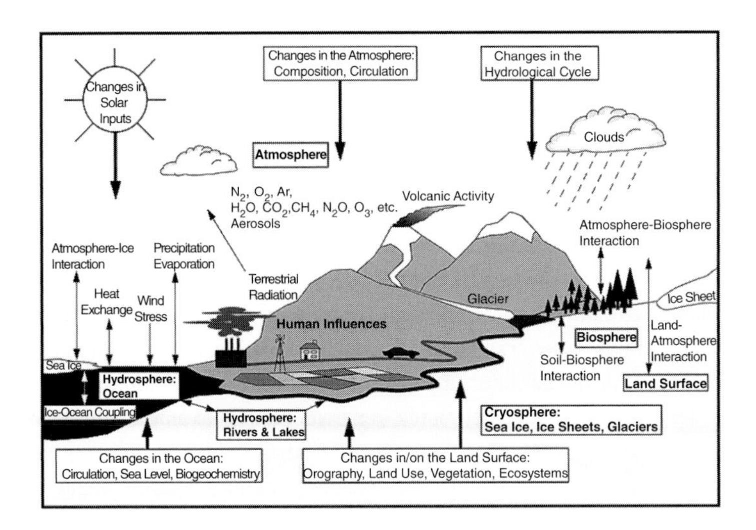

Figure 1.1 Schematic view of the components of the global climate system (bold), their processes and interactions (thin arrows) and some aspects that may change (bold arrows). Source: IPCC Third Assessment Report, Working Group 1, The Scientific Basis.

according to the time scale of the memory[1] of the component. In the atmosphere, this time scale is on the order of two weeks (Kalnay, 2002, p. 31). The ocean's memory is longer than that of the atmosphere, and the other climate system components generally exhibit longer time scales.

The atmosphere is the gaseous envelope extending from the surface to several hundred kilometers. Its characteristics, including temperature, density, composition, and movement, exhibit variations on a wide range of time and space scales. The troposphere and the stratosphere, which make up the lowest 50 km of the atmosphere, are of the greatest significance to our topic and will be the focus here. The troposphere extends from the surface to about 12 km and is characterized by a general decrease of temperature with height; the stratosphere begins just above the upper boundary of the troposphere, the tropopause, and extends up to roughly 50 km. The troposphere and stratosphere exhibit relatively strong coupling[2] with each other and with the surface.

Clearly not all variations of the atmosphere are part of the climate. Time and space scales can separate weather from climate quite effectively, except near the land surface as discussed in this chapter. The atmosphere is a fluid insofar as its variations in

[1] By memory we mean the time scale over which the influence of the initial state endures. For example, the state of the atmosphere at a given time is strongly influenced by its state a few hours before, less so by the state several days before, and little if any by the state a month earlier. A reasonably complete definition of memory in a system can be derived from the behavior of a trusted forecast model: It is the time scale at which predictions beginning from indistinguishable, but not identical, initial conditions diverge sufficiently to be effectively uncorrelated.

[2] Coupling is the linkage between two entities, such as components of the climate system, whose behaviors are influenced relatively strongly by each other. The components of any system are coupled to one another, which provides a useful definition of system: a collection of coupled entities.

temperature, density, and movement or flow are concerned, and its behavior can be described by a set of equations (Panofsky, 2014). Atmospheric composition changes result from interactions with the surface (vegetation, animals, and human society), the solid earth (volcanoes, dust), the ocean (evaporation), and the sun (ultraviolet radiation from the sun creating ozone in the stratosphere). All of these variations can be climate or weather, depending on the scales, and a clear demarcation is not always possible.

In this book, we will consider time scales longer than the memory of the atmosphere, about two weeks, to characterize climate; shorter time scales will be weather. The spatial scales of climate are more difficult to settle upon. In the free atmosphere, above the lowest few hundred meters, climatic scales are relatively large, but near the surface, where large changes in the nature of the surface–atmosphere interaction can occur on very short length scales, microclimates are evident. An example might be where an asphalt parking lot borders woodland; a large diurnal temperature range characterizes the parking lot while the wooded area next to it will exhibit a much smaller range. Such microclimates are found everywhere near the surface, created by terrain, vegetation, and built structures.

The ocean/sea is conventionally defined as the contiguous body of salty water that covers a bit more than 70% of the surface of Earth. For the purposes of this book, ocean will refer to any permanent body of water that is large enough to interact with the atmosphere so as to influence the large-scale climate. Thus land-locked bodies of salt or fresh water like the Caspian Sea, the Great Salt Lake, and Lake Superior are part of the ocean for our purposes. Rivers, transient water bodies, smaller permanent bodies of water, and other land surface water such as swamps will be considered as elements of the land surface.

By far the largest component of the ocean is the conventionally defined World Ocean. While both water and air are fluids and can be simulated using mathematical models, they differ in profound and significant ways. As a liquid, water exhibits defined upper and lateral surfaces, while the gaseous atmosphere does not. Water is also very nearly incompressible, and so the vertical structure and dynamics of the ocean are very different from those of the atmosphere.

In the climate context, it is important that the ocean represents the lower boundary for much of the atmosphere, and that humans rarely reside on or in the ocean for long periods. The surface of the ocean is also more uniform than the land surface. These facts make it convenient for our purposes to ignore most small-scale phenomena in the ocean unless they impact the atmosphere on climatic time and space scales.

Both observation and modeling of the oceans are less advanced and complete than for the atmosphere. This is particularly true for the deep ocean, below the thermocline. The thermocline is the lower boundary of the upper ocean layer that interacts more significantly with the atmosphere, and below which water temperatures vary more slowly with depth. The deep ocean is thought to be extremely important for climatic variation and change on multi-decadal and longer time scales but has been very poorly observed until quite recently.

The cryosphere consists of the persistent ice and snow on Earth's surface. It includes both sea ice and ice and snow on land. Highly transient ice and snow, such as drifting icebergs or short-lived snow cover on land, are not included in the cryosphere, but the seasonal changes in sea ice and snow cover generally are.

The cryosphere, like the ocean, is a part of Earth's surface that is treated in its own category because of its important role in interacting with other components of the climate system and because of its relative homogeneity – otherwise it would be simply included with all other aspects of the land surface. The qualities that distinguish the cryosphere from other land surface elements are temperature, always freezing or below, and albedo. The albedo is the fraction of incoming solar radiation that is reflected from a surface, and snow and ice are characterized by very high values, ranging from 0.5 to as high as 0.9. Few other land surfaces have albedos above 0.4, and so the cryosphere is a very unusual and important element of the climate system due to the role of ice–albedo feedback (to be discussed in the next section).

The cryosphere is made up of a number of different elements, including sea ice, land-based glaciers and ice sheets, and seasonal snow cover. Sea ice can be either transient on a seasonal time scale or persistent, lasting through the warm season but changing on longer time scales. The albedo of sea ice is related to its age – persistent sea ice is typically less reflective – and to the amount of surface melting and fractures, both of which reduce reflectivity. A fresh snow cover greatly increases sea ice albedo. Sea ice is less dense than seawater and therefore floats and moves with both ocean currents and near-surface winds.

Glaciers and ice caps are permanent ice on land, differentiated mainly by size and the nature of movement. Glaciers originate at high altitudes, relative to their surroundings, and flow[3] through valleys to lower altitudes. They terminate at the point where melting and other ablation processes balance the flow. Ice caps or ice sheets are different mainly in that they are thick enough, and extensive enough, that they are not contained within a basin at high altitude. In either case, the climatic significance is the same: They are cold and have a relatively high albedo. Just as with sea ice, albedo tends to decrease with the age of the ice, which, particularly in the case of glaciers, is closely correlated with the distance from the source region. Snow cover increases the albedo of surface-based ice just as it does for sea ice.

Snow that lasts through the winter and melts during the summer is seasonal snow cover. This snow has a strong impact on seasonal climate: The albedo in visible light[4] of freshly fallen snow is the highest of any substantial natural land type, and so a fresh snow cover immediately reduces the amount of solar heating in a region. The albedo of snow depends on its age and thickness. Other characteristics of snow affect the climate in various ways: by insulating the ground below, cooling the air immediately above, and by increasing the thermal inertia of the land surface and delaying the onset of the warm season.

[3] Ice, while solid, is plastic and both deforms and flows over relatively long time periods. Large masses of ice will tend to take on the shape of the underlying and surrounding terrain just as water does, but over much longer time spans.

[4] Albedo can differ for different wavelengths of light. For the most part, the albedo in the visible spectrum is of greatest importance for climate because it controls the heating due to solar forcing.

The land surface component of climate includes everything that contacts the atmosphere and is not included in the ocean and the cryosphere. The land surface is by far the most heterogeneous major component of the climate system, and it is the most difficult to observe and model. It is characterized by abrupt boundaries with sharp gradients in climatically important properties. The most crucial characteristics of the surface with respect to climate are albedo, heat capacity, which is the ability of the surface to absorb heat from the atmosphere or sunlight, and roughness, an abstract parameter that summarizes the degree to which a surface area affects the wind flow over it. These parameters are significant for the ocean and cryosphere as well as the land surface but, as will be discussed in the next section, must be discussed and handled very differently.

The land surface is a mosaic of different elements that have varying roles in the climate. In observations and modeling of the land surface, every resolved element, or pixel,[5] is described by a set of typical values for the climatically significant parameters. Since much of the land surface is covered by vegetation, 10–20 or more vegetation types are often used, in addition to types that describe buildings and other built environments. Unvegetated land, wetlands, and other surface types are also included. Of course, even the most casual inspection of the land surface shows that it is not composed of uniform elements at any spatial scale, and so observing, modeling, and understanding the role of the land surface in the climate system require developing methods for estimating the average impact of relatively large areas with great internal diversity.

1.3 Interactions among Components

Interaction, or coupling, occurs throughout the climate system, both directly and through intermediating processes. Many of the interactions among the components of the climate system occur at the interfaces: atmosphere–ocean, atmosphere–cryosphere, and atmosphere–land surface are particularly important, but every interface plays a role. These interactions are mechanical and thermal in nature, with one component causing change in the movement or temperature of the other; usually each component affects its partner. A particularly important type of interaction is known as feedback, in which a change in some characteristic of the climate leads to a change in another, and the second change leads to a further change in the first characteristic. Feedback can be negative, in which case the initial change leads to subsequent changes in the opposite direction, tending to reduce the first change. On the other hand, a positive feedback leads to a situation in which the first change begins a chain that leads to a further increase. An example of the latter is the ice–albedo feedback in high latitudes, in which an increase/decrease in snow and ice cover causes an albedo change that leads to cooler/warmer conditions and a further increase/decrease in snow and ice cover. A positive feedback tends to amplify a small perturbation, while a negative feedback dampens it.

[5] Pixel is short for picture element and refers to the smallest resolved area in an image. The term originated in the early days of computerized image processing.

Other important interactions are less direct. For example, each of the surface components emits radiation whose intensity and spectrum are related to its temperature and physical characteristics. Some of this radiation is absorbed within the atmosphere above the surface, at locations that depend both on the properties of the atmosphere – temperature, water vapor content, clouds, and other constituents – and the characteristics of the emitted radiation. The atmosphere in turn radiates to the surface, where its effects depend upon the surface properties. This radiative interaction is one of the two principal indirect interactions of climate system components.

Other indirect interactions occur through the movements and phase changes of water. Water in the form of gas, water vapor, is a significant component of the atmosphere and, when in liquid or solid form, constitutes clouds and precipitation. The oceans are essentially all water, and the cryosphere is the solid form of water, or ice. Water vapor and clouds strongly affect radiation within the atmosphere in various ways, and heat is absorbed or released when water changes phase.[6] Movement of water in any form from one climate system component to another is an important aspect of climate variability and change, and it must be observed and understood if we are to understand the workings of the system as a whole.

The roles of radiation and water in climate system interactions are both central and complex, and good observations are vital. Obtaining these observations is difficult, because many of the needed values are nearly impossible to measure directly and must be deduced from observations of related phenomena. For example, direct measurements of clouds are not possible except in limited situations, and so cloud properties must be estimated from observations of atmospheric radiation by satellites. Models are used to complement observations to understand hydrological and radiative processes, but they also suffer from significant limitations.

One additional important agent that plays a role in the climate system is the biosphere. For the most part, we will incorporate observations of the biosphere into our discussion of vegetation on the land surface and the impact of phytoplankton on optical properties such as color in the ocean. Larger animals will not be discussed, with the exception of humanity, which has a number of significant impacts on the climate by changing the characteristics of the land surface and the atmosphere. In that sense, humanity is a part of the climate system and interactions involving humans should be included, but in this book we will take the perspective that humanity is one of the forcing agents for climate, as discussed in the next section.

1.4 Forcing Agents

We have discussed the components and interactions that figure in the climate system, but no system can be fully appreciated without an understanding of the external influences that affect it. In the case of the climate, these are the agents that cause variability

[6] A change in phase, or state, is a change from any of the solid, liquid, or gaseous states to any of the others.

and change but which are not themselves affected significantly by the direct response of the climate. We will consider three main forcing agents of the climate system: radiation from the sun, the geology of Earth, and humanity.[7]

The sun provides nearly all of the energy that drives the climate system on Earth. Nuclear fusion within the sun releases a very large amount of energy that supports a solar surface temperature of nearly 6000K.[8] Like any object, the sun emits thermal radiation with a spectral profile and intensity determined by its surface temperature. This radiation is strong enough that at the average distance of Earth's orbit the solar constant, a measure of the solar radiation at all wavelengths, is about 1361 W/m^2. Two things affect the amount of solar energy available to the climate system: fluctuations in total solar irradiance (TSI)[9], resulting from changes in the sun itself, and changes in the amount of radiation that impinges on Earth at a given point due to changes in its orbit and orientation over time.

Changes in TSI affect the entire planet in a consistent way, increasing or decreasing the amount of energy available at the top of Earth's atmosphere. Variations in TSI could not be measured very accurately until the advent of observations using instruments on Earth-orbiting satellites; since 1978 observations from a number of different satellite instruments have found that the systematic variation associated with the 11-year sunspot cycle is about 0.1%, which is associated with a change in global mean temperature of about 0.2K from solar maximum to solar minimum (Camp and Tung, 2007). Cycles in TSI with much longer periods have been hypothesized based on proxies for TSI, but strong evidence for climatic impacts of such cycles has not been found. Over very long time scales on the order of billions of years, solar evolution has led to an increase in intensity that is expected to continue, but these time scales are far too long to concern us here.

Since Earth is roughly spherical, the amount of solar energy that falls on a given area changes with latitude, being greatest in the tropics when the sun is nearly vertically overhead and least in the polar zones where the sun is near the horizon. Since Earth is tilted with respect to the plane of its orbit around the sun, the amount of sunlight hitting any latitude changes with the season, and because the orbit is not a perfect circle, the intensity of sunlight changes through the year as well. The impact of the tilt is substantially greater than the effect of the elongation (ellipticity) of the orbit in any given year, and while both have substantial impacts on the climate, those impacts manifest themselves as the annual cycle; climate variability and change take place relative to that annual cycle.

Over longer time spans, Earth's spin axis precesses, changing the tilt relative to the timing of the orbit, and the ellipticity of the orbit fluctuates as well, changing the timing

[7] The activities of humans clearly affect the climate on small scales and large, and so treating humanity as a forcing agent is easily justified. We recognize that, conversely, changes in the climate affect humanity, and so it would be possible to consider humans as a climate system component. In this book, however, we choose to limit our discussion to the impacts of humans as a forcing.

[8] K stands for Kelvin, the standard unit of temperature measurement above absolute zero. The other temperature unit we will use is C, or degrees Celsius; 0° on the Celsius scale corresponds to about 273K.

[9] Total solar irradiance is the total amount of radiation emitted from the sun across all wavelengths.

of perihelion, when the Earth is closest to the sun. These Milankovitch Cycles,[10] while small, lead to changes in the seasonal timing and distribution of solar insolation that are thought to determine the timing of glacial/interglacial cycles over the past million years or so.

The geology of Earth is an important influence on the climate system. The distribution of the continents determines the locations of oceans and seas, and mountain ranges and other features of the land surface influence atmospheric circulation and associated weather; all of these contribute to the characteristics of the climate. These features of Earth's geology do not themselves vary on time scales relevant to this book except on spatial scales smaller than we will consider. On much longer time scales, continental drift changes the position of landmasses in ways that modify oceanic and atmospheric circulations and is thought to be one of the factors in the transition between ice-free and glaciated conditions.

While most geological events and changes do not lead to climate variability on time and space scales of interest here, there is one that does. Volcanic eruptions can produce extremely large volumes of climatically relevant gases and small particles that can each have significant climatic effects. Volcanoes emit a wide range of gases, among them carbon dioxide, water vapor, and sulfur dioxide. Carbon dioxide and water vapor act as greenhouse gases,[11] being largely transparent to solar radiation in the visible spectrum and opaque to terrestrial outgoing radiation in the infrared. Individual volcanic eruptions are not thought to produce enough greenhouse gasses to impact the climate, but eras with extremely active volcanism may do so.

Sulfur dioxide has the reverse radiative effect: in the atmosphere, it reacts with water vapor to form sulfuric acid that condenses to sulfate aerosols in the stratosphere. These sulfate aerosols have a high albedo and tend to reflect a significant amount of solar radiation. Since the stratosphere has limited vertical interchange with the troposphere, once there the aerosols tend to have relatively long residence times, leading to a climatically significant reduction in solar heating and a decrease in the global mean surface temperature. The same eruptions that produce sulfate aerosols tend to generate particulate aerosols that also produce cooling, but these have shorter residence times.

The climatic effects of volcanoes depend both on the eruption size and on the number of eruptions in a given time period. There is general agreement that the effect of individual large eruptions on global mean surface temperature is detectable. Some recent studies have suggested that a series of large eruptions over a relatively short time led to a much longer lasting planetary-scale cooling (Miller et al., 2012).

There is no doubt that humans have had detectable impacts on the climate. Changes in the land surface on small spatial scales, such as pavement or changing vegetation, have obvious effects. Since humans do similar things everywhere they live, these effects

[10] After Milutin Milankovitch, a Serbian astrophysicist who calculated the impact of these factors on the amount of solar energy reaching Earth at various latitudes and seasons.

[11] The reference is from a purported analogy to greenhouses, in which visible solar radiation enters generally unimpeded while the infrared radiation of the warm surfaces within the structure is mostly retained. In actuality, greenhouses are warmer because they are enclosed and warm air cannot convect away, but the term has become entrenched.

may aggregate to detectable levels. The variety of ways in which humanity and civilization affect the climate system is large, although the impacts on the global scale remain relatively subtle.

Two of mankind's earliest activities, agriculture and animal husbandry, appear to have had measureable climatic impact. Agriculture always leads to a significant change in vegetation on the surface by replacing indigenous plant life with a less diverse population. The albedo of a cultivated area is nearly always different from its uncultivated state, and the exchanges of water and heat over such surfaces are generally quite different as well. Many of the activities associated with agriculture, such as plowing and harvesting, are also accompanied by aerosol production. All of these factors can have impacts on local climate, changing the microclimates dramatically. When extended to large amounts of land, as in deforestation of entire regions, global scale impacts have been suggested, although not yet clearly demonstrated in data.

Animal husbandry is another of humanity's ubiquitous activities and might impact climate in at least two ways. Many of the animals that humans raise for food are responsible for methane emissions, and because methane is a strong greenhouse gas, a warming effect might occur. Another evident impact of husbandry is through its impact on vegetation: forests are replaced by pastureland to facilitate raising animals. This changes both the albedo and the local fluxes of water and heat, and if extended over large enough areas it can have climatic impacts. A mechanism related to these effects may have led to desertification and drought in western Africa (Charney et al., 1977).

The use of chlorofluorocarbon (CFC) gases as solvents, refrigerants, and fire suppressants expanded rapidly from the mid-twentieth century due primarily to their very low toxicity and generally low reactivity. Their use in air conditioning systems, fire extinguishers, and as aerosol propellants led to significant and growing releases into the atmosphere. CFCs do not interact strongly with other atmospheric constituents and do not condense at atmospheric temperatures and pressures, and therefore they have a long residence time. Their presence in the free atmosphere was first demonstrated in the early 1970s, and shortly thereafter the possibility was raised that significant amounts of CFCs would be found in the stratosphere, where the presence of energetic solar ultraviolet radiation could allow them to play a more active role. It was found fairly soon afterward that CFCs were indeed catalyzing the destruction of stratospheric ozone, leading to significant decreases in ozone, particularly over Antarctica during the Southern Hemisphere winter, and to increases in ultraviolet radiation at the surface with potential associated increases in health effects. CFCs are also a strong greenhouse gas, but the Montreal Protocol, an international treaty that led to large reductions in the use of long-lived CFCs to reduce impacts on health, has limited the impact of CFCs on climate.

Carbon dioxide, or CO_2, is released by the burning of vegetation and is also a principal product of the combustion of fossil fuels. Carbon dioxide is another greenhouse gas, transparent to visible radiation from the sun but opaque to infrared radiation from Earth's surface, and so changes in its atmospheric concentration can impact global climate. While biomass burning, the term applied to combustion of vegetation, releases substantial amounts of CO_2, the climatic effects are partially balanced by the fact that the preceding growth of those same plants had reduced the carbon dioxide in the atmosphere by approximately the same amount. However, fossil fuels contain carbon that was taken from the atmosphere in the distant past, and so their combustion increases the total concentration of CO_2 and affects the climate.

CO_2 varies naturally on time scales of thousands of years and longer (e.g. Pearson and Palmer, 2000). Time series of atmospheric CO_2 derived from ice cores for the past 400,000 years indicate that values varied from approximately 180–280 parts per million (ppm) until the past several hundred years, with the highest values occurring during interglacial periods when the ice caps were the smallest. Earth is currently in such a period, and the pre-industrial revolution baseline value for atmospheric CO_2 is generally thought to be 280 ppm.

Regular measurements of CO_2 began in the late 1950s, led by Ralph Keeling of the Scripps Institution of Oceanography in La Jolla, California. He and his colleagues found that CO_2 exhibited a consistent year-to-year increase as well as a clear annual cycle. The annual cycle results chiefly from the growth/decay of vegetation on land in mid- and high northern latitudes, where the growing season is relatively short. The time series of atmospheric CO_2 is one of the iconic climate observations and is referred to as the Keeling Curve. The overall increase in mean CO_2 values has continued to the present and is consistent with the amount of fossil fuel burning (IPCC AR5). Global values now exceed 400 ppm, and continued increases are inevitable as long as fossil fuel use continues at the current level.

Theory predicts that increasing the concentration of any greenhouse gas in the atmosphere will result in an increase in global mean temperature, due to the stronger absorption by the atmosphere of infrared radiation emitted from Earth's surface. Observing the mean surface temperature of the globe is more difficult than it may sound (details will be discussed in Chapter 6), but the expected increase has been detected (IPCC AR5).

1.5 Cycles in the Climate System

The elements, interactions, and forcings discussed earlier in this chapter result in the collection of phenomena that characterize the climate system. For the most part, the behavior of the climate system is undetectable to humans because it takes place over time scales much longer than our senses and brains integrate – nearly everything that we observe directly is weather, not climate. However, humans are very good at pattern recognition and recognized long ago that weather was characterized by cycles or oscillations[12] over longer time scales: Those cycles are the most readily observed behaviors of the climate system.

[12] The climate system exhibits a number of large-scale phenomena that recur in similar form, like the seasonal cycle. In some case, like the seasonal cycle, the phenomenon is strictly periodic, while in others it is recurrent but at irregular intervals. These phenomena are referred to as cycles or oscillations. There is a tendency to use cycle for the periodic phenomena and oscillation for others, but that usage is not universal.

Observing climate is challenging. Usable observations depend on reliable record keeping, something that is not trivial even with current technology, and the ability to combine observations from different locations and times so that the patterns that characterize climate emerge from the large variations due to weather and to noise.[13]

A critical characteristic of these features is that they recur, as implied by the label – a phenomenon that happened only twice would not be recognized as a cycle in the way that several successive ones would. This repetitive nature, characterized by a similar sequence of events recurring at somewhat regular intervals, makes cycles in the climate system predictable to some degree, often to a useful degree. While everyone reading this book is fully aware that some seasons are warmer or wetter than others, due to the annual cycle, the first people who discovered and used this fact found it produced extremely useful predictions.

Cycles play a vital role in observing and understanding the climate system above and beyond their practical applications. They are valuable tools for understanding how the system works, and for checking the accuracy of our observations, derived analyses, and simulations. These cycles can be grouped into three broad categories, each of which has a different role in enhancing our understanding of the climate. Forced cycles are oscillations in climate system behavior that result from periodic external forcing; process cycles describe the reservoirs and exchanges that characterize some crucial element of the system; and phenomenological cycles are oscillations that emerge from interactions within the system. We discuss examples of these three groups in the following sections.

1.5.1 Forced Cycles

Forced cycles result from periodic external forcing of the climate system. The most prominent and obvious of these is the annual cycle, which is the most important form of climate variation virtually everywhere on Earth. The annual cycle is caused by the changes in solar heating that result from the combination of Earth's revolution around the sun and the inclination of the axis of rotation to its orbit. The annual cycle is so strong that atmospheric climate variations from other causes are often nearly impossible to discern without first removing the mean annual cycle from observations.

Earth's axis of rotation, defined by the line connecting the north and south poles, is tilted relative to a line perpendicular to the plane of its orbit around the sun by about 23.4°. This axial tilt, or obliquity, results in a substantial change in the height, relative to the horizon, of the noontime sun through the year. If Earth's axis were vertical to the orbital plane, the noontime sun would reach the same height every day, with the value determined strictly by latitude: At the equator, the sun would be directly overhead and at the poles it would be precisely on the horizon. The axial tilt produces a change of 46.8°

[13] Noise in this context refers to the bias and random variations in observations introduced by the observing system, the human interaction (if any), and the recording and storage mechanisms. In addition, weather variations themselves are sometimes referred to as noise in the context of climate, since they can be thought of as high frequency variations about the more slowly varying climate signal.

in elevation above the horizon between the winter solstice, when the axis is directed away from the sun, and the summer solstice when it points toward the sun. The solstices are opposite in the Northern and Southern Hemispheres: Winter solstice is roughly December 21 in the Northern Hemisphere and June 21 in the Southern Hemisphere, and the summer solstice occurs on the same dates but in the alternate hemisphere.

Solar radiation more effectively heats the surface when the sun is closer to overhead, and so the summer hemisphere in general is warmer than the winter hemisphere. In high latitudes, greater than about 45°, this is by far the most prominent feature of the annual cycle – during winter there is little sunlight, that sunlight is weak due to the low sun angle, and the long nights permit greater radiation of terrestrial heat to space. However, even in high latitudes the annual cycle is far from simple, and in lower latitudes other factors, like the distribution of ocean and land and elevation, greatly complicate the picture.

The annual cycle results from changes in solar radiation through the year, and at Earth's surface it is characterized by substantial temperature changes. Since the moisture content of the atmosphere near the surface of the ocean tends to be in balance with the ocean surface temperature, the warmer seasons are more humid. These changes in surface temperature and moisture are accompanied by substantial changes in atmospheric circulation. A more extensive discussion of the annual cycle is found in Chapter 4.

The diurnal cycle[14] is a result of the rotation of Earth on its axis every 24 hours. For the most part, events with a life span of one day would not be considered part of the climate. However, because the diurnal cycle forces a nearly identical pattern of solar heating day after day, there are many locations where the resulting weather patterns repeat in ways that lead us to consider them aspects of climate. The most common of these occur through the daily heating by the sun of Earth's surface and the atmosphere.

Heating of the land surface following sunrise can lead to cloud and shower formation during the day under certain conditions. This diurnal convection is a pronounced feature of the climate in many tropical areas and in midlatitudes during the summer season. Similarly, in coastal regions, the sun heats the land surface more quickly than the adjacent waters. The warm air over land rises and the cooler air over the water flows landward. This sea breeze is observed very regularly in many areas at many times of year, and it forms an important part of the climate in those regions. At night, quicker cooling of the land surface relative to adjacent waters can lead to a land breeze over coastal waters.

The patterns of differential heating driven by surface features extend into the atmosphere, but with lesser amplitude as altitude increases up to the tropopause. At very high altitudes in the stratosphere, some atmospheric constituents, such as ozone and water vapor, absorb sunlight, leading to very large-scale heating during daytime and cooling at night. These global-scale diurnal changes in heating lead to

[14] Diurnal cycles are daily patterns of variation, particularly referring to phenomena that are completed within a 24-hour period and repeat every 24 hours.

very large-scale low-intensity diurnal circulations in the stratosphere that vary with the season and are sometimes referred to as atmospheric tides. This terminology can lead to confusion between the heat-driven features and tides forced by gradients in gravitational attraction between Earth and the moon or the sun. The gravitational tides arise because gravitational attraction on Earth of the moon is greater on the side of the planet closest to the moon and weaker on the opposite side. The solid earth deforms slightly in response to this gradient, but both the oceans and the atmosphere, being fluid, deform much more, leading to very evident variations in height. These gravitationally forced lunar tides are a very prominent feature of coastal ocean variability and are nearly diurnal, but with an approximately monthly cycle due to the revolution of the moon around Earth superimposed. Solar gravitational tides are smaller than lunar tides, but are exactly diurnal. Gravitational tides in the atmosphere have much lower amplitude than the changes forced by the diurnal cycle of solar heating.

Changes in solar radiation over time create climate variations, as discussed in Section 1.4. The most prominent such phenomenon is the sunspot cycle, which is the common term for a comprehensive suite of solar variations characterized by changes in the number and location of sunspots and by changes in the intensity of total solar irradiance of about 0.1%. This amount of variation was too small to be detected until satellite observations were available, and has not yet been unequivocally linked to specific climate variations. The sunspot cycle itself varies over longer time scales as well, and it has been suggested that extended periods of low sunspot number, and presumably intensity, like the Maunder Minimum (1645–1715) might be associated with widespread cold conditions, but detailed analyses have not confirmed this hypothesis. Sunspots are discussed further in Chapter 5.

Earth's orbit itself changes over much longer time periods in ways that are thought to produce climate variations. These changes produce the Milankovitch Cycles (see Section 1.4), and analyses of data related to past ice ages have shown that these orbital changes are reflected in the onset and end of glacial epochs, particularly in the Northern Hemisphere. The total solar energy reaching Earth over the course of a year is not affected by the Milankovitch Cycles, but the distribution in time and space during the year is. In some phases, winters are warmer than the long-term average, while summers are cooler, which is thought to lead to less summer melting of snow and ice in high latitudes. Over time, the ice–albedo feedback amplifies this effect and leads to an ice age.

1.5.2 Process Cycles

Process cycles are important tools in helping us to understand the functioning of the climate system. A process cycle in the climate system is a description, preferably one that includes both the physics and the measurements, of how some quantity behaves. Useful process cycles are those for which the quantity in question is neither created nor destroyed during the functioning of the climate system. The cycles most relevant to the climate system are water, energy, and carbon.

The water cycle is a crucial part of the climate system because of both the ubiquity of water on Earth and its significance to humanity. While the amount of water in the climate system is essentially constant, its distribution varies dramatically in time and space. Water is found in the atmosphere, mostly in the form of water vapor, but also in liquid and solid form in clouds. It constitutes nearly all of the oceans, is present in permanent solid form as ice caps in high-latitude locations like Antarctica and Greenland and nearby seas, and in liquid form in rivers and lakes on land. It is also present below the surface in land areas, and it plays a role in climate in all of those reservoirs. The exchanges of water among these reservoirs are enormously, and obviously, important in climate variations like droughts and floods. Precipitation[15] and evaporation[16] are important both because they are movements among reservoirs and because of the implications of their phase changes for the energy cycle.

The energy cycle has less obvious impacts on human life but is a vital tool for understanding how solar radiation, which is the only significant external (to Earth) source of energy for the climate system, moves through the system. The energy from the Sun is unevenly distributed in time and space, with much more available in the tropics than in high latitudes and more in summer than winter in middle and high latitudes. Even a completely uniform planet would experience energy flows to redistribute these imbalances. These inhomogeneities are exacerbated by the very different responses of different substances to solar heating. Bodies of water experience small increases in temperature over large volumes, while land surfaces warm more over a much smaller volume, leading to larger surface temperature increases. Vegetated areas absorb sunlight in the process of photosynthesis, storing some of the energy in complex chemicals. Some energy is absorbed in the atmosphere, with important effects on clouds and in the stratosphere.

The carbon cycle describes where carbon is found and how it moves between reservoirs. The element carbon is key for life on Earth because of its role in organic compounds, but in the climate system its most important role is as a constituent of two of the most important greenhouse gases (discussed in Section 1.4): carbon dioxide and methane. Carbon dioxide in particular is strongly linked to global temperature variations as illustrated by time series based on ice cores from Greenland and Antarctica. Human use of fossil fuels beginning with the Industrial Revolution has led to a significant increase in the amount of carbon dioxide in the atmosphere and to an increase in the amount of energy absorbed within the climate system. While measurements of fossil fuel burning are available, improved understanding and observation of the carbon cycle is essential to determine where the released carbon ends up and how future releases will affect atmospheric concentrations.

[15] Precipitation refers to any of the processes that transfer water in either solid or liquid form from the atmosphere to the surface.

[16] Evaporation is the process through which surface liquid water, from water bodies or land, enters the atmosphere as vapor. For convenience, it is generally used to include sublimation, in which ice transforms to vapor. Evapotranspiration is the term used for evaporation facilitated by plants in which water is transported from the roots to the leaves and evaporated there, thus enabling nutrient transport.

1.5.3 Phenomenological Cycles

Phenomenological cycles are emergent behaviors[17] of the climate system. They are not periodic, as many forced cycles are, nor are they heuristic devices in the manner that process cycles are. Phenomenological cycles are distinctively oscillatory, in that each instance is observed to possess a life history that resembles previous ones, and instances recur at intervals that are somewhat regular. These phenomena are sometimes referred to as quasi-periodic, and a spectral analysis (see Appendix A) of the time series of any of the associated observations will exhibit a broad but significant peak. The sunspot cycle, discussed earlier, is an emergent phenomenological cycle of the complex system that is our Sun.

The most studied and widely known such phenomenon in the climate system is the El Niño/Southern Oscillation, or ENSO. Aspects of ENSO have been known for centuries, but the observations and theory necessary for a reasonably complete understanding have been available only since the 1970s. ENSO is a coupled oscillation primarily involving the upper tropical Pacific Ocean and the tropical troposphere. The ocean surface temperature oscillates between a state with relatively cold (compared to the rest of the equatorial ocean) water near the equator in the central and eastern Pacific, and a state with equatorial temperatures more nearly uniform across the Pacific. Atmospheric changes in lower and upper tropospheric winds accompany the temperature changes. While the active coupled oscillation is a feature of the tropical Pacific Ocean, the occurrence of either the warm, El Niño, or cold, La Niña, phase has impacts on climatic events over a much larger region. Warm and cold events generally alternate, but the recurrence period varies substantially, and on occasion multiple events of the same sign can occur in succession.

The Madden-Julian Oscillation (MJO) and Quasi-Biennial Oscillation (QBO) are also well-observed phenomena for which the beginnings of successful theories are available. Both are primarily atmospheric oscillations that involve reversals in wind direction over time, and both are most clearly observed in the tropics. The QBO is a reversal of winds in the stratosphere that has a period of close to, but not exactly, two years. Its cause is now understood to be vertically propagating waves originating at lower altitudes in the tropical atmosphere, and climate models are beginning to successfully simulate it. The QBO appears to have relatively little impact on surface climate. The MJO is an oscillation in tropical near-surface winds, precipitation, and tropospheric circulation that propagates eastward. This complex of winds and precipitation passes over a location over a period of several weeks to a few months, and a quiescent period generally separates one manifestation from succeeding ones. While some aspects of the MJO are observed to propagate continuously around the globe, the most prominent wind and precipitation signatures are found in longitudes from the Indian Ocean to the central Pacific. The large-scale changes in upper tropospheric winds

[17] Emergent behaviors are complex phenomena that arise from the interaction of components of systems, but are not observed in the individual components in isolation.

that occur during passage of a MJO may induce extratropical anomalies as well, making modeling and prediction of the MJO an important goal.

Many other such phenomena have been identified by analysis of atmospheric and oceanic observations. An important distinction among them is the degree to which potentially predictable coupled physics are responsible, because predictive potential depends upon our ability to model the oscillation and to predict future states based on available observations. For example, ENSO predictions based on models of the coupling between the tropical atmosphere and the upper tropical Pacific Ocean have achieved some skill on seasonal time scales.

1.6 Outline of the Book

Climate analysis is the science and practice of creating and examining complete and comprehensive datasets to describe, understand, and predict the current state and evolution of the climate system. Our first three chapters provide an overview of the climate system, describe how climate analyses are performed, and discuss the observations, instruments, and datasets that form the basis for climate analysis. Chapters 4 and 5 summarize the nature of climate variability and change. Chapter 6 is focused on the considerations required in constructing climate datasets using atmospheric temperature over land as an example. Chapter 7 discusses how *in situ* and remotely sensed observations are combined to produce climate datasets using precipitation as an example. These first seven chapters concentrate on the atmospheric component of the climate system. In chapters 8, 9, and 10, we focus on the ocean, cryosphere, and land components of the climate system, respectively. Climate models as sources of data and as analysis tools form the basis for Chapter 11. The final chapter describes how the current state of the climate is monitored.

Suggested Further Reading

Burroughs, W. (Ed.), 2003: *Climate into the 21st Century*, Cambridge University Press, 240 pp. ISBN 0 521 79202 9 (A review of climate variability, climate change, and impacts for non-experts.)

Peixoto, J. P., and A. H. Oort, 1992: *The Physics of Climate*, American Institute of Physics, 520 pp. ISBN 978-0883187128. (A broad survey of the physics of climate for those with knowledge of calculus.)

Questions for Discussion

1.1 Describe ways in which humanity has affected the climate system as a forcing agent.

1.2 In what ways could humanity be considered an interactive component of the climate system?

1.3 Which of the cycles discussed here (forced, process, and phenomenological) exhibit a specific stable temporal period, and why?

1.4 We describe three process cycles: water, energy, and carbon. Can you identify the common characteristic of these that allows us to measure and analyze these cycles?

2 Climate Analysis
Goals and Methods

2.1 Introduction

At the mid-point of the twentieth century the American Meteorological Society commissioned the Compendium of Meteorology (Malone – Editor, 1951) to document the then current state of the understanding across all aspects of the science including a chapter on climate. In the chapter on climate it states that:

> *... there has been a woeful tendency to use of the bones of bare statistics and mean values without the flesh of physical understanding ... climatology as presently practiced is primarily a statistical study without the basis of physical understanding, which is essential for progress."* (p. 967)

Much progress has been made since that statement was written. In this chapter and the ones that follow we strive to provide some practical guidance for continued movement from the "bones of bare statistics" toward the "flesh of physical understanding."

Climate analysis is the name we give to the process aimed at improving our physical understanding. It has three associated but independent meanings: (1) the use of data to create a complete set of maps and/or time series of climate information from both regularly and irregularly distributed observations, (2) the production of datasets for archive and further use, and (3) the examination of climate data and information to describe and understand climate including its variations and changes.

From early in the twentieth century, meteorologists had called the process of creating maps of weather features, such as temperature and pressure, by drawing contours based on plotted data from individual stations "analysis." This process was eventually automated and referred to as "objective analysis." Objective analyses are used to create the gridded data fields, also called analyses, required to initialize numerical weather prediction, and used by forecasters, scientists, and the public to describe weather and climate.

In order to assess progress in climate analysis since the publication of the Compendium, a Climate Diagnostics Workshop, the first in a continuing series of annual meetings, was convened in 1976 by National Oceanic and Atmospheric Administration (NOAA). The Workshop featured descriptive studies of climate variability based on gridded datasets as well as based on more traditional station observations at specific points (Ropelewski and Arkin, 2017). The Proceedings from these annual Workshops serve to illustrate the progress in analysis techniques and availability of data over the years. From the mid-to-late 1970s forward, satellite and other data began to be employed for climate analysis. The purposeful development of climate analysis datasets, including atmospheric circulation, sea surface temperature, snow cover, sea ice, and others, accelerated in the early 1980s, motivated by studies of climate phenomena, climate change, and the El Niño/Southern Oscillation (ENSO) in particular.

Throughout the 1980s datasets based on climate analysis techniques expanded and were used more extensively to study climate phenomena, such as ENSO and the North Atlantic Oscillation (NAO), among others, on monthly to interannual time scales. Decadal to centennial variability such as the Pacific Decadal Oscillation (PDO) and the Sahel drought were studied and datasets based on observations and capable of depicting global climate changes were developed and made available for further analysis.

Many climate studies during the 1980s and early 1990s relied on semi-monthly and monthly gridded data fields obtained by averaging data fields produced by numerical weather prediction models in initializing operational weather forecasts. Such fields, while useful for climate studies, exhibited variations whenever operational models changed in efforts to improve the forecasts. To eliminate these sorts of artificial variations from the datasets, a number of efforts to develop modern analysis systems more appropriate for climate data began. These systems, which are described in more detail in Chapters 3 and 11, are applied to historical data to produce reanalyses, and to current data to yield analyses for real-time climate monitoring (Chapter 12). The reanalysis datasets resulting from the combination of the two have proven immensely useful in atmospheric climate studies. At present, the need to develop analysis procedures that use all available data to specify most accurately the true state of the global atmosphere at a series of times is clear, and efforts to create a true climate analysis capability are underway.

In this book, we illustrate how climate analyses are used to characterize climate variability and change, to identify and describe phenomena within the climate system, to validate and improve climate simulation and forecast models, and to support operational climate monitoring and prediction. We also describe the development and application of analysis procedures to create extended time series of a variety of climate variables, including those in the atmosphere, land, and ocean.

2.2 The Basics of Climate Analysis

In this chapter we view climate analysis as the process used to describe the state of the climate system and its evolution over some period in order to identify and understand systematic patterns of climate variability and the character of climate change. The process of climate analysis must adhere to a set of well-established basic steps if the analysis is to be useful in furthering deeper understanding of climate variability and change. These basic steps, introduced here, serve as a theme that is woven through the text and Questions for Discussion. The basic steps of climate analysis are:

Clearly identify the purpose for the analysis
Select or develop data appropriate to that purpose
Apply the appropriate analysis tools
Critically interpret the results of the analysis

The basic steps of climate analysis are defined in more detail in the discussion that follows.

2.2.1 Step I – Identify the Purpose of the Climate Analysis

The purpose of the analysis strongly influences the data and tools that will be used in the analysis. The question "Why are we performing this analysis?" should guide the work that follows. The climate may be analyzed for a variety of reasons, ranging from the need to validate or disprove a hypothesis (SST anomalies in the equatorial Pacific are causally related to winter weather anomalies in the United States) to entirely pragmatic (what are the heating and air conditioning requirements over a new building's design lifetime at a given location). Is the analysis for exploratory research aimed at getting a better understanding of climate variations or is the analysis to guide some activity, e.g. water resource management or agricultural practice? While the reasons for performing climate analyses are innumerable, we can describe a few broad justifications.

To Describe the Current State of the Climate

Many practical activities are influenced by climate and climate variability. Obviously, descriptions of mean or current climate conditions in the context of the recent past are the simplest and most familiar kinds of climate analysis. They address the everyday questions encountered when traveling to a new place, for example "Do I need to pack a sweater and umbrella?" The informal use of words like "normal" and "typical" often used in these inquiries hides a mountain of complexity, and great care is needed to understand the purpose behind the question asked. Is the question based on idle curiosity or to satisfy the need for some vital piece of information that may fit into a larger decision-making framework? Often when a person asks what is "normal" or "typical" they are really asking, "What can I expect?" The usual response is to provide an average of basic weather conditions such as temperature or rainfall over some time period, for example the current month over the past 30 years. For simple curiosity, and some applications, the statistical expectation, or average, is sufficient. However, the mean, or median for that matter, isn't necessarily informative. Although many users of climate information, including researchers, ask for the average, what they really want and need is some sense of the relative frequency of occurrence, the histogram.[1] There are many situations where the histograms show that the climatological data are not symmetrically distributed with respect to the mean. Such skewed distributions (Figure 2.1) limit the usefulness of the mean or median. Precipitation at a location is an example of a climate variable that most often has a skewed distribution. Considerations of the frequency distribution of the historical data are no less important in more theoretical and in exploratory analyses.

In addition to information about the average or "typical" climate over some period there is often interest in knowing the characteristic spatial patterns (described in Chapter 4) and evolution of the climate over time. Describing the evolution of the climate system adds to the complexity of the analysis. Once again, the types of analyses and the data to be used are necessarily guided by the specific questions being asked. Very often descriptions of climate are expected to include a description of the mean

[1] Histogram, a graph of the number occurrences of a certain value, or within a certain interval, on the vertical axis, versus values of the data on the horizontal axis. See Appendix A.

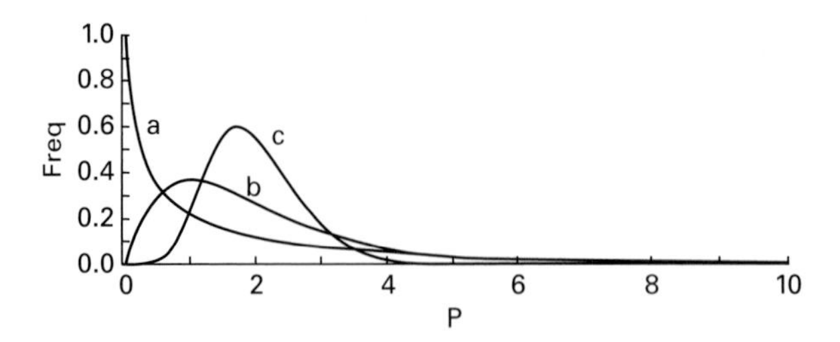

Figure 2.1 Sketch of typical gamma function distributions based on empirical fits to the data. Note that the distinctly different distributions, curves "b" and "c," have the same mean value (2 units), illustrating that two locations that might have, for instance, the same mean precipitation may experience different climates. Note also, that even though curve "a" has a mean of 1 unit both curves "a" and "b" have about the same probability of exceeding 5 units.
(Adapted from Ropelewski et al., 1985).

seasonal cycle of temperature and precipitation at a location. Other applications may depend on the probability that some meteorological variable such as maximum or minimum temperature might exceed a threshold.

To Identify Patterns of Climate Variability and Change

When the analysis is expanded to include describing climate variability and change, the challenges are significantly increased. Common experience tells us that the coming winter will differ from the last, and few if any of the days, weeks, months or seasons are likely to fall exactly on the mean value given by climatology.[2] Climate variability most often refers to differences between the mean for a given month or season and the long term, for instance 30- year, historical mean. The climate varies everywhere for a variety of reasons, as discussed in Chapter 4. Variations about the monthly and seasonal mean include the variability associated with day-to-day weather not tied to any particular climate pattern. These random weather variations may, for instance, result in monthly mean temperatures that differ from year to year. Thus a more complete description of the climate is given by measures, like histograms, or statistical measures like the standard deviation (Appendix A), that include some indication of the typical magnitudes of the spread of year-to-year variation in the monthly or seasonal means. Interactions among components of the climate system, as with ENSO (Chapter 4) may temporarily shift the entire histogram at some location (Ropelewski and Halpert, 1989; Ropelewski and Bell, 2008) in different phases of the phenomenon. Further variability may be introduced by volcanic eruptions that inject aerosols and gases into the atmosphere and by changes in the concentrations of greenhouse gases.

The purpose of this kind of climate analysis is to quantify the relative magnitudes of the variability related to different sources and to identify systematic patterns, or modes, in the variability. The need to separate random variability from systematic climate

[2] Climatology is the term used for an estimate of the mean climate. It is most often based on the average of observed data over some recent extended period.

patterns, such as variability associated with air-sea interactions like ENSO, imposes requirements on the spatial and temporal resolution, as well as the duration, of the data used in the analysis.

To Test Hypotheses of Variability and Change

Within the past 30 years or so the real-time monitoring of climate and climate variability has become routine within the weather services of several nations as well as at academic and research institutions. This monitoring, focused primarily on monthly and seasonal climate, has led to the production of large amounts of data that, in turn, have provided the opportunity to form and test hypotheses that attempt to explain the sources of variations in the observed climate. Very often these hypotheses are based on statistical relationships among components of the climate system or on observations taken from case studies. A crucial role for climate analysis in these kinds of studies is to move beyond simple statistical inference toward examination of the physical links among observed phenomena: for example, between larger than usual seasonal increases in the equatorial Pacific Ocean sea surface temperature and precipitation in the United States. Only through testing hypotheses based on the underlying physics using independent observations or model simulations can climate analysis advance our understanding of the climate system.

To Compare Observations against Model Simulations or Forecasts

Numerical models of the climate system have become more and more sophisticated. There is continuing progress in increasing the spatial and temporal resolution of these models together with advances in including more components of the climate system (e.g. atmosphere, ocean, land surface, sea-ice and snow cover, atmospheric constituents) and slowly improving parameterizations of sub-grid scale processes.[3] Climate analysis has increasingly been used to perform comparisons between observations and models. The purpose of these comparisons is to help establish the capabilities and limitations of the models in replicating and predicting past climate and, by implication, help establish model abilities to forecast future climate events.

2.2.2 Step II – Select or Develop the Appropriate Data

Until the latter part of the twentieth century climate data were often difficult to obtain and analyze. Researchers, private companies, and many national weather services considered their archived data as proprietary. Limited data were exchanged and made available through international agreements brokered by the World Meteorological Organization (WMO – see Box 2.1). In many cases obtaining even these limited

[3] All such models calculate parameters at a grid of points or using a spectral representation that imposes some implicit spatial resolution. Atmospheric and oceanic phenomena with scales smaller than the grid spacing cannot be simulated directly by the model, and so must be calculated using some other approach. This is referred to broadly as parameterization, and usually relies on the use of a formula that uses local values from the full model and statistical relationships derived from observations to calculate the required quantities.

amounts of data required paying the costs of reproduction and delivery, which, for large datasets, could be prohibitive. In many cases, the scientist or organization paying the reproduction and delivery costs was required to agree not to share these data with other entities.

Over recent decades, this situation has improved to some degree, with significant amounts of data available for most, but not all, regions of the world, as the wider community has become aware of the advantages of data sharing. In addition, many derived datasets based on the original proprietary data are shared freely among members of the research community. While the current situation is not ideal, freely available archives of analyzed climate data can now be easily found through Internet searches. A partial list of the organizations and websites for these data sources appears in the References section of this book.

In addition to the traditional "weather station" data, remotely sensed satellite observations now provide global, or near global, data coverage for a number of climate variables. These include estimates of sea surface temperature, sea-ice, vegetation, snow cover, and an increasing number of variables described in Chapters 6 through 10 as well as estimates of temperatures through the depth of the free atmosphere and precipitation. Some satellite observations now extend back for several decades and are discussed in detail in the following chapters. Our understanding of what constitutes climate data has been expanded to include the retrospective analyses that assimilate historical data using global climate models. In this usage one can think of the numerical models as tools for sophisticated data interpolation where the interpolations are constrained by the physics contained in the model. Several versions of this kind of model-based reanalysis are available and continue to be generated (see Chapter 3 for more details).

Box 2.1 Data Availability and Access

Historically, the individuals, institutions, and nations who had kept climate records held these records very closely for real or perceived scientific, economic, or military advantage. It was common practice for these entities to present analyses based on climate records but not provide access to the climate records themselves. In 1995 the World Meteorological Organization, sensing the growing importance of the need for wider collaboration in the exchange of weather data, passed Resolution 40 (URL in the reference list) calling for free and open exchange of at least some climate data that are based on weather observations. Around the same time, the US Government adopted a policy of free and open data availability for data gathered with the aid of government funding or archived by government agencies. In this case "free" has been interpreted to mean that the agencies holding data could only charge for the cost of reproduction and/or transmission of the data. Many climate datasets held by US Government data centers are available via the Internet without charge.

In the United States, observational climate data can be obtained from several sources, including the National Centers for Environmental Information (NCEI), the National Center for Atmospheric Research (NCAR), and others. Many of the climate

Box 2.1 *(cont.)*

indices discussed later in this book can be obtained from the web sites of the institutions mentioned in this section as well as from NOAA's Climate Prediction Center (CPC), the Physical Science Division (PSD) of the NOAA Earth System Research Laboratory, and the International Research Institute for climate and society (IRI). Numerical model data may be accessed from the Program for Climate Model Diagnosis and Intercomparison (PCMDI), as well as from the NASA Distributed Active Archive Centers (DAACs). Many of the sources of model data/analyses provide the capability to download geographic or temporal subsets of data. Additionally, many of the organizations hosting climate datasets have found it more efficient to provide analysis and display capability on their own computer facilities to avoid the need for large data transfers.

The contemporary climate analyst may be overwhelmed by the availability of model-analyzed data, remotely sensed data, and large numbers of individual station observations. One of their first tasks, once the purpose of the analysis is well defined, is to examine these existing datasets to determine whether one or more of them are appropriate for the problem being studied. If none are available, the research activity will need to include the development of the necessary datasets. Alternatively, the scope and purpose of the analysis may have to be modified given the realities of the data constraints. The spatial and temporal characteristics of the data must be compatible with the purpose of the analysis; for example, studies of monthly or seasonal temperature variations at a location or region cannot be conducted solely using the annual mean global temperature time series.

Data Selection – Compatibility between the Temporal Characteristics of the Data and Phenomenon Being Examined

As discussed previously, there are many open sources of climate data. These data need to be examined carefully before they are adopted to ensure that available archived data are consistent with the purpose of the analysis. The temporal aspects of the data are characterized by their period of record (calendar dates), duration, completeness, continuity, and sampling frequency. It is imperative to carefully study the documentation associated with the data to identify these characteristics and assess the advantages and limitations of a dataset in the context of the analysis at hand. The analysis will be greatly strengthened if more than one dataset with the appropriate characteristics is included in the study. If none of the available datasets have the necessary characteristics the analysis may require using two or more datasets that have complementary advantages and limitations. For instance, the precipitation record at a location may have temporal gaps that can be filled using statistical relationships between satellite-derived estimates of precipitation and precipitation at that location.

Even though the routine observation, recording, and archiving of data seem to be straightforward tasks, it has proven extremely difficult, maybe impossible, to maintain a gap-free dataset for all of the locations in a climate observing network for any time period. Instruments, data loggers, and communication systems and their backup systems will occasionally fail, and human observers will sometimes become ill or miss an observation for some other reason. Thus, in a time series of monthly mean temperature, for example, it is likely that some monthly observational records will have fewer daily observations than the number of days in the month. If the average daily temperature for every day of the month were absolutely necessary, the number of locations available for analysis would be very small, vanishingly so if averages over large areas are required. Thus, as a practical matter, some number of daily observations is established as the minimum required for estimating monthly values consistent with the purpose of the analysis. Typically, data centers and other sources of these data arbitrarily allow no more than 10% of the temperature record to be missing, so for example three missing days for a monthly estimate. A further complication, in this example, is that some of the daily data, recorded before or after the missing data, may have come from interpolations or statistical estimates. In any case it is prudent to estimate the uncertainty introduced by incomplete data. One way of estimating this uncertainty is to compare monthly values from complete datasets to monthly values formed by randomly withholding N-days of data.

The lack of appropriate data to study phenomena of interest has been a principal motivation for the design and implementation of field experiments and development of new observing networks. These activities are generally far beyond the capabilities of individual researchers or research groups. Many observational networks and past field experiments were not designed with climate analysis as their primary focus, but observations from such activities can be useful with appropriate processing.

Data Selection – Compatibility between the Spatial Characteristics of the Data and Phenomenon Being Examined

Once it has been established that the data under consideration have the required temporal characteristics for the proposed analysis they must also be examined to ensure that they are spatially consistent with the analysis. The spatial aspects of the data are characterized by spatial extent or domain, data density,[4] uniformity in the observing instruments, and spatial homogeneity[5] over a geographic region. These characteristics often vary as a function of the period of analysis and may limit the ability to obtain unambiguous conclusions.

Many datasets are derived from surface meteorological stations or other point observations by averaging or interpolating the observations into bins defined by latitude and longitude boxes or grid areas.[6] While processing the observations in this manner to form more manageable datasets has many advantages for analysis, it also masks the spatial and temporal characteristics and uncertainties in the data within the grid areas. Some gridded datasets provide quantitative information relating to these

[4] Data density, here, refers to number of observations per unit area.

[5] Data homogeneity refers to the (uniform) distribution of the observations within the area.

[6] This process is one form of objective analysis. See discussion of analysis methods in Chapter 3 for more details.

issues that should be consulted when interpreting analyses based on these data (for example Jones et al., 1986a, b).

Gridded datasets based on satellite observations are common. Some of the longest periods of satellite data extend back more than 40 years and have been reprocessed for use in the analysis of past weather events and climate studies. In analyzing these and all remotely sensed datasets it is important to be aware of changes in instrumentation, orbit characteristics, and data processing over the lifetime of the dataset. The consequences of changes in observational satellite data are discussed in the following chapters on temperature, precipitation, and land surface characteristics.

Numerical weather forecast models are also a ready source of gridded datasets created in the processing of observations to create initial conditions (this process is called data assimilation and is discussed in detail in Chapters 3 and 11). The assimilation systems of several atmospheric models have been run retrospectively using historical observations to provide estimates of the three-dimensional structure of the atmosphere, and in some cases, the upper ocean, for periods dating back several decades. These datasets are influenced by changes in the observational data being assimilated. The inclusion of satellite data into forecast model data assimilation starting in 1979 has a documented impact on the model-based reanalyses for periods before and after inclusion of these data (Kalnay et al., 1996). While it is tempting to treat the retrospective model outputs as observational data, it is important to remember that these are in themselves model-based analyses with limitations and uncertainties inherent in the input data as well as limitations in the model's abilities to replicate the real world.

2.2.3 Step III – Apply the Appropriate Analysis Tools

Climate analysis often begins with a search for appropriate data that have the longest period of record for the station(s), area, or regions of interest. Implicit in the choice of appropriate data is that they have been, or can be, corrected for biases and errors, as well as changes in instrumentation, location, observational practices, and the sampling environment. Simple visualizations, such as time series, scatter diagrams, or maps, should be employed as a first step in any analysis of climate data. Often these will identify errors and artifacts that, if left unattended, will lead to confusing and/or misleading analyses. For example, Figure 2.2 illustrates an uncorrected bias probably due to a change in location or instrumentation, and Figure 2.3 exhibits an outlier that could represent random error, or a rare event. These simple plots will provide some insight and "feel" for how the data behave and might provide hints of where and what to look for in the subsequent analysis. They should also occasionally be redone to identify the influence of any subsequent data processing or data filtering as the analysis proceeds.

The total observational period and sampling frequency of the data will place limitations on the choice of the appropriate analysis tool and on the interpretation of its output. The statistical significance (see Appendix A) of analysis results depends on many things, including the number of independent samples in the dataset being examined. This is true when computing simple correlations between a pair of time series, as well as in the application of more complex statistical tools, such as spectral analysis

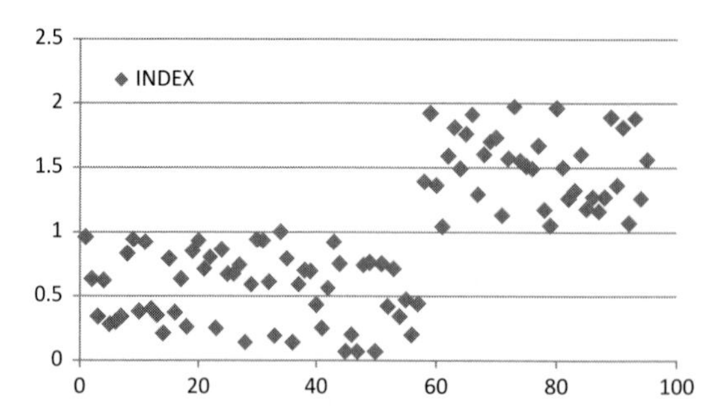

Figure 2.2 This time series of a hypothetical climate index that experienced a "step-jump" because of a change in instrumentation, location, or elevation is an example of an artifact in the data identified through simple graphical plots.

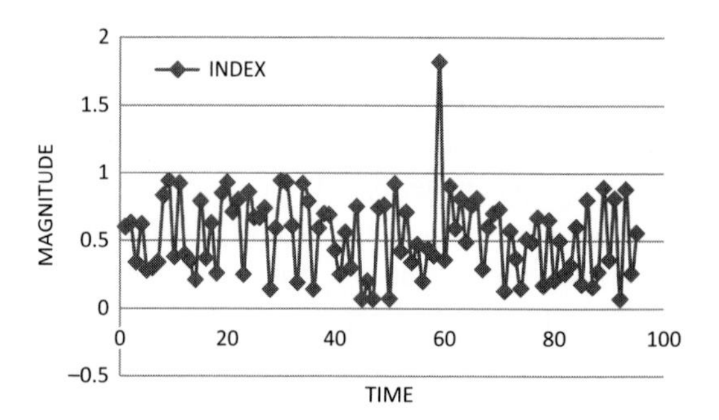

Figure 2.3 This is an example of an outlier in the time series that is 2.5 standard deviations from the mean. The outlier could be an error or an extreme event. In either case, outliers like this are closely examined in routine examination of datasets.

or Empirical Orthogonal Functions (EOFs), both discussed in the following and, in more detail, in Appendix A. Spectral analysis, for example, is often used to identify periodic or quasiperiodic behavior. The spectral analysis will, in principle, identify periods ranging from the total length of the time series being analyzed to twice the period at which the data were sampled. In practice, resolved frequencies will depend on exactly how the sample spectrum is computed. For example, a spectral analysis of 30 years of monthly data will try to resolve periodic behavior with periods ranging from 30 years to 2 months. In this hypothetical case, spectral analysis can identify the annual cycle because in 30 years of monthly data there are 30 independent samples. It will provide some indication of biennial variability based on 15, or fewer, independent samples, but it cannot provide much reliable information on decadal or longer variability since there are only three or fewer 10-year samples in the 30-year period. Thus, spectral analysis of a 30-year dataset is not an appropriate tool for studies of decadal variability. Spectral analysis of monthly data is also not appropriate for studies of phenomena with periods of less than two months, such as the intra-seasonal variability discussed in the next chapter.

The limitations discussed in this section are easily seen in the spectral analysis example given here, but they apply just as much to other analyses of a dataset, including simple correlations, the whole family of linear decompositions such as EOF analyses and Principle Component Analysis (PCA), described in Appendix A, and all other analyses no matter how sophisticated. Any-and-all statistical tools can produce results that may not have any physical significance. The interpretation of statistical analyses may also be limited because in any climate dataset the number of independent samples is likely to be less that the number of data points because adjacent data points are often highly correlated with each other. (See Appendix A for a further discussion of this and other statistical topics.) The fewer the number of independent samples the larger the uncertainties in the analysis. Once there are only a few independent samples, as in the case of decadal variability in the example above, there is little choice but to analyze the data as a collection of case studies.

Likewise, the spatial dimensions of the phenomenon being studied relative to the number of spatially independent samples in the domain covered by the data must be considered. For instance, the mean sea surface temperature (SST) on spatial scales of a couple of degrees of latitude and longitude over much of the world's oceans may be correlated over distances of hundreds of kilometers in large areas of the ocean (Reynolds and Marsico, 1993). Thus, an analysis of monthly mean SST that covers a spatial domain of a few adjacent 2.5° latitude/longitude grids in these areas will not contain spatially independent data over the relatively small domain.

In addition to the statistical tools discussed in this section, many climate studies use composite analysis, also referred to as superposed epoch analysis (see, for example, Reed and Recker, 1971; Rasmusson and Carpenter, 1982), in which averages over several samples are keyed to indices or occurrences of events. For example, averages of rainfall in an area may be keyed to some threshold value of a sea surface temperature index for some location. If the analysis is to further our understanding it is essential that the index chosen to form the composite be consistent with the physical mechanisms hypothesized to be associated with the phenomenon.

Much contemporary climate analysis is performed using tools found in various commercially available or institutionally developed software packages. A decreasing number of climate researchers continue to write climate analysis programs from scratch. Proficiency in computer programming, once a requirement for such work, is rapidly becoming a rare skill. Some proficiency with programming is, nonetheless, advantageous to ensure that the climate analyses are not artificially constrained by the limitations of the tools in the available software.

2.2.4 Step IV – Critically Interpret the Results of the Analysis

The purpose of climate analysis is to better understand how the climate system, or some part of the climate system, works. Continuing analyses provide suggestions that point the way toward more complete understanding and thus move climate science forward. It is important that the analysis include discussions of any implicit assumptions, as well as uncertainties in the analysis and alternative interpretations.

The implicit and explicit assumptions in the analysis constrain the extent to which a hypothesis may be tested with a particular dataset. These assumptions relate to the representativeness of the data, the ability of the instrument, or the model output, to provide realistic values of the climate variable(s), and the suitability of the spatial and temporal domain.

Uncertainties in the data include those related to accuracy and precision in the instrumentation as well as those associated with bias corrections, adjustments in the data associated with changes in the location of instruments, changes in instrumentation and observing practices, and in the sampling environment. If numerical model data are being analyzed, then changes in the data being assimilated by the model, the model parameterizations, as well as the model domain and grid size (discussed further in Chapters 3 and 11) are also important. If satellite data are analyzed, changes in satellite instrumentation, orbital characteristics, pixel and/or grid size, and in the algorithms applied to the raw data should be noted and, as with the model data, the magnitude of the influence on the analysis documented or considered.

All analysis tools are associated with a set of explicit formal assumptions. These assumptions should be interpreted in the context of the phenomenon being studied and the data being analyzed. Statistical analysis tools, for instance, usually include assumptions about stationarity.[7] In climate change studies, on the other hand, the analysis may be aimed at documenting the extent to which the data are nonstationary. While a complete discussion of statistical techniques and specific guidance for interpretation of these analysis tools is beyond the scope of this book, Appendix A has a general discussion of the analysis tools and factors to consider before application of some of the more common statistical tools.

In addition to reading the documentation and becoming familiar with the reference material associated with the analysis tools, many experienced researchers will test the analysis packages to ensure that they perform as expected. Testing the analysis software on data for which the outcome is known is one method to become familiar with what the software does. For example, a spectral analysis program can be expected to identify the annual cycle from several years of daily or monthly data.

Statistical tools will do what they were designed to do. EOF, Principal Component Analysis, Factor Analysis, and the like will produce coherent spatial patterns even from an analysis of fields of random data. Off-the-shelf analysis tools may be easy to use but the proper use of the tools is often more difficult (Wheelan, 2014). Any time-series can be decomposed into sines and cosines but these periodic functions may or may not have physical significance. Identifying a pattern of periodicity is not enough. Spatial and temporal patterns found through sophisticated use of analysis tools may account for a very small percentage of the total variation in the dataset and, thus, not have much practical significance. A more detailed account of the capabilities and limitations of data analysis tools is found in Appendix A.

[7] A stationary dataset is one where the statistics of subsets of the data are the same as those of the full dataset. For instance, the mean and standard deviation of one 30-year sample in 100 years of data will be essentially the same as any another 30-year sample.

It is relatively easy to find and use statistical analyses packages leased or maintained at most research institutions. In addition, open access sources of statistical routines are fairly common. Thus, we are often relieved of the burden of programming and testing our own computer programs. This does not relieve us of the responsibility for testing and understanding the analysis tool in the context of the problem at hand. After reading the documentation that accompanies a particular analysis tool a good practice before application of that tool is to expend some effort understanding its characteristics. A relatively easy way to do this is through construction and analysis of synthetic data whose characteristics are known. For example, a time series generated using one or more sine functions plus some random noise should yield no surprises from a spectral analysis of that series. Insight on the characteristics of an analysis package could also come from analysis of a computer-generated pseudo-random time series compared to a number of pseudo-random time series that have various levels of lag-one autocorrelations. The testing exercise also provides experience in interpretation of the output from the statistical or other analysis package.

It is as important to understand the characteristics, advantages, and limitations of the statistical or other methods being employed as it is to understand the limitations of the data being analyzed. Simple testing of the analysis techniques itself, as discussed earlier, provides some guidance on whether that particular technique is appropriate to the problem at hand. A good rule of thumb when looking at the first results from an analysis scheme is to check for mistakes or errors. Most researchers have at some time during their careers been extremely enthusiastic about some analysis results only to find on closer inspection that the results were in error or a misinterpretation. Often, the temptation to go directly to interpretation of the data at hand is too great. Interesting results are no excuse to skip the testing stage, and analysts do this at their own peril.

2.3 Analysis of Data from a Single Location

The analysis of data from a single location or a group of observing stations may be appropriate for many variously motivated climatological studies. For example, studies intended to characterize the typical weather and climate at a location to support planning and design of some new activity. In some analyses, data from a single location are used to characterize the climate of an extended area. In such cases, the local environment of that station must be considered. Some locations are particularly sensitive to relatively small changes in location. For example, a few tens of meters can make a big difference in the microclimate in mountainous terrain or near bodies of water. Very often the stations that have the longest and most complete data records are not optimally sited to characterize the climate for the entire region. In these cases, the longest period of record from a single station can be compared to data from other stations with shorter records. In this way data from several stations with less than complete data records might be useful in regional and global climate analysis. Data from single stations can also be integrated with data from other sources, such as estimates derived from satellite or other remotely sensed data, to get a better description of the spatial patterns of the climate in a region.

The oldest station-based instrumental temperature data are likely those for central England dating back to the 1600s (Manley, 1974). Interpretations from tree rings, ice cores (discussed further in Chapter 9), corals, and other noninstrumental records provide temperature and precipitation proxies dating back hundreds to thousands of years. These records provide invaluable information about past climates at a location and may provide some information about the climate over larger regions. Interpretation of these single location records requires great care to separate the variations associated with artifacts in the local measurement due, for example, to growth of a human settlement near the site where the data are gathered from those associated with variations in climate. Since this is not always possible, uncertainties in interpretation of single station climate records are common. Before embarking on the climate analysis of data from a single location it is necessary to examine the metadata.[8] accompanying the data and, if the metadata are found wanting, to systematically examine the record for artifacts that may have arisen from a variety of sources not related to climate. Sources of non–climate-related variations in the record include changes in location, instrumentation, or environment, discussed in the next section.

2.3.1 Changes in Instrument Location

Observations from weather stations are often used in climate analyses. Individual weather stations are usually identified by name and/or number. The World Meteorological Organization (WMO) lists all stations that are available for regional or international data exchange by both a name and a unique number; for example, Atlanta Hartsfield International Airport is 722190. (Note that in the WMO numbering system the first two digits refer to the country where the station is located.) The name and identification number for a station on the WMO list, or any other similar list, do not, however, guarantee that the station has remained in the same exact location for the entire period of record. It is common to find that instruments have moved to accommodate practical and operational changes at airport locations, for instance (See Box 2.2). The instrument location changes are often documented, but may not be clear in the data record, underlining the need to examine the metadata describing how the data were taken and any changes associated with the station. Examination of the metadata and interpretation of how changes may have biased the climate record can be tedious and time consuming, especially if data from several stations (or hundreds of stations) are being considered. Fortunately, many of the station datasets available from archival centers, for example at the US National Centers for Environmental Information (NCEI), formerly the National Climatic Data Center (NCDC), have been examined in detail to

[8] Metadata provide information about the data. This may include the latitude, longitude, elevation, type of instrumentation, observational practices, and any changes in these over the period of record as well as other information that may be relevant for interpretation of the data.

Box 2.2 Large-Scale Influence of Undocumented Changes in Station Location

The bias that can arise in a climate analysis from incorrect instrument location and height can be illustrated by an example in the late 1980s when routine real-time monitoring of monthly climate anomalies was first attempted. During those early years one of us (CFR) was tasked with developing maps of global land surface temperature anomalies, based on monthly data exchanged over the Global Telecommunications System (GTS). It was soon noticed that the monthly mean temperatures from the South Atlantic showed large positive anomalies compared to the historical mean temperatures that were being used as the climatological base period.

There are only a handful of islands in the South Atlantic, and thus very few permanent observing sites. The 1980s observations, which were thought to be from St. Helena (15.9°S, 5.7° W, 436 m elevation), were the only observed data available in real time to represent the temperature over a large area of the South Atlantic. Real-time monthly temperature anomalies were based on the long-term mean from St. Helena.

These anomalies influenced the analysis over a large area, at the same time that scientists were becoming aware that the global climate might be changing on human timescales. The question arose as to whether the island observations were signaling an early and large Southern Hemisphere warming. Metadata were not easily found, since responsibility for the observations on the island changed several times. Responsibility for the observations was shared by the United States and Royal Air Forces during the World War II era, then passed to NASA in the 1960s, reverted to the British during the Falklands War. Since then, these observations have been taken by temporary contract employees.

After some effort, it was discovered that the real-time observing site was actually at Wide Awake Field, on Ascension Island, 7.9°S, 14.4°S, elevation 79 m, with irregular observations available from the 1950s to 1980 and more regular observations from 1981 forward. Since the mean atmospheric temperature in a moist tropical boundary layer decreases with height at a rate of several tenths of a °C per hundred meters, a difference in observation height of 357 m between the actual and assumed observation site led to a systematic increase in the observed temperature of about 2.5°C at Wide Awake Field compared to the mean from the St. Helena record. It is likely that other changes occurred at this observation site, limiting the usefulness of its data for climate analysis. These early estimates of global and regional surface temperatures were based solely on land observations. Since then sea surface temperature observations over the South Atlantic Ocean and other ocean basins have been incorporated into the global surface temperature analysis through use of marine surface air temperature based on ship, buoy, and satellite observations. Without these observations, our estimates of global temperature variability and change are considerably limited.

identify station location changes and their effects on the observational record. Documentation that accompanies the data provides details of any corrections applied, allowing users to assess whether to accept the archive center's corrected data or to embark on their own analysis using uncorrected data. Very often the station data from the archive centers are also made available in their "raw" form, giving the opportunity to deal with station changes as appropriate for a specific analysis.

2.3.2 Changes in Instrumentation

Meteorological instruments have changed quite significantly since the middle of the nineteenth century, when observations of weather started to become somewhat routine. Even the instruments for the most basic of climatological measurements, daily temperature and precipitation amounts, have evolved, with the most rapid changes occurring within the past 50 years. Initially, the daily maximum and minimum temperatures, along with the temperature at observation time, were read once each day using alcohol or mercury in glass maximum/minimum thermometers. While many volunteer (cooperative) observers in the United States and elsewhere continue to make these observations in this manner, temperature observations by many national weather services are now based on thermistors that provide continuous data throughout the diurnal cycle. Daily precipitation observations have likewise evolved from a simple "bucket and measuring stick" to automated tipping buckets and laser-based technologies. The introduction of new instrumentation provides more accurate and precise measurements but may also introduce biases and discontinuities in the climate record. Most modern instruments are designed to optimize daily weather observations to support the operational needs of aviation, agriculture, water management, recreation, manufacturing, and business as well as the general public, and not necessarily with a priority for producing a stable climate record. The data archive centers generally strive to provide metadata documenting instrument changes and, when possible, estimates of biases in the climate record associated with instrument changes. Even if the magnitude of a bias introduced by an instrument change has not been documented, awareness of when changes in instrumentation occurred could help to avoid errors in interpretation of the climate record based on data from these instruments. The importance to the analysis of the bias introduced by an instrument change should be assessed in the context of the magnitude of the changes important in the purpose of the analysis.

2.3.3 Changes in the Instrument's Environment

Increases in population, changes in land use, and other factors often affect an instrument's environment. These processes can occur slowly and are, perhaps, more difficult to quantify than the changes in location and the instrument itself discussed earlier. Questions like "When did that tree grow to be large enough so that its shade started to influence the observed temperature?" or "When did those buildings start to influence the air flow around the rain gauge?" are not easy to resolve. Comparisons between the data from the observing site in question and another nearby site can provide some estimates of local environmental influences, provided, of course, that the comparison site doesn't have changes. Whether the changes in the local environment render the data inappropriate for climate analysis depends on the goals of the analysis. Climate change since the nineteenth century may be difficult to establish from data taken at a site that has experienced significant changes in the local environment over the period of record. For example, the site with one of the longest temperature records in Melbourne, Australia, sat on a small traffic island at a major intersection in the downtown area, making it difficult to use data from that instrument for detailed studies of century-long climate change. On the other hand, that site may be ideal for studies of urban climate variability compared to rural sites and to contrast the urban climate over recent decades with the climate in the early years of the record when the site was relatively free from urban influence. As with changes in location and instrumentation, metadata describing major changes in the local environment are extremely important aids in the interpretation of the climate record. Even if the biases associated with environmental changes can't be quantified directly, knowing when in the record these changes took place may serve as markers for where to look for data shifts that may not be associated with climate change.

2.4 Analysis of Data Fields

The earliest climate studies could only rely on data from fixed observing sites over land and from ship data taken mostly from well-traveled routes on the world's oceans. It is a testament to the skill and insight of eighteenth- and nineteenth-century scientists that they were able to deduce many of the major features of the global wind patterns from these limited observations. The advent of routine satellite observations and development of numerical model-based analyses, starting in the second half of the twentieth century, now provide incredibly rich data sources for climate studies. These datasets have revolutionized climate analysis, but they are subject to their own set of limitations. Considerations for the use of data fields generated by models and satellite observations in climate analysis are discussed in the next section.

2.4.1 Climate Analyses Based on Numerical Models

The earliest successful numerical (computer) model of the general circulation of the atmosphere in a climate framework was developed by Phillips (1956) and implemented at the Institute for Advanced Study of Princeton University, based in part on his Ph.D. dissertation work at the University of Chicago. Much of the development of numerical models in subsequent years was concentrated on improved short-range weather forecasting. Simulation of climate, however, continued at the Geophysical Fluid Dynamics Laboratory (GFDL), also at Princeton (Manabe et al., 1965) and at the National Center for Atmospheric Research (NCAR; Kasahara and Washington, 1971). As discussed in Chapter 3, one application of such a numerical model of the atmosphere is to act as a sophisticated analysis system that is constrained by the physics of the atmosphere as represented by the equations constituting the model.

Model-based analyses are appropriate for investigations of large-scale regional and global climate phenomena. Their utility for studies of climate for a specific location is limited, however, since the smallest area represented by a model analysis is tens to hundreds of square kilometers. Several of the large climate centers such as GFDL, NCAR, and ECMWF have long-term goals to eventually reduce the model grid scale to the order of a few kilometers or less. No matter the size of the model grid the values produced by the model are fundamentally different from those taken at any specific location, e.g. a weather station, within the grid area. Whether it is appropriate to use the analysis from a model to study climate depends on whether the aim of the study is to characterize the area or whether the climate at a specific location is the goal. Interpretation of model analyses in regions with grid sizes too coarse to resolve changes in topography within the grid can be especially challenging. For example, the observed mean elevation of the snow line at a specific ski lift site will not be the same as the single grid value for snow elevation inferred from a model. Some reanalysis-based climate studies determine statistical relationships between the values at a model grid and stations within the grid. These statistical relationships can be used for estimating local values in forecasts or for estimation of values where local data are no longer available or not available for real-time climate monitoring. A detailed discussion of model-based analysis and reanalysis can be found in Chapter 11.

2.4.2 Climate Analyses Based on Satellite Observations

Earth-orbiting artificial satellites provide routine global data in support of numerical weather prediction and have become very useful for the analysis of the global climate system. Daily and higher frequency estimates of sea surface temperature and sea ice over the world's oceans and precipitation over nearly the entire globe (discussed in Chapters 7 and 8) provide unprecedented climate analysis opportunities. Satellite data have also been used to characterize vegetation and other land features as well as tropospheric temperature and moisture. Many datasets derived from satellite data are archived and available for climate analysis from NOAA and NASA centers listed in the References.

Considerable efforts have been made to ensure that artifacts in the record due to changes in instrumentation, orbital characteristics, and sampling strategies have been removed from the archived satellite data record, but some uncertainties remain. Documentation describing the bias and other corrections applied to the satellite data record is often available and should be consulted when using these datasets. While we routinely refer to satellite "observations" of, for example, sea surface temperature, moisture, and vegetation, the basic observations are from instruments that measure reflected or emitted radiation in some part of the electromagnetic spectrum. Thus, all the satellite-derived climate variable observations are estimates based on algorithms that transform the observed radiation into a weather or climate variable.

Satellite data share a characteristic with the numerical model data in that they both represent averages over some area or samples within a grid area. On the one hand, this makes comparisons between numerical models and satellite data straightforward, but, on the other hand, the satellite estimates are not directly comparable to *in situ* observations at a location, sharing the same limitations as the model-derived data discussed earlier. The spatial resolution of the satellite instrument sensor defines the raw satellite data. Observations from different sensors are handled differently, but in general a usable data value, or pixel, is formed from one or a number of individual data points. Often the satellite data pixels are averaged into grid areas, comparable in size to the grids of many models. In some cases, rather than averaging the pixels to form a grid mean, data within a grid area were sampled over a time period to avoid cloud contamination over an area. In these cases, the satellite observation has a higher resolution than the nominal size of the observational grid but the location of the representative pixel within the grid is unclear. Thus, successive satellite observations for the same grid may be from different locations within the grid.

Satellite data provide a valuable resource for climate analysis from 1979 forward. Some of the earlier satellite data starting in the 1960s may also be useful for some climate studies but require more care and interpretation than the later data.

2.5 Budget Analyses and Other Climate Studies

In addition to the basic observed quantities, climate analyses may also include investigation of atmospheric momentum, energy, radiation, and water budgets. Budget studies investigate the climate system from the standpoint of conservation of mass and energy (Piexoto and Oort, 1992; Fasullo and Trenberth, 2008a, 2008b). Budget analyses, ideally, account for all sources and sinks of a quantity. For example, global water budget studies attempt to track precipitation, evaporation, river discharge (streamflow), and changes in storage in the atmosphere, in lakes, ponds, and rivers as well as groundwater below the surface. The domain of the budget computations need not be global, but then fluxes of quantities into and out of the analysis boundaries need to be included (Rasmusson, 1967; Ropelewski and Yarosh, 1998).

Budget analyses may reveal insights into the detailed behavior of the climate system, but are often limited by the lack of observational data for all components of the budget. While budget calculations based on model output are complete, they often reveal more about the workings of the model than of the actual climate. Budget calculations based on observations, on the other hand, are rarely complete, and usually (at least) one quantity must be estimated as a residual.[9] (See Appendix D for an example of the calculations used in water budgets.), The residual in many budget computations is often large relative to the measured quantities, which are, in themselves, uncertain to varying degrees. Budget calculations are often very helpful in providing estimates of quantities that are difficult or impossible to measure, especially for large areas. Examples of such quantities include vertical motion in the atmosphere, evaporation over the oceans, and evapotranspiration over land.

[9] A residual in a budget calculation is the estimated value of one term, given observed values for all the others.

New and innovative methods to analyze climate variations are likely to emerge as higher spatial and temporal resolution data become available for longer durations. Daily, and for some datasets hourly, data can free climate analysis from the artificial constraints in the historical record that traditionally have limited analyses to the calendar month and seasonal boundaries. As these higher resolution datasets become available we can expect future climate analyses to more realistically include climate phenomena that span these artificial boundaries. Daily data can be used to investigate the ways in which probability distribution functions, or histograms, of daily weather may be modulated by some modes of climate variability. For instance, ENSO is related to wetter and drier seasonal rainfall in different parts of the world, but for most locations influenced by ENSO we have not yet investigated the changes in the daily frequency and intensity of rainfall. In other words, is a seasonally wetter (drier) region wet (dry) because it rains more (less) frequently or because the intensity of the rainfall is (less) greater when it rains? Similar studies of the modulation of daily weather by other modes of climate variability are likely to have a great practical utility.

2.6 Summary

This chapter outlines the fundamental principles of climate analysis that will be referred to, illustrated, and reinforced in the chapters that follow. Just as "[T]he purpose of computation is insight not numbers" (Hamming, 1986), the purpose of climate analysis is improved understanding of the climate system and not mere documentation of statistics. The goal of climate analysis is to uncover the underlying signal in very noisy data and thus contribute to a better understanding of how the climate system works and how we might arrive at better climate forecasts. Climate analysis is not simple because the climate system is not simple, but adherence to the principles discussed here provides a frame of reference that will aid in furthering understanding. Following the basic steps introduced and discussed in this chapter,

- clearly identifying the purpose for the analysis;
- selecting or developing data appropriate to that purpose;
- applying the appropriate analysis tools; and
- critically interpreting the results of the analysis,

will not guarantee a successful outcome of the analysis, but will help to minimize errors, quantify uncertainty, and increase the probability of achieving an improved understanding of the climate system.

Suggested Further Reading

UCAR Climate Data Guide – Available on line at https://climatedataguide.ucar.edu (Comprehensive data discovery tool, listing 200 datasets used in climate analysis, including detailed dataset descriptions and further discussions of climate analysis basics.)

Wheelan, C., 2014: *Naked Statistics: Stripping the Dread from Data*, 282 pp., W.W. Norton and Company. (A book on statistics for the nonexpert).

Wilks, D. S., 2011: *Statistical Methods in Meteorology*, 3rd edition, Academic Press, Elsevier, 704 pp. ISBN-13: 978–0–12–385022–5. (A standard reference book for the use of statistics in meteorology and climate studies).

Questions for Discussion

2.1 How would you go about examining the observational record for changes in instrument location at, e.g., the local airport?

2.2 How would you search for changes in bias associated with changes in instrument location, type, and environmental factors?

2.3 If you were traveling to a location would whether the historical occurrence of rainfall looked like curve (a) versus curve (b) in Figure 2.1 (where the horizontal axis is taken to be rain amount and vertical axis frequency of occurrence) influence whether you pack for wet or dry conditions?

2.4 What are the characteristics of the data that would be required to examine variability in the mean diurnal cycle of precipitation in January and July at one location? For the entire United States?

3 Climate Analysis
Atmospheric Instruments, Observations, and Datasets

3.1 Introduction

We have described the nature of the climate system, including the components of which it is comprised (Chapter 1), and the methods scientists use to examine and understand the system (Chapter 2). In this chapter, we will describe the atmospheric observations, many of them obtained in support of weather forecasting, used in climate analysis, along with the instruments used and the methods employed to process the observations into global datasets. We emphasize that until very recently no coordinated atmospheric observations were made specifically for climate, and even now the majority of observations of the atmosphere are taken for other purposes. Observations made specifically to monitor the climate and the datasets derived from them will be described in later chapters: temperature and precipitation in Chapters 6 and 7, respectively, ocean observations in Chapter 8, cryosphere in Chapter 9, and land surface in Chapter 10.

Many applications of climate analysis require datasets that include a complete depiction of the atmosphere, including full spatial coverage of the basic atmospheric parameters extending upward to a height sufficient to describe the important weather and climate variability and covering the time period of interest. We refer to these as four-dimensional datasets, in which horizontal position, altitude, and time are needed to locate the data values. The separation of the data locations in space and time is referred to as the resolution. For weather and climate applications, the parameters required are motion (wind), both horizontal and vertical, temperature, pressure, geopotential height, and moisture. It is, of course, not possible to observe each atmospheric parameter at all the locations that would be needed, so methods have been developed to combine the observations that we do have into more complete datasets.

Atmospheric datasets, in combination with mathematical models based on the governing equations of the atmosphere, can be used to forecast future weather, and were prepared by meteorologists long before any effort to describe and understand climate variability and change began. Since observations of weather parameters were initially available only at the surface, and only where weather stations were located, and since complete maps of at least some variables are needed for forecasting, procedures for estimating values in other locations were developed. These techniques resulted in contour maps of the atmospheric data that were of greatest utility to the forecaster. For example, a map showing lines connecting locations where the atmospheric pressure is the same (such lines are called isobars) can be used to identify areas of relatively high and low pressure. A sequence of such maps can depict the movement of those pressure systems, and extrapolation of the movement can provide a forecast. As introduced in Chapter 1, the process of creating contour maps of atmospheric pressure and other variables was, and still is, called analysis. Initially, maps were prepared by manually drawing lines on paper base maps, but eventually digital datasets, with values on regularly spaced grids, became the standard.

Increasingly accurate and complete datasets, or analyses, are necessary for making improved forecasts, and so the effort to obtain more plentiful, more widespread, and more accurate, observations and to develop better analysis techniques has been a continuing theme throughout the history of meteorology and weather forecasting. We will discuss the history and current status of meteorological observing systems in Sections 3.2–3.4. Then Section 3.5 will address the history of the development of atmospheric analysis techniques and describe the current state of atmospheric climate datasets, including reanalyses, which are created by applying an unchanging analysis system to a collection of observations covering an extended time period.

Observations of atmospheric parameters presumably began in concert with the development of agriculture, when humans first began to live in fixed locations and noticed that weather changed in interesting and relevant ways. The history of weather observations is one of continually improving instruments, capable of observing more parameters more accurately in more locations, and greater reach, enabling observers to collect information in more and more locations, including higher in the atmosphere. These advances were driven firstly by the need to produce better manual forecasts, which benefitted from more accurate and complete maps of current conditions, and then by the need to define the initial state of the atmosphere for mathematical models, which eventually proved to provide superior forecasts.

3.2 Direct Surface Observations

The earliest weather observations were made at the surface of the Earth, and surface observations continue to be important. Air temperature, precipitation, atmospheric pressure, winds, and humidity are all observed at many sites and are used in many ways: to provide current weather conditions for the public, to help with very short range forecasting for the local area, and to contribute to atmospheric analyses that serve as initial conditions for forecast models. Such surface observations contribute to climate data sets in two principal ways: through statistics calculated for a single location, and as inputs to analyses covering larger areas.

Most observations made at surface stations are on land, and one important concern with their use in climate datasets is that they are influenced in varying degrees by the presence of the land surface (Chapter 10). This is less of an issue over water, but fewer observations are taken there. Climate datasets in the form of statistics calculated from a single station are generally used without adjustments for factors such as elevation or land surface characteristics, but changes in station location or instrumentation or urban development must be taken into consideration when using such data in climate studies;

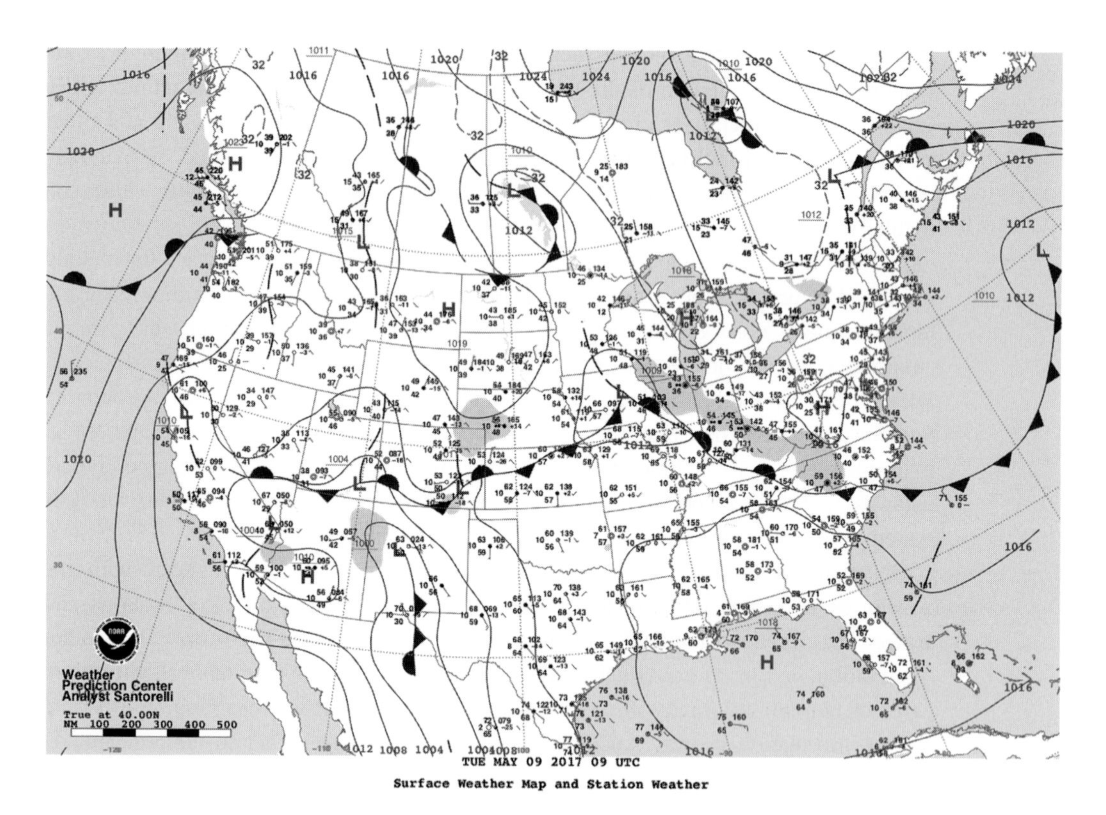

Figure 3.1 An example of the surface weather map showing isopleths of sea level pressure, frontal systems, and observational data at US National Weather Service (NWS) sites for May 9, 2017 at 09 UTC.
Source: National Weather Service/NOAA.

this is particularly important for surface air temperature data (see Chapter 6 for more details). Procedures have also been developed to adjust surface observations to make them more consistent with free atmospheric values in order to facilitate their use in modern analysis methods.

Temperature, precipitation, atmospheric pressure, horizontal winds, and humidity are all observed at surface stations and used to create surface-only analyses (Figure 3.1). Temperature and precipitation analyses are discussed in greater detail in Chapters 6 and 7. Station pressure is observed using barometers and converted to sea level pressure using an equation that requires the station temperature, thus making possible the creation of maps of sea level pressure that show high and low pressure systems using consistent elevation values. Early sea level pressure maps enabled weather forecasts to be produced by extrapolating the movement of low-pressure systems associated with storms. However, modern atmospheric models and data assimilation systems (discussed in Section 3.5) often use a vertical coordinate called sigma, defined as the ratio of the pressure to that at the surface.

Because systematic meteorological measurements and forecasting began first in Europe in the middle latitudes, surface wind measurements proved less essential than pressure observations. This is because the large-scale wind field, which is more relevant than is local wind to forecasts, is strongly constrained to flow along isobars in middle (and high) latitudes, making maps of sea level pressure more useful than maps of surface wind speed and direction, which are often influenced by the local topography. Observations of surface winds are important, however, since surface wind forecasts are widely used for many applications and observations are used to develop statistical corrections for model forecasts.

Surface winds are measured using instruments called anemometers that can take various forms. Most direct applications of surface wind observations or of wind forecasts make use of the speed and direction, which may be combined to form the velocity. The velocity can be easily converted to two values that represent the components along two perpendicular directions, usually east–west, referred to as the zonal wind component, and north–south, referred to as the meridional component, and can be used in and forecast by atmospheric models. Other decompositions of the winds are discussed in Chapter 4 and Appendix B. Atmospheric models are more skillful in predicting winds above the surface, and both observations and forecasts of surface winds are used in conjunction with secondary models, generally referred to as boundary layer models, that include the effects of surface friction on the air motions near the surface.

Humidity is a measure of the amount of water vapor in the atmosphere and is quantified in various ways. Absolute humidity is the mass of water vapor in a specified volume in the atmosphere. Atmospheric models make use of the specific humidity, which is the mass of water vapor in a unit mass of moist air. The most familiar measure to most people is relative humidity, which is the ratio of the observed specific humidity to the saturation specific humidity at the observed temperature and pressure. Water vapor is a relatively minor atmospheric constituent, but is significant for a number of reasons: It is highly variable, with concentrations ranging from very near zero at high altitudes and in cold conditions to as much as 4% over the warmest ocean waters, and through evaporation, condensation, and precipitation it plays a crucial role in the development and life cycle of thunderstorms and other important weather phenomena.

Humidity is important for many purposes but is very difficult to observe, analyze, and predict. In a laboratory, the absolute humidity can be determined by carefully measuring the temperature and pressure of a volume of air and then extracting all of the water vapor from the volume, usually through condensation. Both specific and relative humidity can then be calculated from the absolute humidity and the observed temperature and pressure. However, this laboratory procedure is impractical for weather- and climate-observing systems that require frequent observations at many locations. Instead, humidity is most often calculated from observations of dewpoint (or frost point for temperatures below freezing). The dewpoint is the temperature at which a parcel of air at constant pressure reaches a relative humidity of 100% and moisture begins to condense, as often happens at the surface on cold nights to form dew or frost. Manual observations have generally relied on an instrument called a psychrometer, which measures the air temperature and the "wet-bulb" temperature to which evaporation cools a collocated thermometer. Dewpoint is then calculated from the air temperature and wet-bulb temperature. Modern automated weather stations use a chilled mirror to directly determine the dewpoint temperature at which the water vapor condenses.

All humidity measurements near the Earth's surface are strongly influenced by the surface characteristics and other factors, and so analyses of surface dewpoint or relative humidity must be used carefully in climate studies. Statistics of humidity at individual stations can be useful in understanding long-term changes in near-surface atmospheric moisture, but the observations are very sensitive to changes in the local environment and instruments used.

3.3 Direct Upper Air Observations

While surface observations are important, observations of conditions in the free atmosphere well away from the surface are vital to understanding and predicting weather and climate. Forecasters learned in the first half of the twentieth century that changes in the movement and strength of weather systems could be understood much better when viewed from the perspective of winds and pressures in the mid and upper troposphere.

The need for upper air measurements was understood before there were practical methods to obtain them. Early observations taken on mountains showed that the atmospheric temperature and pressure decreased with increasing elevation. The decreases of temperature and pressure with height in the free atmosphere were confirmed in early manned balloon observations. It was, however, not until the mid-twentieth century that observations in the free atmosphere became routinely available. Before that time limited observations were made from aircraft, instrumented balloons, and kites. The earliest kite observations date to the late 1700s, but routine observations did not take place until a century later. Regularly scheduled kite observations of atmospheric temperature reaching to 4 km and above were taken at a handful of sites, primarily in North America and Europe (Ewen et al., 2008; Strickler et al., 2010). Recording instruments were attached to the kites and read after the flight was completed.

Upper air wind observations using un-instrumented balloons have been made in the United States and Europe since the early part of the twentieth century. These pilot balloons continue to have some specialized applications but have not been used extensively in climate analysis. The wind direction and speed are determined by changes in the viewing angles of optical tracking scopes, called theodolites, combined with the assumption that the balloons rise at a constant rate. The balloons do not carry instruments with them and thus do not observe wind values directly. Tracking with a pair of theodolites is often required to remove ambiguities in interpretation of the data. These shortcomings limit their potential use in climate studies.

Routine upper-air observations had to wait until the advent of instrumentation that was small enough, could be powered by lightweight batteries, and could transmit the observations back to the ground. Cost and simplicity were also factors. By the 1930s instrument packages called radiosondes were designed to meet these requirements

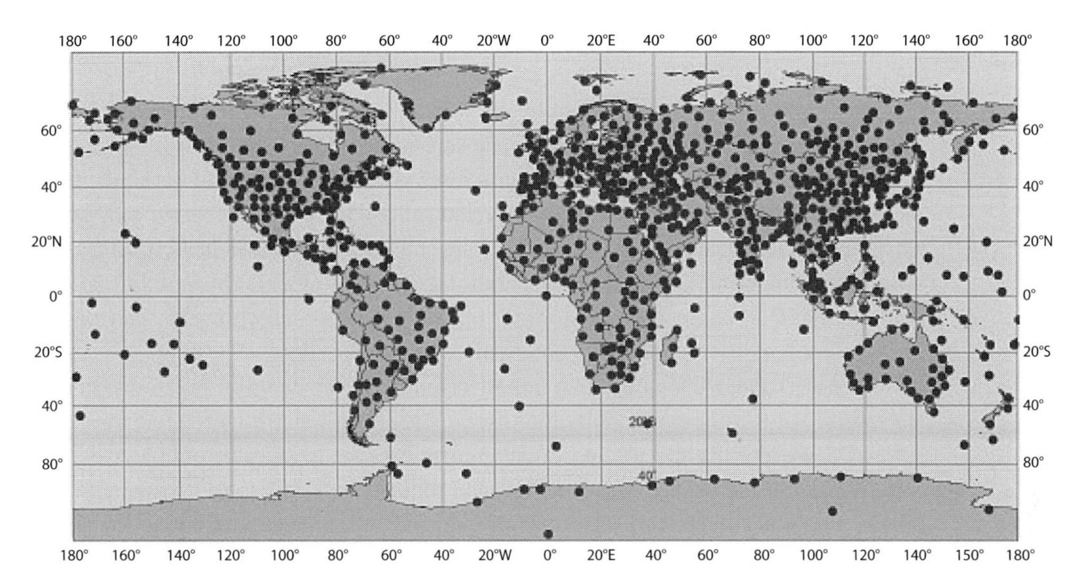

Figure 3.2 Location of radiosonde launch sites that form part of the Global Observing System. Source: National Weather Service/NOAA.

(see the website www.weather.gov/upperair/factsheet). Regular radiosonde temperature observations were inaugurated in the United States by the Weather Bureau in 1937. The addition of wind observations in 1942 was made possible by using radar to track the instrument package; sondes that observed winds in addition to temperatures are generally called rawinsondes. By the late 1940s a global network of sondes was established by international agreement under guidance of the World Meteorological Organization (WMO). This network has evolved considerably but remains the primary source of real-time *in situ* upper-air data, including moisture, wind speed and direction, pressure, and, of course, temperature. The development of instruments that can observe location directly using GPS (Global Positioning System) has made possible radiosondes that need not be tracked by radar. Sonde observations are taken twice daily (00 and 12 UTC) at 69 locations in the coterminous United States and at roughly 800 sounding sites globally, to support weather forecasts (Figure 3.2).

A global network of upper air observations, coordinated by the World Meteorological Organization, has been in place since the end of the 1950s. This network continues to operate and provides data of vital importance to weather prediction efforts around the world. Climate datasets based on radiosonde observations, like those derived from surface station observations, exist and are quite useful, but, again like surface observations, changes in instrumentation and gaps in data at many locations limit their utility for global climate studies.

Instrumentation has evolved over the years, introducing some time-varying biases in the historical records. The uncertainties associated with these biases are compounded by the use of sondes from different manufacturers in different parts of the network. Sondes transmit their data back to a receiving station while ascending into the stratosphere with a target level of 10 hPa (about 31 km) or higher. When the balloons burst at the top of their ascent the sonde instrument package descends on a parachute. The USA's instrument package contains instructions on how it may be returned to the National Weather Service for re-use. About 20% of the sondes are recovered.

Other upper air direct observing systems have been and continue to be used. All of these have been developed for special circumstances for which the basic radiosonde network is inadequate for some reason. For example, rocketsondes have been used to observe the upper atmosphere, since radiosondes cannot ascend above about 35 km. Aircraft flying through tropical storms or severe weather environments can deploy dropsondes that observe locations unreachable by radiosondes.

Piloted aircraft have long been an important source of upper air observations. In fact, the earliest regular upper air observations were made by pilots who were commissioned by the US Weather Bureau to carry an instrument package to altitudes of at least 13,000 feet. While the radiosonde network replaced those efforts, pilots of commercial and private aircraft continue to provide manual reports of upper air conditions for use in forecasting. Since 1979, automated systems that make and report observations of temperature, wind, location, and altitude have been available for commercial aircraft (Moninger et al., 2003). These observations have greatly increased in number since the early 1990s, and provide a very helpful complement to the radiosonde observations, since they are available along air routes over the ocean, where radiosondes are very scarce, and they provide very frequent sampling near major airports. They have not been used to form standalone climate datasets.

Another recent development in upper air observations is the implementation of remotely piloted aircraft, commonly referred to as drones. Drones have become very capable and affordable, and can be fitted with radiosonde-type instrument packages to take observations in situations of interest. They are not yet used in the creation of climate datasets, although proposals have been offered to create a network of high capability remotely piloted vehicles that would perform observations on a regular schedule over the entire globe (MacDonald, 2005).

3.4 Remotely Sensed Observations

All of the systems discussed so far observe values in their immediate vicinity. Such observations are often referred to as *in situ*, or on site, in contrast to remotely sensed observations, which are radiation measurements used to derive meteorological parameters. *In situ* observations are usually more accurate and precise than remotely derived parameters, since the latter require assumptions regarding atmospheric or surface radiative properties. The advantage of remote sensing is the ability to observe much more broadly than is possible using *in situ* instruments. Human senses provide an analog: touch and taste are *in situ* observations, requiring close proximity to whatever is being observed, while sight and hearing are remote sensing, greatly expanding our knowledge of the environment in which we live, but subject to greater uncertainty and potential error.

Atmospheric remote sensing systems are nearly all located on the surface of the Earth or on orbiting satellites. In this section, we will focus on remote sensing that provides observations that can be used in data assimilation systems to improve our knowledge of the climate of the atmosphere.

3.4.1 Surface Remote Sensing

The first surface-based remote sensing systems were people using their eyes to observe clouds, sunshine, winds, and precipitation, and those observations continue to be useful for weather nowcasting and forecasting. Such observations are difficult to quantify for use in climate analysis, and to use in data assimilation systems. Additionally, the unaided human eye has limited range and precision, and more powerful remote sensing systems have been developed. Cloud observations provide an example of the challenges associated with subjective observations. Regular weather observations all over the world included cloud type and amount until fairly recently when automated observing systems came into use. Datasets derived from these observations have been used to describe the climatology and variability of cloud type and amount (Warren et al., 2015). Surface observation-based cloud datasets are available for the period 1971–2009 under Surface-Based Cloud Observations listed in the References under websites. Changes in weather observing practices, particularly resulting from the advent of automated weather observing systems such as ASOS beginning near the end of the twentieth century, have made it difficult to extend the cloud datasets, and their utility for climate analysis is viewed as limited (Dai et al., 2006).

Radar (from RAdio Detection And Ranging) was originally developed during World War II to detect aircraft, and that remains a principal application. Radars transmit a beam of radio frequency radiation and detect the reflected energy. By pointing the beam in different directions and at different elevations and by measuring the strength of the reflection and the time delay, a three-dimensional picture of reflecting phenomena can be produced. The initial development and use of radar was motivated by the need to detect enemy ships and aircraft. It became quickly apparent that falling precipitation produced a detectable return, and following the war, research led to the development and use of radars that were optimized for the observation of precipitation. Observations of the shift in frequency of the radiation reflected by precipitation (or anything else) can be used to calculate the velocity toward or away from the radar, making wind estimates possible as well. Different frequencies of radiation are sensitive to different atmospheric parameters, and so radars that can detect clouds rather than precipitation, or winds more directly by measuring the reflections from small atmospheric eddies, have been developed.

At present, many countries have networks of surface-based radars and profilers, which are vertically pointing radars that measure updrafts and downdrafts through the depth of the atmosphere above the instrument. These networks provide estimates of precipitation intensity and location, and wind speed and direction. They contribute to the production of analyses of accumulated rainfall that contribute to climate studies (Chapter 7); but the wind observations must be used in conjunction with data assimilation systems to make them useful for climate analyses. So far, no large-scale climate datasets have been based solely on surface remote sensing systems.

3.4.2 Satellite Remote Sensing

Earth orbiting satellites have proved to be very well suited for remote sensing of the atmosphere and surface, both land and ocean. This came as no surprise – observations from towers, balloons, and rockets had shown the benefit of an elevated point of view, and early efforts to launch satellites included weather-related experiments. The first satellite optimized for weather observations was TIROS-1 in 1960, followed by several experimental satellites that tested a variety of instruments and then operational[1] satellites that provide data used in regular weather forecasting. Currently, a number of both operational and research[2] satellites are in orbit and providing data that can be used to describe the Earth's atmosphere and surface. Examples related to the observation and analysis of precipitation are described in Chapter 7, analysis of sea surface temperature in Chapter 8, and analysis of land surface properties in Chapter 10.

Satellites that support meteorological and climate activities fall mostly into two groups: low earth orbit (LEO) and geostationary orbit (GEO). The typical orbits of LEO and GEO satellites are illustrated in Figure 3.3. LEO operational satellites are often in orbits called sun-synchronous, which are at an altitude and an inclination[3] that allows them to travel close to both Poles and pass over the equator at the same local time on each orbit. This timing tends to lessen the impact of the diurnal cycle on the observations, making them easier to use in data assimilation systems. LEO orbits tend to be at altitudes of less than 1000 km, enabling their instruments to make fine resolution observations. These satellites are also referred to as polar orbiting satellites.

GEO satellites orbit at an altitude of 35,880 km. Their orbital period[4] is identical to the rotational period of the Earth's surface at the equator, and so when placed over the equator satellites in geostationary orbit make one complete orbit in a day. This has the effect of making the geostationary satellite appear to remain in the same location relative to the surface, making it easy to infer the motion and development of cloud systems.

Most satellite observations of the Earth are made with radiometers that measure the amount of radiation coming from the atmosphere and surface in different parts of

[1] Operational in the context of weather forecasting refers to regular, generally periodic, activities that are associated with forecasts issued to the public. Since such forecasts are time-sensitive and must be issued as quickly as possible, operational activities are often referred to as real-time, meaning as quick as possible.

[2] Research satellites are those that carry experimental instruments, usually to investigate specific hypotheses, and that are not intended to be part of a series of identical satellites supporting routine monitoring and forecasting. While their data are often used in weather forecasting and other ongoing applications, those are not the primary motivations.

[3] Inclination is the angle that the satellite's orbit makes with the equator. An orbit with an inclination of 0° would remain over the equator, while one with a 90° inclination passes over the North and South Poles.

[4] The orbital period is the length of time it takes a satellite to complete one orbit. It increases as the altitude increases; for LEO satellites it is between 1 and 2 hours, while at geostationary altitude it is 23 hours and 56 minutes.

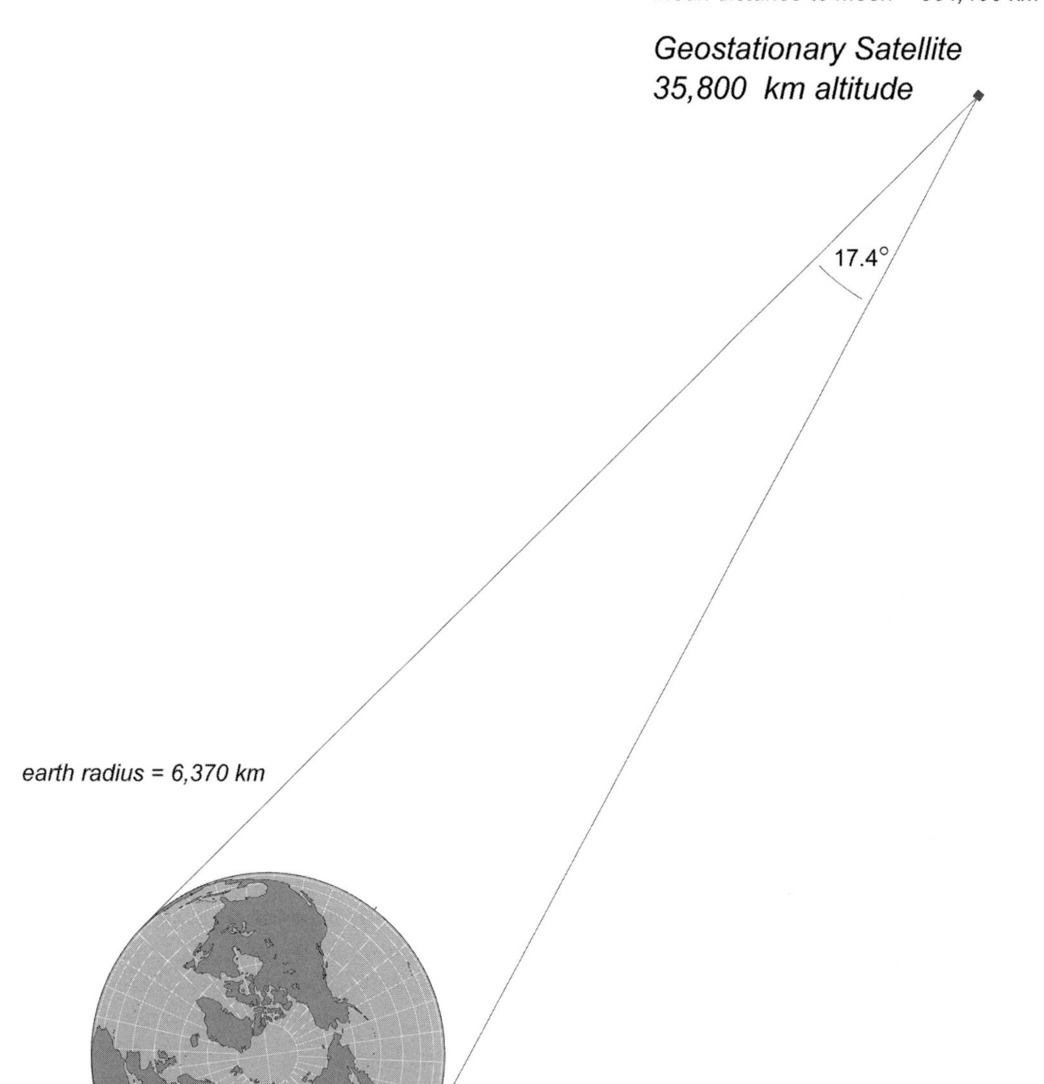

Figure 3.3 Comparison of the views of the Earth from a geostationary satellite and a polar orbiting satellite. Note that the Geostationary Orbiting Satellite field of view includes the entire hemisphere but foreshortening will limit its abilities to view features at high latitudes. In this drawing the Polar Orbiting Satellite is shown as it crosses the equator. The typical field of view of this satellite in relatively small but can provide higher resolution images over the entire globe. Source: Research Applications Program, NCAR, courtesy of David B. Johnson.

the electromagnetic spectrum. The measurements of radiation as a function of wavelength are referred to as radiances and can be used in several ways. The earliest and still most frequent use is as input to algorithms to infer physical properties of the atmosphere or surface. These algorithms can be categorized in a number of ways, according to the source of the radiation, the location of the quantity sought, or the nature of the algorithm used. The radiances may also be directly assimilated into mathematical models of the atmosphere and ocean. In assimilation, observed radiances are compared to those calculated from atmospheric models and used to compute corrections to improve the initial state in the models, resulting in better forecasts.

Satellite-based observations are exceptionally valuable for initializing and validating weather forecast models, since they are available in areas where few other observations are made. Their utility for creating independent climate datasets is somewhat limited by both their relatively short length of record and by the continuing evolution of the instruments used.

Since satellites that orbit the Earth must be outside virtually the entire atmosphere in order to maintain a stable orbit, they only observe radiation that reaches their instruments. In most cases, these instruments passively observe radiation that is emitted by the Earth itself, or reflected solar radiation. The range of properties and parameters that can be inferred from these observations is impressively large, including the altitude, geographic location and movement of clouds, the vertical variation of temperature and moisture in the atmosphere, the surface temperature and winds over the oceans, and the reflectivity of the land surface.

The Earth, like any physical body, emits radiation with a spectrum that depends on its temperature and its properties. An object that absorbs all the radiation that falls on it (a blackbody) emits radiation with a specific spectral profile that depends only on its temperature. The Earth is not a blackbody, and so reflects some of the radiation, principally from the sun, that reaches the surface. The amount and spectrum of reflected radiation, as well as the differences between the radiation emitted by the Earth and an ideal blackbody spectrum, vary tremendously from place to place and reveal a great deal about the properties of the surface. For example, snow and ice reflect a large fraction of the solar radiation while ocean water reflects much less.

Of course, radiation emitted by the Earth as well as solar radiation that it reflects must pass through the atmosphere before reaching satellite instruments, and atmospheric gases are not perfectly transparent. At most wavelengths of radiation, the atmosphere absorbs nearly all the radiation that reaches it and so does act like a blackbody – in those spectral regions a satellite instrument can only observe the effective temperature of some high level in the atmosphere. However, there are some parts of the spectrum in which the atmosphere is more nearly transparent (spectral windows) thus permitting the surface radiation to reach satellite instruments with little modification. At wavelengths in such window regions the surface temperature and reflectance can be estimated quite well, making possible estimates of ocean temperature (see Chapter 8) and sea ice cover (see Chapter 9), for example. Some surface properties such as vegetation reflect different proportions of solar radiation at different wavelengths, a phenomenon that we perceive as color, and change over the course of a growing season, making possible estimates of vegetation characteristics (see Chapter 10).

The most useful and prominent window region is in what we called the visible[5] spectrum, which corresponds to the type of radiation that humans see and which we call light. While the Earth is too cool to emit radiation in the visible range, solar radiation is most intense at visible wavelengths and so satellite instruments that detect reflected solar radiation and its spatial, temporal, and spectral variations in the visible spectrum are possible and observations have been made since the 1960s.

Atmospheric gases are almost entirely transparent to visible light. However, the atmosphere often contains particles called aerosols that are large enough to either reflect or absorb visible light, and satellite instruments can observe their effect. The most prominent and important of these are clouds, which are collections of water droplets and ice particles that are small enough to remain at an approximately constant altitude, unlike precipitation, which is formed of particles that fall and either evaporate or reach the surface. Most clouds reflect visible solar radiation quite strongly and so are readily detected by appropriate instruments. Of course, clouds also obscure the Earth's surface where they occur, leading to sampling challenges for surface datasets. Other aerosols that can be usefully observed by satellite instruments include dust raised from the surface by winds, ash from volcanic eruptions, and smoke from natural and manmade fires.

Two other spectral regions are useful for weather and climate observations – the radio frequency (RF) and infrared (IR). The IR region is adjacent to the visible window, and it is reasonable to consider the entire range from roughly 400 nm (violet) to 14 μm (far infrared) to be a single window interrupted by bands in which absorption is high. The "pure" windows, wavelengths in which atmospheric absorption is negligible, are used to detect surface properties such as reflectance and temperature. The bands of higher absorption result from absorption and reemission by atmospheric gases, and can be used to detect the location and density of gases such as water vapor and carbon dioxide that vary in space and time. Wavelengths at the spectral edges of the absorption bands are particularly valuable since observations in those bands can be used to estimate the temperature and humidity of the atmosphere at different altitudes.

RF observations are extensively used in astronomy, but somewhat less so in observing the Earth from space. As discussed in Chapter 7, observations at microwave frequencies can be used to estimate water and ice concentrations in the atmosphere and on the surface, and radars can be used on satellites to estimate precipitation and cloud properties. Other surface properties can be inferred from microwave measurements as well, such as ocean wind speed and direction, and sea ice concentrations. Active microwave observations are generally more sensitive and accurate than passive observations, but are much more limited in coverage. As with visible and IR remote

[5] All animal vision detects radiation in the visible spectrum, which for humans corresponds to radiation with wavelengths from about 400–700 nanometers. Color vision requires the ability to detect radiation at different points in the visible spectrum.

sensing, vertical profiles of atmospheric properties can be derived; the spatial resolution is coarser due to the longer wavelengths of microwave radiation and consequent limits on antenna size, but such observations are less impacted by clouds and precipitation than are visible and IR observations.

Other observations of the Earth and its near-space environment are available and at least potentially useful in completing the total picture of the climate system. We introduce some of these in this chapter, but we do not intend this discussion to be comprehensive.

The International Satellite Cloud Climatology Project (ISCCP) uses data extracted from satellite imagery to create a large number of datasets of cloud properties and related climate information. ISCCP was established by the World Climate Research Programme in 1982, with data collection beginning on July 1, 1983, and continues to collect observations and produce datasets. A great deal of research into the role of clouds in the climate system has been carried out using ISCCP datasets (Rossow and Schiffer, 1999). However, the cloud properties diagnosed from ISCCP datasets and from cloud atlases derived from surface cloud observations are quite different, and understanding their relationship is challenging (Hahn et al., 2001).

The Earth's total radiation budget is also determined best using satellite observations. The large-scale balance between incoming solar radiation and outgoing radiation from the top of the atmosphere is a crucial element of the climate system. Small imbalances between incoming and outgoing energy can lead to very large changes: for example, the oscillation between glacial and interglacial conditions during the past million years is thought to be a consequence of small changes in the annual cycle of incoming solar radiation in high latitudes of the Northern Hemisphere that result from variations in the Earth's orbit around the sun (Huybers, 2006; McGehee and Lehman, 2012). Observations of the components of the radiation budget at the top of the atmosphere: incoming solar radiation, albedo, and outgoing radiation, began to be attempted with satellite instruments designed for that purpose in the 1970s. In 1984, the Earth Radiation Budget Experiment used instruments on three satellites to estimate the spatial distribution of the components of the radiation budget, and similar instruments have been flown on succeeding satellites to the present. Climate studies using top of the atmosphere radiation budget datasets together with measurements of the total amount of solar radiation reaching the Earth have proven very useful in understanding the changes in climate associated with volcanic eruptions and increasing greenhouse gas concentrations (Trenberth et al., 2009).

Two novel additional examples are worthy of mention: gravity measurements and radio occultation observations. GRACE (Gravity Recovery and Climate Experiment) is a two-satellite joint mission of NASA and DLR in Germany launched in 2002 (Tapley et al., 2004). Very accurate measurements of the changes in separation between the two satellites permit the measurement of changes in Earth's gravity in space and time, and those can be used to infer changes in groundwater and ice sheets over time. Details about the GRACE mission may be found under the heading GRACE in the references under websites. Occultation of Global Positioning System radio waves can be used to derive vertical profiles of temperature and moisture in the atmosphere (Nishida et al., 2000). While these data are not easily incorporated into climate datasets in their original form, they are useful in improving global analyses of the atmosphere as discussed later in this chapter.

3.5 Atmospheric Climate Datasets

As discussed in Section 3.1, the initial atmospheric analyses were based on contour lines manually drawn by a meteorologist on a map on which observations of some parameter were plotted. These analyses were used in their original form by forecasters, but eventually were converted to digital datasets of values at locations that represented a regular grid. Such gridded datasets could be used by objective analysis schemes (algorithms that derived contour lines from the collection of gridded points). This made possible both rapid production of contour maps, since the computer programs fairly quickly became quicker than human analysts, and the creation of initial conditions for numerical weather forecast models. In this section, we describe the historical development of atmospheric analyses and reanalyses, and their applications to climate analysis.

3.5.1 Atmospheric Analyses

Studies of atmospheric climate became feasible in the 1950s. Initially, two quite different types of analyses were available for use in atmospheric climate research: those based on datasets of radiosonde observations and those used in operational weather forecasting. The analyses based on radiosonde observations were statistical in nature, and were limited by the restricted spatial coverage of radiosondes. Ocean areas, and the Southern Hemisphere in particular, were poorly sampled, and the first such analyses used for climate purposes were limited to the extratropics of the Northern Hemisphere. The general circulation project led by Professor Victor Starr at the Massachusetts Institute of Technology created the first usable archives of atmospheric observations and derived analyses of the climate of the general circulation (e.g. Oort and Rasmusson, 1971). Such observation-based analyses dominated atmospheric climate studies through the 1960s and 1970s and were still relevant into the 1980s (see, for example, Angell and Korshover, 1983; Angell, 1988).

Analyses created for operational weather prediction were not considered useful for studies of atmospheric climate until the 1970s. By that time, 15–25 years of analyses had been produced and archives that could be used to calculate statistics were available. By the mid-1980s two facts had become clear: operational forecast model-based initial condition analyses were vastly superior to analyses derived from radiosonde and other observations alone for comprehensive studies of the atmospheric climate, and the continual changes in the operational analysis systems, required to ensure the best model forecast results, were causing significant problems for climate analyses. The number of studies based on global analyses from the National Meteorological Center (NMC, later NCEP), and later the European Centre for Medium Range Weather Forecasts (ECMWF), grew dramatically in the following decade, and in many cases one of the clear results was that atmospheric climate variability was challenging to understand in the context of frequent changes in the analysis system itself. Often a seemingly innocuous change in how observational data were handled, while impossible to detect in individual analyses, would result in substantial changes in monthly means.

The community of atmospheric scientists conducting studies of the climate of the general circulation was eager to find and implement a solution to this problem, and while the solution – to conduct a long time series of analyses using a single consistent system – was obvious, implementation was difficult. The operational weather forecast centers that created the analyses were unable to stop changing their system, since their mission demanded continual improvement in both the analysis procedure and in the kind and amount of observational data used. On the other hand, the computer resources required to conduct a separate analysis of all available data were not available to individual scientists. A series of publications and proposals were put forward (see, for example, Bengtsson and Shukla, 1988) and, combined with an accelerating interest in global climate variability and change, resulted in the initiation of the first reanalysis efforts.

As discussed earlier, weather models strive to constantly adapt to take advantage of observations from new technologies in order to produce the best possible real-time analyses and short-term weather forecasts. On the other hand, data assimilation in climate models strives to include the most consistent set of observations. This limits the kinds of data that can be assimilated without introducing artificial changes in the modeled climate. This limitation then imposes limits on the period of record that may be studied by a climate model with a given data assimilation system. Reanalyses using climate models that assimilate satellite data are limited to the period starting in the late 1970s and reanalyses that require only radiosonde data are limited to the period starting in the mid-to-late 1940s. Reanalyses that require only surface pressure data have been extended back to the mid-nineteenth century (Compo et al., 2011). Even during the relatively short periods that use either radiosonde data exclusively or satellite and radiosonde data the models must be examined for artificial climate variations introduced by changes in the instrumentation since both radiosondes and satellites instruments have experienced considerable changes over their lifetimes.

3.5.2 A Brief History of Global Reanalyses

The logistical challenges of actually performing reanalyses were significant. The operational numerical weather prediction (NWP) analysis/forecast systems in place in the mid-1980s were among the most demanding in the computational world in terms of computer resources required. All of them were operated by government agencies or by consortia of governments, and only one, the NASA Data Assimilation Office (DAO, later the GMAO), had a research mission that might be construed to justify a reanalysis for climate purposes. Even the DAO, however, was focused on weather prediction problems.

At about the same time, major national and international climate research programs were beginning. In the United States, for example, the multi-agency Global Change Research Program (GCRP) brought significant funding to climate research, and reanalysis was one of its early areas of emphasis. Interactions between the climate science community and the various NWP operational centers, along with the DAO, led to an accommodation in which the NWP centers provided the computing and scientific manpower resources for reanalyses while the climate research programs provided incremental financial support to ensure successful production and exploitation by the climate science community.

The DAO and ECMWF carried out the first reanalyses for short periods as proofs of concept during the 1980s. The DAO version (Schubert et al., 1993) covered the period from 1985 to 1989 and became available for use in 1994. While too short in duration to permit most climate studies, it provided a different perspective from other early reanalyses in that it focused more on the "quality and utility of the assimilated data" (Schubert et al., 1993) than on forecast accuracy. The initial ECMWF reanalysis was labeled ERA-15 and covered the period from 1979 to 1993. It was released in 1996. As with the DAO reanalysis, the 15-year duration was insufficient for most climate studies. However, both of these pioneering efforts provided invaluable experience for the scientists and centers involved and contributed to the success of follow-on projects.

The first reanalysis to create a data time series long enough for most climate variability research was completed by the US National Centers for Environmental Prediction (NCEP) in the mid-1990s (Kalnay et al., 1996). This reanalysis began in 1948, and continues to the present, providing time series of almost 70 years as this is written. The analysis/forecast system used for this NCEP/NCAR[6] Version 1 reanalysis was based on the operational system circa 1994, although with slightly coarser spatial resolution. While the length of the time series available from this reanalysis is impressive, the methodology, which dates from the early 1990s, increasingly limits its value given the availability of more advanced analytical techniques. An NCEP/NCAR Version 2 (Kanamitsu et al., 2002) covering the time period from 1979 to present was created to fix some problems in the original and to use a more modern analysis/forecast system to improve accuracy during the period with satellite observations.

Two other major NWP centers produced long reanalyses that became available early in the twenty-first century. ECMWF conducted a reanalysis called ERA-40 that began in September 1957 and extended through August 2002. That reanalysis was shorter than the NCEP/NCAR versions, but used a more advanced analysis/forecast system and a finer spatial resolution. The Japan Meteorological Agency (JMA) produced JRA-25 covering the period from 1979 to 2004, when it was superseded by a more modern version. JRA-25 used a more current data assimilation system than ERA-40 (2004 versus 2001), but was calculated on a coarser spatial resolution.

[6] While the development and data processing were carried out at NCEP, the data archive and distribution were handled by the National Center for Atmospheric Research (NCAR).

All three of these reanalyses, NCEP/NCAR Versions 1 and 2 (which are generally viewed as 2 versions of a single product), ERA-40, and JMA-25, have been extensively used by the climate research community. NCEP/NCAR in particular received extremely wide distribution and has been used and referenced in thousands of scientific publications. As with any first-generation effort, of course, opportunities for improvement were found. Detailed examinations found problems that could be fixed, and of course assimilation/forecast systems and computing power continued to advance.

The clear need for improved atmospheric reanalyses and the proven benefits to both climate research and operational weather prediction activities have resulted in a relative plethora of ongoing reanalysis efforts. Three operational NWP centers, NCEP, ECMWF, and JMA, have continued to engage in reanalyses using a variety of approaches. The NASA Goddard Space Flight Center (GSFC) Global Modeling and Assimilation Office (GMAO), the successor to the DAO, has continued to experiment with reanalysis aimed at improving the use of NASA research satellite data. The NOAA Earth System Research Laboratory (listed in References under websites) has developed a unique approach that enables a global atmospheric reanalysis to be calculated using only surface observations, which are available for a much longer time period than are upper air and satellite data.

JRA-25 ended at the end of 2004, but the identical assimilation system continues to be applied to observations to create the JMA Climate Data Assimilation System (JCDAS) as a continuation. JRA-25 and JCDAS data together enable near real-time monitoring of climate variations in the context of historical variations during the period of JRA-25, which began in 1979. As with every first-generation reanalysis, JRA-25 suffered from errors and deficiencies that were uncovered during exploration of its data, and an increasingly outdated data assimilation system. In 2009, JMA began production of JRA-55, a new reanalysis that used the current data assimilation system and created analyses beginning in 1958. Retrospective production was finished in 2013 and the dataset was made available to the scientific community. As with the earlier version, JRA-55 continues to be extended as a tool for climate research.

The first-generation ECMWF reanalysis, ERA-40, was completed in 2002. As with each of the first-generation products, the dataset contained errors of consequence, particularly in the tropical hydrological cycle (Trenberth et al., 2001). So the ERA project, rather than extending ERA-40 in real-time, chose to use an updated analysis system, circa 2006, to perform a reanalysis back to 1979 and to continue that in real-time as a climate monitoring system. This version is called ERA-interim, and uses an assimilation system that considers observations available in a window of time (12 hours in ERA-interim) to create the analysis at a given time. This approach is technically challenging but is thought to enable improved weather forecasts at least in some situations (Lorenc and Rawlins, 2005). While its benefits for climate analysis are unknown, it certainly could provide a superior initial guess for a starting point. ERA-interim includes land surface properties as well as ocean waves, although the ocean surface temperature is specified. It continues to be produced with data products available with a roughly two-month lag. While this lag prevents general use in real-time climate monitoring, ERA-interim products are excellent material for diagnostic and retrospective studies due to their high spatial resolution (about 80 km) and advanced analysis system.

The NCEP/NCAR Versions 1 and 2 reanalyses both continue to be produced by the NOAA NWS Climate Prediction Center. While they are no longer based on state-of-the-art technology, their long record and prompt availability make them uniquely valuable for real-time climate monitoring. NCEP completed a more recent reanalysis called the Climate Forecast System Reanalysis (CFSR) in 2010. The CFSR used an atmospheric assimilation system based on the NCEP operational NWP system circa 2009, and provided data products with a spatial resolution of about 50 km. The CFSR atmospheric analysis was coupled with interactive analyses of the ocean, land surface, and sea ice, with specified variations in atmospheric composition, aerosols, and solar radiation. The CFSR dataset ended in March 2011 at the same time that the Climate Forecast System v2 became the operational NCEP climate analysis and forecast system. The CFSv2 analysis, which produces fields at six-hourly intervals as initial conditions for the forecast model, is similar (although not identical) to the CFSR analysis system and its output can be used to extend the length of the CFSR. NCEP does not archive or provide CFSR data, but does facilitate archives provided by NCAR and the NOAA National Centers for Environmental Information (NCEI, formerly NCDC). All three of the NCEP reanalyses (NCEP/NCAR 1 and 2, and CFSR) continue to be produced and monthly output products are available within a few days of the end of the month.

The NASA GMAO has produced the Modern-Era Retrospective Analysis for Research and Applications (MERRA) using the 2008 version (Version 5) of the Goddard Earth Observing System Data Assimilation System (GEOS-5). MERRA begins in 1979 and continues to the present, with updated datasets available a few months after the end of a month. The spatial resolution is approximately 50 km, and the analysis is computed at six-hourly intervals. The focus of MERRA is on the impact of satellite observations on both climate analyses and forecasts.

All of the reanalyses discussed in this section were derived from a common origin as systems intended to create initial conditions for weather forecast models. Over time, each of the ongoing efforts has morphed into more of a climate-oriented process, with the goal being more to define the state of the climate system rather than to improve the resulting forecast of upcoming weather.[7] One common characteristic of these reanalyses is that they tend to begin either in 1979, when the satellite observing system became well established, or sometime in the 15 years following World War II, when the radiosonde upper air observing system became established. This is because the analysis procedures used rely on a reasonably dense collection of observations in the free atmosphere, away from the surface. This reliance also means that such a reanalysis

[7] These two goals are certainly not mutually exclusive, but a choice of either will tend to lead to differing decisions on, for example, when and how to incorporate new observations.

can only practically be conducted at a major forecasting and/or research center because of the need to ingest and quality control the very large number of observations used. Summary descriptions of the products available from many of the reanalyses discussed in this section may be found online in the NCAR/UCAR Climate Data Guide listed in the references.

A research group led by Drs. Gil Compo and Jeff Whittaker of NOAA ESRL has developed and implemented a new approach for climate reanalysis that can be applied over a much longer period of time using a much more limited set of observations. Their analysis uses an ensemble Kalman Filter, which produces a large set, or ensemble, of realizations,[8] for each time period – the mean and statistics of the individual realizations are both used. Their product is referred to as the 20th Century Reanalysis (20CR), and the current version 2 begins in 1870 and continues to the recent past, with annual updates. The spatial resolution of 20CR is 2° in both latitude and longitude, with six-hourly temporal resolution. The most unique element of 20CR, and the reason that such a long time-span is practical, is the use of surface observations of atmospheric pressure and sea surface temperature without any other observed data. 20CR products have proven their value in diagnostic studies covering long time-spans, but are less accurate and detailed than other reanalyses in more recent years due to their limited input data.

3.5.3 Regional and Special Purpose Reanalyses

Reanalyses can be performed on limited regions, and a couple of regional reanalyses have the temporal coverage required to be considered climate analyses. The North American Regional Reanalysis (NARR) was developed from the initial NCEP/NCAR Reanalysis effort. NARR used an analysis/forecast system that was quite similar to the global reanalysis, but with a nested higher resolution grid, roughly 32 km, over the North American continent. The time period covered is 1979–present, and the analysis/forecast system used was current in 2003. A unique aspect of NARR was the assimilation of observed precipitation, which was derived from an analysis of rain gauge observations over land and from an analysis of satellite-derived precipitation estimates over the oceans. The precipitation assimilation uses hourly-observed precipitation, which is largely interpolated in time from coarser observations, to constrain the diurnal cycle in the associated atmospheric circulation fields. As with the other NCEP reanalyses, the NARR continues to be produced as a climate monitoring tool.

The Arctic Ocean and surrounding land regions, referred to together as "The Arctic," appear to be undergoing rapid and significant changes in their climate. Scientists who specialize in studies of the Arctic consider global reanalyses to be inadequate for their purposes, and a project called the Arctic System Reanalysis (ASR) is intended to provide improved climate reanalyses of the region. The ASR is a (so far) unique example of an ongoing climate reanalysis effort that is supported purely by national

[8] The process generates a large number of possible atmospheric states that are consistent with the input data and the model. Each state is called a realization, and the set of all realizations is referred to as an ensemble.

research funds and executed by academic scientists. It is based on a polar version of a regional atmospheric model with an associated state-of-the-art data assimilation scheme. The ASR has completed production of 30 km/3 hourly resolution analyses for the period 2000–2012, and is currently processing a 15-km version. Extension of the ASR into the past and future is intended, but depends on both financial and institutional support.

3.5.4 Reanalysis of Other Climate System Components

Reanalysis of other climate system elements is important for climate research and monitoring as well. As discussed in Chapter 1, the ocean, land surface, and cryosphere are all important components of the global climate system, and each can be analyzed in the same way that the atmosphere is – by combining observations, and sometimes model output, using a statistical process to create the most accurate complete depiction possible. The requirement for reanalysis is most acute in the atmosphere, since the analyses produced for NWP applications are the only other ones available, but the concept is relevant for other climate system components.

Atmospheric analysis/forecast systems, whether used for NWP or reanalysis, require analyses of some aspects of the land and ocean surface as boundary conditions. Sea surface temperature (SST) is a strong influence on the atmospheric boundary layer, and SST analyses have been used in NWP for more than 30 years. Certain aspects of the land surface, such as moisture content, vegetation, and roughness,[9] also interact strongly with the atmospheric boundary layer, and analyses of these properties have become essential for analysis/forecast systems. Snow on land, and ice on lakes and oceans, have very large impacts on atmospheric analysis/forecast systems in several ways: They change the surface albedo, they change the heat transfer between atmosphere and surface, and they change the roughness of the surface. Reanalyses of SST and sea ice are most often executed by specialists in those data, and used, generally without adjustment, in atmospheric reanalyses. Reanalyses of other land surface characteristics are highly dependent on the land surface models (discussed in Chapter 10) used in a given analysis/forecast system, and are generally done in conjunction with each new atmospheric reanalysis.

Atmospheric analyses have a long history that is driven by weather forecasting needs and capabilities. On the other hand, ocean analyses and reanalyses, aside from sea surface temperature, have emerged from the oceanographic research community. A recent listing of ocean reanalyses by the Integrated Climate Data Center at the University of Hamburg contains 44 global or near global reanalyses,[10] all of which

[9] Roughness refers to the characteristics of the surface that affect the friction experienced by near-surface winds. The parameters used to represent roughness do not correspond precisely to any conventional measurements, but do change in reasonable ways: for example, a forest has larger roughness values than a field of grass.

[10] The oceanographic community often refers to reanalyses as *syntheses*. Some members of the oceanographic community tend to view the problem of completing the four-dimensional data field given a set of observations as a matter of treating all four dimensions in the same manner, while in the atmospheric

are useful in varying degrees for studying the climate. Two are particularly useful for climate research and monitoring: the ECMWF Ocean Reanalysis System 4 (ORAS4), and the NCEP Global Ocean Data Assimilation System (GODAS). Both of these consist of reanalyses (beginning in 1957 for ORAS4 and 1980 for GODAS) as well as current analyses at six-hourly intervals that are used as initial conditions for each center's operational coupled climate forecast system.

3.5.5 Ongoing Reanalyses and Future Prospects

The concept and history of atmospheric analysis and reanalysis is one of constant change and directed evolution, as scientists continually attempt to improve the capability of the analysis systems to provide users with improved atmospheric (and other climate component) datasets. We expect that the earlier parts of this section will provide our readers with the background necessary to understand the context for the datasets they are likely to use, but progress on these efforts is rapid and specific details are certain to require investigation of the most recent efforts. Here we offer the most up-to-date technical details as of the time of writing, not so much for their specifics as for the background and context that we hope will enable the reader to locate and understand the most current information.

It remains true at the time of writing that full global atmospheric and oceanic reanalyses, as well as real-time analyses, are only possible at operational NWP centers and major research centers with close links to the NWP centers. At present, only NCEP, JMA, ECMWF, and GMAO have completed such reanalyses, and all continue to extend their products with lags ranging from a few days (NCEP) to a few months (the other three). After some initial reluctance, all of these centers have incorporated the reanalysis efforts into their ongoing operational mission to some degree, and plans for future improvements and enhancements are in place or under consideration.

GMAO has completed retrospective processing for MERRA-2, which is a reanalysis of the same nature as MERRA-1, focused on the use of research satellite observations, but using a significantly upgraded data assimilation and forecast system circa 2014. The upgraded data assimilation system permits the use of observations from newer satellites, unusable in MERRA-1, and improves the use of other observations. The forecast model contains significant improvements as well. MERRA-2 has replaced the earlier version as an ongoing climate analysis.

JMA has already gone through one replacement cycle, as described in this section. The JCDAS and associated JRA-55 reanalysis are quite thoroughly incorporated into JMA operations, and continuation and eventual replacement with a further enhanced system seems likely. One augmentation available in JRA-55 is especially noteworthy – two additional versions, JRA-55C and JRA-55AMIP. These products are well suited

community there is more of a tendency to treat time as different from the spatial dimensions. The result is that the ocean analyses are optimizations of all the analyses at once, while the atmospheric analyses are treated as forward (or backward) sequences of optimizations.

to illustrate the impacts of changes in the observing system on reanalysis products. JRA-55AMIP is a dataset created solely from the atmospheric model and datasets that specify things like sea surface temperature and sea ice, without the assimilation of any observations. It should be valuable in identifying model biases and other characteristics. JRA-55C is a full assimilation but using only non-satellite observations. Since the great increase in satellite observations during the past 50 years has been shown repeatedly to lead to substantial inhomogeneities in time series derived from reanalyses, JRA-55C, which uses a much more uniform set of observations, will provide a very useful baseline with which to assess long time scale variations in atmospheric climate. JRA-55C and JRA-55AMIP are available through the websites listed in the References.

The ECMWF has begun the production of the next generation ERA-5 reanalysis, which uses the most current version of the Centre's operational forecasting system and runs at a much higher spatial resolution than ERA-interim, 31 km horizontally and 137 layers in the vertical, extending from the surface to about 80 km. ERA-5 retrospective processing from 1979 forward is expected to be complete by the end of 2018, at which point it will replace ERA-interim as the ECMWF operational reanalysis. More details are available in Hersbach and Dee (2016).

ECMWF is conducting additional reanalyses as part of the ERA-CLIM and ERA-CLIM2 European Community projects. These are three-year research, development, and applications efforts involving 16 organizations from 9 countries in addition to the multinational ECMWF and European Organization for the Exploitation of Meteorological Satellites (EUMETSAT). ERA-CLIM2 began in 2014 "to apply and extend the current global reanalysis capability in Europe, in order to meet the challenging requirements for climate monitoring, climate research, and the development of climate services." The reanalysis component of ERA-CLIM2 is being carried out by ECMWF. ERA-20C, which is similar to JRA-55C and 20CR in that it covers a long time period (1900–2010) and uses a limited set of observations, has been completed and data are available from their project website listed in the references under ERA20c. The assimilation system used is based on the 2012 operational ECMWF version with a spatial resolution of about 125 km and only surface pressure and surface marine winds are assimilated. It is complemented by ERA-20CM, which uses the same model run over the same period but with only sea surface temperature and sea ice boundary forcing together with observed radiative forcing.

The most recent ECMWF long-period reanalysis is CERA-20C, which like ERA-20C uses a limited set of observations, but which uses a coupled ocean-atmosphere assimilation and forecast system. Details on CERA-20C can be found in Laloyaux et al. (2017), and a broad range of datasets can be obtained from the project website listed under CERA-20C in the website references. According to its web site, ECMWF "periodically uses its forecast models and data assimilation systems to 'reanalyse' archived observations," which implies that it will continue to produce new global reanalyses as assimilation/forecast system capabilities advance and computing resources permit.

NCEP global reanalyses, NCEP/NCAR Versions 1 and 2, and CFSR will continue to be extended for the foreseeable future, as will the regional NARR. At present, NCEP has no published plan to conduct a new global reanalysis. The extension of CFSR is an ancillary outcome of the initialization of the operational CFSv2; as assimilation/forecast system improvements are developed and eventually implemented, this process will lose the uniformity that characterizes reanalyses, and the CFSR time series will effectively come to an end. NCEP and its parent organization, the National Weather Service (NWS), have not committed to conducting further climate reanalyses.

Other reanalyses have been produced, including 20CR and ASR, but long-term, ongoing commitments from the scientists and organizations responsible are not in place. Unlike the NWP centers, the relatively small groups of researchers responsible for 20CR and ASR can only continue to develop and extend their reanalyses as long as they are successful in securing financial support, always an uncertain element. The NOAA scientists responsible for 20CR intend to continue to extend it roughly annually, and plan to replace it as new methods and computing resources become available, but no specific details have been published. The producers of ASR intend to extend it into the future, and possibly further into the past than at present, but this also depends on the availability of financial and computing support. GMAO, which supported the execution of MERRA 1 and 2 with the help of NASA research funds, is somewhat more likely to be able to continue with reanalyses because of the broader institutional commitment provided by NASA and the Goddard Space Flight Center, but the general thrust of NASA's research could certainly change.

Reanalyses of ocean circulation will continue to be the object of research efforts, and the increasing trend toward coupled climate prediction models will ensure atmospheric reanalyses will often be accompanied by ocean reanalyses. Reanalyses of other climate system elements that serve as boundary forcing for atmospheric models, such as the land surface, sea surface temperature, and sea ice, will be produced as new methods and observations become available.

3.6 Summary

This chapter introduces the observations, instruments, and datasets that are commonly used in analyzing the climate of the atmosphere. These include traditional *in situ* observations as well as remotely sensed and numerical model-based datasets. One of the themes running through the chapter is a realization of the tension between the requirement that datasets that have long enough records to capture climate means and variations, versus the reality that, very often, observations, instruments, and models change over the time-spans of interest. This is because there are a limited number of sources of climate data that come from instruments and observations whose primary purpose is to monitor climate.

Most modern atmospheric datasets used for climate research and applications contain a significant level of information from mathematical models of the atmosphere. The representation of atmospheric parameters that represent its dynamical behavior, including winds, pressure, and temperature, is significantly more accurate, meaning more consistent with good observations, than that of the parameters related to water, such as clouds, precipitation, and humidity. The hydrological variables are both more difficult to observe on large scales (necessary for the global datasets required for climate studies) and more challenging to model due to the complex physics involved. These facts have important implications for the users of atmospheric climate datasets, as we discuss further in Chapter 11.

Suggested Further Reading

Hamblyn, R. 2001: *The Invention of Clouds: How an Amateur Meteorologist Forged the Language of the Skies*, Farrar, Straus and Giroux, 403 pp. (The popular history of how Luke Howard developed and popularized the current convention of naming clouds types and clouds.)

WMO Guide to Instruments and Methods of Observation – Available online at www.weather.gov/media/epz/mesonet/CWOP-WMO8.pdf (A comprehensive guide to meteorological instruments, observing systems, and quality assurance.)

Questions for Discussion

3.1 Why are direct observations of the atmosphere alone inadequate for assessing the climate?

3.2 What are the strengths and weaknesses of remote sensing observations in observing the climate?

3.3 What are the strengths and shortcomings of data assimilation in producing climate datasets?

3.4 Describe the strengths and weaknesses of the reanalysis efforts led by NOAA NCEP, NASA GMAO, ECMWF, and NOAA ESRL. Which products would be best in assessing decadal-scale variability? Which is best for assessing variations in water and energy cycles?

3.5 The introduction of satellite data in 1979 impacts the use of the NCEP/NCAR reanalysis. Discuss ways that despite these limitations the (1948–present) reanalysis can be used in climate change and decadal variability studies?

4 Climate Variability

As discussed in Chapter 1, the climate exhibits preferred patterns of variability in space and time. Some of these are the result of external forcing, such as the characteristics of the earth's orbit around the sun (the annual cycle), while others appear to be the result of internal interactions among or within the components of the system (the El Niño/ Southern Oscillation, or ENSO, is one example). One characteristic that all such patterns share is a detectable spatial and temporal coherence,[1] which is obvious in some cases, but which can be revealed by a wide variety of statistical analyses of the kind discussed in Chapter 2 and Appendix A. In the first part of this chapter (Sections 4.1 and 4.2) we describe the mean seasonal cycle of the atmospheric component of the climate system, including a discussion of some of the early attempts to describe and explain global wind patterns. We concentrate on the atmospheric circulation, sea level pressure, surface temperature, and precipitation. In the remainder we describe the principal patterns of climate variability on sub-seasonal to multi-decadal time scales.

While we cannot offer a rigorous definition of climate variability as distinct from climate change, a usable heuristic explanation can be provided. The climate phenomena that we discuss here are those with anomalies exhibiting consistent spatial patterns and temporal evolution that recur often enough to be identifiable as part of the climate of the Earth. The duration of these phenomena is longer than two weeks (to distinguish them from weather events), and we will restrict our discussion to phenomena whose spatial extent is significantly greater than 1000 km.

Understanding climate variability requires some familiarity with the mean climate state, including the general circulation of the atmosphere, as well as familiarity with analyses typically used to illustrate, study, and monitor these phenomena. An overview and introduction to the Earth's climate system can be found in Chapter 1; more complete presentations are available in the texts cited in Suggestions for Further Reading at the end of this chapter. Our more modest aim here is to: (1) Introduce concepts and terminology often used in discussions of climate variability, (2) Briefly review the historical background associated with the study of climate variability, and (3) Describe various examples of climate variability.

[1] Coherence is a term that is used in various ways, and a clear and rigorous definition is not really available. We use it here to describe phenomena that exhibit a range of behaviors that are understood to be associated within the physics of the climate system; these phenomena have manifestations over a substantial area of the earth, and endure in a fairly consistent manner over a period of at least several weeks.

Since the climate involves coupling of all of the components of the climate system, coherent patterns may be detected in any of them. Very often the atmospheric features are the easiest to examine, particularly because atmospheric climate datasets provide far more detail and length of record than oceanic and land surface datasets. The atmosphere also provides the connections between the Earth's surface and the rest of the climate system, forming the bridge between the ocean and the land. The purpose of identifying and describing patterns of climate variability is to advance our understanding of the behavior of the climate by examining the observed relationships among climatic variables, which in turn leads to a better understanding of the underlying physics and eventually to improved modeling and prediction of the phenomena involved. Successful predictions have benefits to society beyond the intellectual satisfaction of enhanced understanding, but consistently successful climate predictions continue to be a challenge.

Climate variations include oscillations that result from periodic external forcing such as the seasonal and diurnal cycles, phenomena that result from coupled oscillations within the climate system, such as the El Niño/Southern Oscillation (ENSO), and features that occur often but for which there isn't a clear physical explanation. The climate variations associated with ENSO, for example, arise from a coupled oscillation of the ocean and atmosphere in the tropical Pacific Ocean with manifestations found both locally and in distant regions. The North Atlantic Oscillation (NAO), on the other hand, is an example of a systematic pattern of fluctuations in the climate system that manifests as changes in the jet stream, storm tracks, temperature, and precipitation in and around the North Atlantic Ocean basin that occurs quite often but without clear coupling with other parts of the climate system.

Patterns of climate variations that have been found to be associated with coupled interactions between two or more components of the climate system, such as ENSO, tend to have longer lifetimes and to lend themselves to more skillful predictions. Those like the NAO without strong couplings seem not to exhibit consistent lifetimes or recurrence intervals. However, the coupling associated with climate variations can be subtle and one goal of climate analysis is to identify and characterize such coupled modes.

Climate variability is defined in relation to a notional unchanging steady state referred to as the background climatology. Thus, we begin by first discussing the mean state, the climatology, of the atmosphere and ocean. As discussed in Chapter 2, the working definition of climatology, by tradition and international agreement, is taken to be 30-year periods and is characterized by statistics of parameters observed at weather stations. These statistics are updated at 30-year intervals and stored at World Data Centers and designated as the official climatological records by the World Meteorological Organization (WMO). In practice, for the last several decades, provisional 30-year climatologies have been updated at 10-year intervals by the NOAA Climate Prediction Center and other centers that monitor the climate.

The original purpose of the 30-year updates was primarily to account for changes in the observing system due to additions to and deletions from the set of observing sites, and changes in their location and instrumentation. The 30-year period also provided ample time for quality control of the data and the computation of monthly and seasonal means, which was quite challenging before the availability of computers and digital archives. These estimates assumed that the background climate varied extremely slowly, and changes in the climatology reflected, for the most part, changes in station location, instrumentation, and observational practices, rather than changes in the climate itself. However, as discussed in Chapter 5, the actual climate is always changing, and, thus the concept of a stationary background climate against which all observations occur is never fully accurate. Nonetheless, the concept is helpful in the study of climate variations with amplitudes much larger, and time scales much shorter, than long-term trends. These traditional climatologies were generally computed for monthly, seasonal, and annual periods. With the improvement in computational capabilities climatologies began to include estimates of daily and weekly means, often derived by fitting curves to the raw data, in addition to the monthly, seasonal, and annual means. Traditional climate records based on weather station data are increasingly supplemented in recent years by data from other environmental networks and by satellite and other remotely sensed data.

In addition to characterizing the mean climate observed at stations on land, there has been a historic interest in the observation of mean global wind patterns, including over the world's oceans. Understanding the global mean wind patterns, generally referred to as the General Circulation, is fundamental to understanding of the climate system and, in fact, numerical weather and climate models were traditionally known as General Circulation Models (GCMs).

4.1 The Mean Climate

4.1.1 Global Wind Patterns – A Historical Perspective

Interest in the mean climate harks back to the earliest civilizations. These societies were dependent on agriculture and animal husbandry and sought ways to determine when, on average, the weather would be favorable for planting their crops, or for changing pastures, based on the mean annual cycle of temperature and precipitation. Early understanding of the timing of the annual cycle in precipitation and temperature was tied to changes in the times and locations on the horizon of sunrise and sunset and observations of star positions in the night sky. By observing the stars, for instance, Egyptian farmers would know the time of year to plant crops in anticipation of the flooding of the Nile delta. Some climatologists would claim that this early interest in the mean annual cycle inspired the development of the science of astronomy, eventually leading to development of mathematics and the rest of the physical sciences.[2]

In the fifteenth and sixteenth centuries, general interest in the mean global wind patterns, particularly over the oceans, was spurred by the competition among several

[2] Of course, the link of the annual rains to the stars also led to the false conclusion that the stars controlled the weather in addition to controlling the destiny of individuals, a belief that, despite the advances in science, is still with us today, and evidenced by the continued popularity of daily horoscopes.

European countries for sailing routes that provided the easiest access to trade and conquest in other parts of the world. At first the individual nations, and groups within nations, jealously guarded any information on the nature of the wind patterns over the routes to destinations of interest. Despite efforts at secrecy, however, the nature of mean wind patterns at the largest scales began to emerge. These included a general tendency for easterly winds, winds blowing from east to west, at low latitudes and winds blowing the opposite direction, westerly winds, at midlatitudes.

The British polymath Halley (1686) is credited with the first integrated descriptions and attempted explanation of the global wind patterns over the oceans. Halley's work was followed by a more complete description and explanation by Hadley (1735). The low latitude easterlies, commonly referred to as the Trade Winds, were explained in these earliest studies as being due to the differences in the mean speed of the earth's surface at various latitudes. (See Box 4.1). While not all the details of Hadley's explanation for the global wind patterns have stood up to modern observations and theory, current literature often refers to the "Hadley cell" as a shorthand for low level easterlies and upper level westerlies, forming a closed mean circulation with rising air in equatorial regions and sinking air at higher latitudes (Figure 4.1).

Further aspects of the general circulation were the subject of study and debate throughout the eighteenth, nineteenth, and early twentieth centuries. (Box 4.2) Among the shortcomings of these early attempts to characterize the general circulation was the failure to account for the complexities due to the configuration and topography of the continents, as well as the lack of understanding of the importance of other parameters, such as moisture, to the behavior of the climate system. A more complete understanding of the general circulation had to wait until the mid-twentieth century when global observations became available and, with them, the development of a theoretical framework that more fully explained the observations.

Box 4.1 Why Are the Trade Winds Easterly?

At higher latitudes air at rest relative to the ground travels a smaller distance to complete one rotation of the Earth than air at rest near the equator. Air at rest at higher latitudes, for example air at 60° of latitude (north or south), is moving relatively slower than air that is at rest relative to the earth's surface near the equator, because both have to make one Earth's rotation in 24 hours. The air at 60° latitude only needs to travel 12,500 miles (approximately 20,000 km) in one rotation, while the air near the equator must move nearly 25,000 miles (40,000 km). Air initially moving toward the equator from midlatitudes will turn westward, producing an "easterly" component to the wind. Likewise, air moving from lower to higher latitudes will turn eastward, resulting in a "westerly" component to the wind. To an observer on Earth these winds seem to be responding to an apparent force. In meteorological textbooks this apparent force is called Coriolis force (Wallace and Hobbs, 2006). The Coriolis force is a consequence of the rotation of the Earth, resulting in the tendency for the winds to blow toward the west when moving equatorward and toward the east when moving poleward.

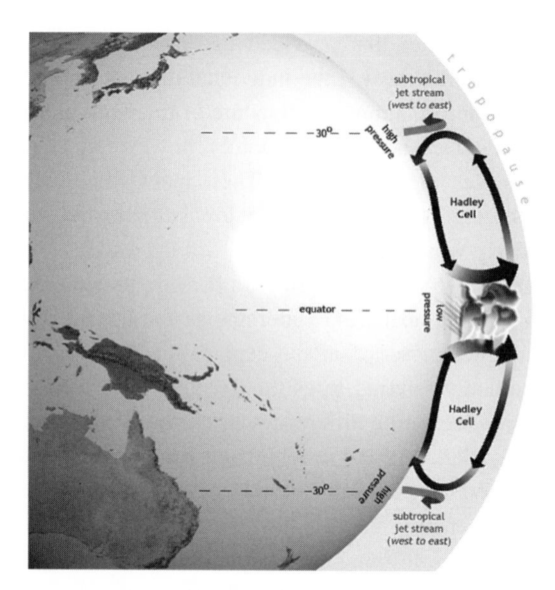

Figure 4.1 Schematic of the Hadley Circulation. The dark arrows indicate the rising air associated with precipitation in the equatorial regions and sinking air associated with dry conditions near 30° north and south.
Source: NOAA/Climate.gov.

Box 4.2 Beyond Halley/Hadley

The Hadley cell was originally conceived as a closed circulation extending from the equator to the poles. While this has been found not to be the case, climate studies often refer to the Hadley cell as shorthand for the zonally (east–west) averaged mean meridional (north–south) atmospheric circulation over tropical and subtropical ocean basins. As more global wind observations became available this conceptual model was modified. Most significantly, a Hadley cell is now taken to describe cells, one in each hemisphere, which extend from near the equator into the subtropics. In this model the descending branches of the Hadley cells occur around 30° north and south of the equator, rather than near the poles, and have an annual variation in mean position and strength. Regions with descending air are associated with dry conditions, thus the descending branch of the Hadley cell is associated with many of the arid and semi-arid regions of the globe, for example, the Sahara and Kalahari deserts.

Once observations revealed that the Hadley cell was neither longitudinally continuous nor did it extend much further poleward than 30 degrees of latitude, nineteenth-century scientists (Maury, 1855; Ferrel, 1889; Thompson, 1892; and others) postulated the existence of a secondary cell (a polar cell) that came to be called the Ferrel cell. Their Ferrel cell extended from the descending branch of the Hadley cell poleward to high latitudes. Subsequent observational data of the mean midlatitude circulation did not support the existence of the Ferrel cell in the mean circulation. It wasn't until the latter half of the twentieth century that observations of the complexity of the general circulation came to be viewed in the context of the entire climate system.

The Hadley circulation is a feature of the mean climate and is not always apparent in day-to-day observations of global winds. Likewise, the strength of the Hadley circulation is not the same at all equatorial locations. It is most easily identified over the Pacific Ocean, and has an annual cycle. The mean annual cycle of the global Hadley cell can be seen in observational data by averaging the wind fields around the globe in latitude bands (Peixoto and Oort, 1992). If the strength of the Hadley circulation were uniform along the equator we would expect to see a continuous band of rainfall, associated with the rising air, along the equator, especially over the open oceans where topography doesn't complicate wind patterns. However, this continuous band of rainfall is not observed, even over the oceans. In the Equatorial Pacific Ocean, we observe that the eastern Pacific tends to be extremely dry, except during El Niño episodes, while the western Pacific contains the largest area of strong convective precipitation on Earth (see, for example, the discussion below and Chapter 7). We can observe the pattern of mean precipitation most easily in data derived from satellite observations but the pattern, based on station observations, was also noted in the early twentieth century (Walker 1924, 1928). Later in the twentieth century the rainfall maximum in the western Pacific and minimum in the eastern Pacific was found to be related to rising air associated with deep convection in the western Pacific and sinking motion in the eastern, approximately in the equatorial plane (Figure 4.2). In recognition of his early studies this circulation generally goes by the name of the "Walker circulation."

In the latter half of the twentieth century, climate research and upper air observations revealed features of the midlatitude general circulation, such as the jet streams (discussed later), that could not be inferred by investigation of the mean surface winds.

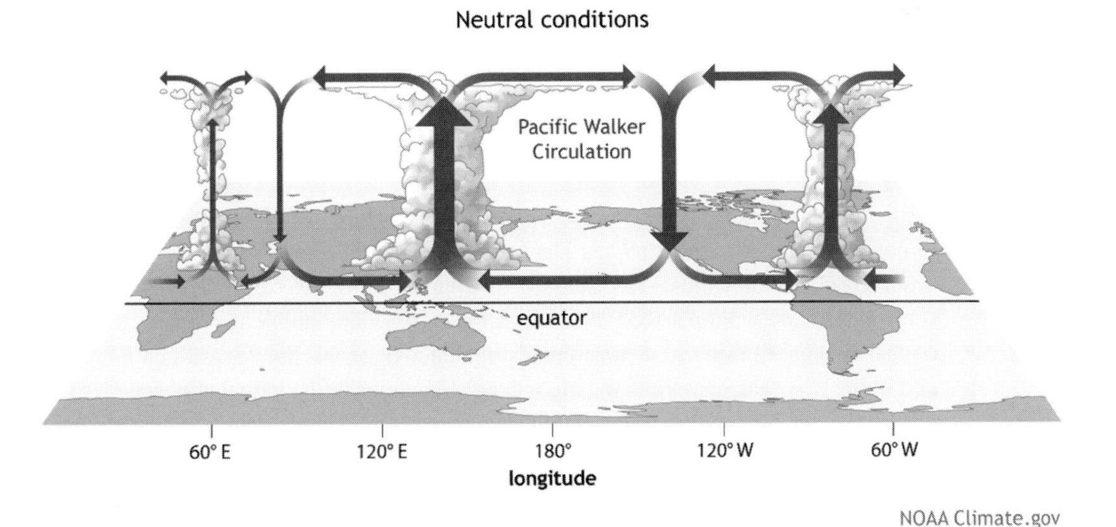

Figure 4.2 Idealized Walker Circulation (December–February). Convection associated with rising branches of the Walker Circulation is found over the Maritime continent, northern South America, and eastern Africa.
Source: NOAA/Climate.gov, drawing by Fiona Martin.

Complete understanding of how the excess of solar energy arriving in equatorial regions gets redistributed to higher latitudes involves not only the mean atmospheric circulation but also the cumulative effects of transient phenomena associated with weather. As emphasized throughout this book, interactions with the oceans as well as complex interactions with clouds, water in its various phases, the global land surface, and high latitude cryosphere also influence the general circulation. Early studies of the major climatic wind regimes have evolved into richer, and much more complex, studies of the entire climate system. Current research is supported by a large array of global observations and use of computers powerful enough to study the general circulation.

4.2 The Mean Seasonal Cycle

The evolution in mean temperature, precipitation, and winds through the year is known as the seasonal cycle, and is the most prominent large-scale manifestation of natural climate variability. The mean seasonal cycle is forced by the changes in solar radiation that result from the Earth's annual journey around the sun, and provides the background against which other climate variations are identified and studied. Observational studies of the global climate were revolutionized in the late twentieth century by the availability of global, or near global, satellite data and the reanalysis of historical station observations using numerical models, discussed in detail in Chapter 3. Here it is sufficient to reiterate that model-based analyses provide a consistent depiction of the climate system that is constrained by the physics encapsulated in the equations that form the numerical model. The data from a reanalysis are limited by the size of the model grid and by the model's approximations of sub-grid scale phenomena, as well as by the coverage and quality of the observational data. In the examples given here we rely heavily on data from the NCEP/NCAR Reanalysis (Kalnay et al., 1996; Kanamitsu et al., 2002).

4.2.1 Atmospheric Circulation

With a few exceptions, early observational studies of global wind patterns were limited to consideration of surface winds. Here we introduce the mean seasonal cycle of the observed upper air winds. As discussed in Chapter 3, by the early twentieth century some systematic attempts were made to sample the winds above the surface through optical tracking of instrumented and uninstrumented balloons, but it wasn't until the last half of the twentieth century that daily wind observations of the global atmosphere from the surface and into the stratosphere were made with instrumented balloons called radiosondes as part of an international effort to support weather prediction. These observations, augmented by satellite estimates of atmospheric temperature and moisture, serve as one of the primary data inputs to numerical weather and climate models including the NCEP/NCAR and other reanalyses.

In the atmospheric sciences, wind and other meteorological fields are usually presented at pressure levels rather than at heights above ground or heights above sea level. In the discussions of various mean wind fields that follows we present mean fields at

925 and 200 hPa to illustrate the mean features of the lower and upper atmosphere. These levels roughly correspond to 0.75 km (2,500 ft) and 11.6 km (36,000 ft) in the US Standard Atmosphere.[3] Height gradients on pressure level plots are equivalent to pressure gradients on constant height plots.

The eastward and westward, or zonal, components of the winds in the upper atmosphere are major components of the general circulation. The mean zonal winds at 200 hPa are strongest in midlatitudes, roughly from 30° to 50° latitude, in the winter hemisphere (Figure 4.3).

The existence of regions of strong upper atmospheric winds, often referred to as the jet stream, was suspected but not confirmed until encountered by high-flying aircraft in the 1940s. In the Northern Hemisphere winter, December through February (DJF), westerly winds with maximum mean speeds of 50 m/s (see Box 4.3) or more extend from northern Africa well into the Central North Pacific. A secondary westerly wind maximum is found along the east coast of North America. The jet streams reflect the mean Asian and North American winter storm tracks, the region with the greatest winter storm activity. Also in Northern winter, a band of mean equatorial easterlies is nearly continuous, interrupted by weak westerlies only in the Eastern Pacific.

Box 4.3 A Note on Units Used to Describe Wind

As with other meteorological variables, the units used to describe the strength of the winds have varied through history. Subjective estimates of wind speed were recorded, mainly by sailing vessels, from the sixteenth century onward. In 1806 the British hydrographer Francis Beaufort, while on a voyage to South America, developed a 14-category list, ranging from Category 1 (calm), to Category 14 (storm), in an attempt to codify and standardize subjective wind estimates. Subsequently, the subjective descriptions were modified several times, and in the modern era by the World Meteorological Organization published a 13 category "Beaufort scale" (WMO, 1970; Huler, 2004). Although the Beaufort scale is still used today, wind speed is most often measured by instruments in knots (aviation, sailing), miles per hour (by the general public in the United States), and meters per second (m/sec) in scientific studies. Note that one knot, or nautical mile per hour, is roughly 1.15 statute miles per hour (mph) and 1 m/sec is roughly 2.237 mph.

Vertical velocity, the speed of the rising or sinking component of the wind, is generally too small to measure directly and is often given in terms of rate of change of pressure, i.e. pascals per sec (pa/sec) as well as m/sec.

In meteorology and climate, the wind direction is defined by the direction from which the wind is blowing, for example, a westerly (northerly) wind is one that blows from the west (north).

[3] A hypothetical vertical distribution of atmospheric temperature, pressure, and density that is taken to be representative of the atmosphere for aviation and engineering purposes. (After the AMS Glossary of Meteorology).

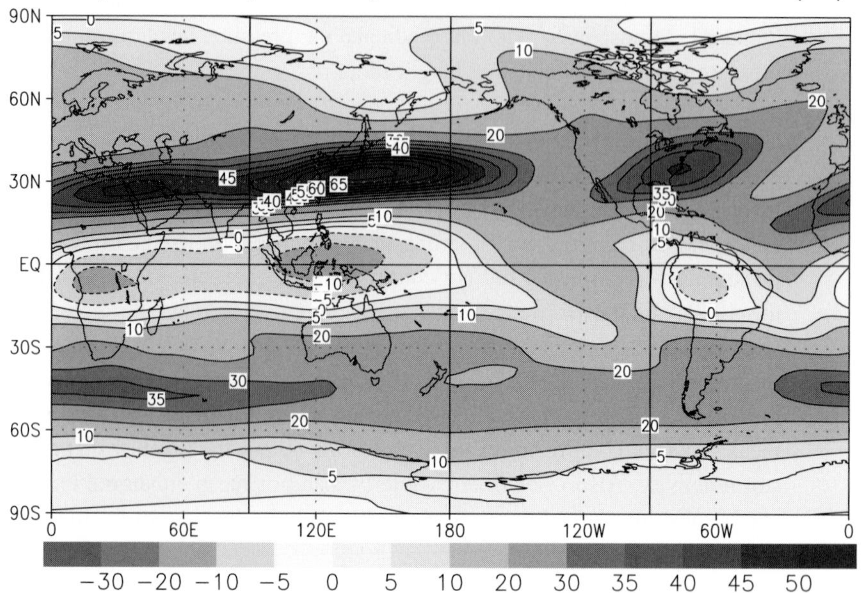

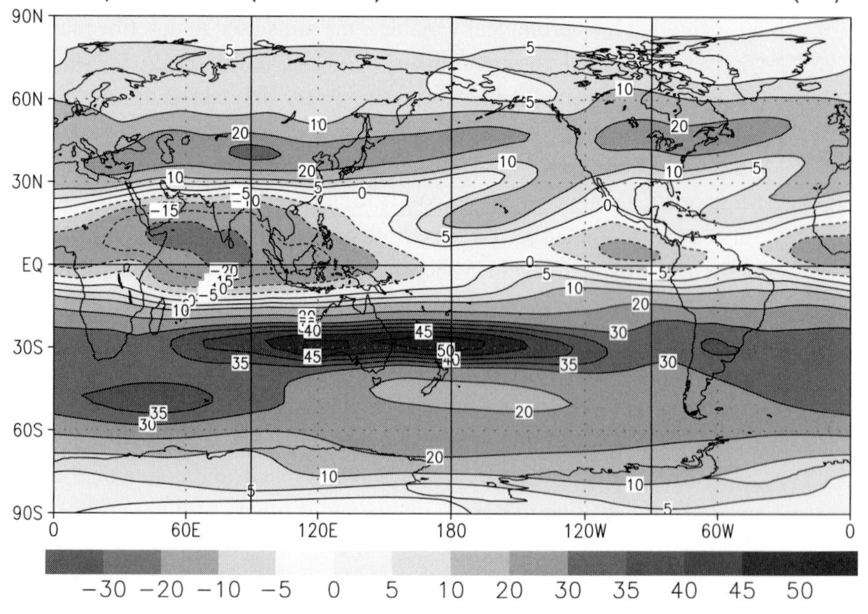

Figure 4.3 The mean zonal wind component at upper levels (200 hPa) for the 1981–2010 period based on the NCEP Reanalysis 2 (Saha et al., 2010), DJF (top) and JJA (bottom). Yellow and red shading denotes westerly winds; blue shading denotes easterlies (m/s). The strongest mean westerlies in the winter hemisphere occur to the east of the Asian, North American, and Australian continents. A black-and-white version of this figure appears in some formats. For the color version, please refer to the plate section.

Figure courtesy of M. Chelliah, CPC.

During Southern Hemisphere winter, June through August (JJA), the band of equatorial easterly mean winds at the 200 hPa level is considerably stronger than in southern summer and extends around the globe without significant interruption. We also note a manifestation of the Monsoon circulation over southern Asia in the dramatic shift of the 200 hPa winds from westerly in northern winter to easterly in northern summer. The geographical extent of the Southern Hemisphere winter jet stream, as well as the mean maximum speed, is generally less than is observed in northern winter. The core of the strongest zonal winds is generally near 30°S latitude during winter and diminishes and shifts poleward during Southern Hemisphere summer.

Similar dramatic seasonal shifts in the zonal winds are observed at 925 hPa, much nearer the surface, as shown in Figure 4.4. Winds at this level generally confirm the surface wind patterns over the oceans discussed by Halley, Hadley, and others. A broad band of easterlies dominates the tropics, with westerly winds at midlatitudes over both the Atlantic and Pacific Oceans in the winter and summer seasons of both hemispheres. Over the Indian Ocean, southern Asia, and parts of the extreme western Pacific a dramatic seasonal shift in the winds associated with the Asian Monsoon is observed. These near-surface winds shift from generally easterly in the northern winter to westerly in northern summer, with especially strong westerly winds across the Arabian Sea from just off the Somali coast to near the Indian subcontinent. These features of the near-surface wind patterns have long been known (Hastenrath, 1991) from wind data gathered by ships. The reanalysis data shown here help put the mean monsoon wind systems into a larger context.

A detailed examination of Figure 4.4, however, shows features that serve as useful caveats for the use of reanalysis, or any other model-based tools. Even though in mountainous regions the land surface is higher than the 925 hPa pressure surface, which is generally between 750 and 800 m above sea level, some models and reanalyses generate winds for this surface. For instance, the analysis shows a small area of westerly winds near the Tibetan plateau and a band of easterlies in the Rocky Mountains of North America on the top panel of Figure 4.4 that are artifacts of this reanalysis. One remedy is to simply "mask out"; that is, do not show, areas with topographic features above the analysis level to help minimize misinterpretation of the model data (Ropelewski et al., 2005).

This discussion of the mean zonal winds at 200 hPa and 950 hPa for the winter and summer illustrates one way of documenting one aspect of the seasonal differences in the global wind fields. Several other analyses of mean fields are commonly used. For example, one can compute and display the differences between the annual mean and seasonal mean fields. Of course, the zonal component of the winds is not the whole story, and the total wind vectors may also be described in a number of ways, as described in this section.

Vorticity and Divergence

Meteorological analyses of the day-to-day weather often include discussions of the winds in terms of vorticity and, to a lesser extent, divergence. The vorticity is a measure

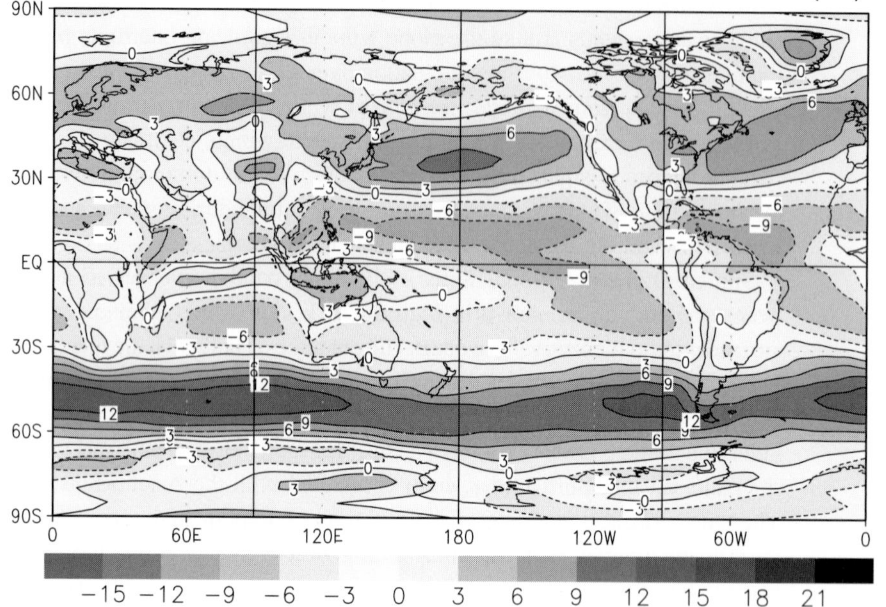

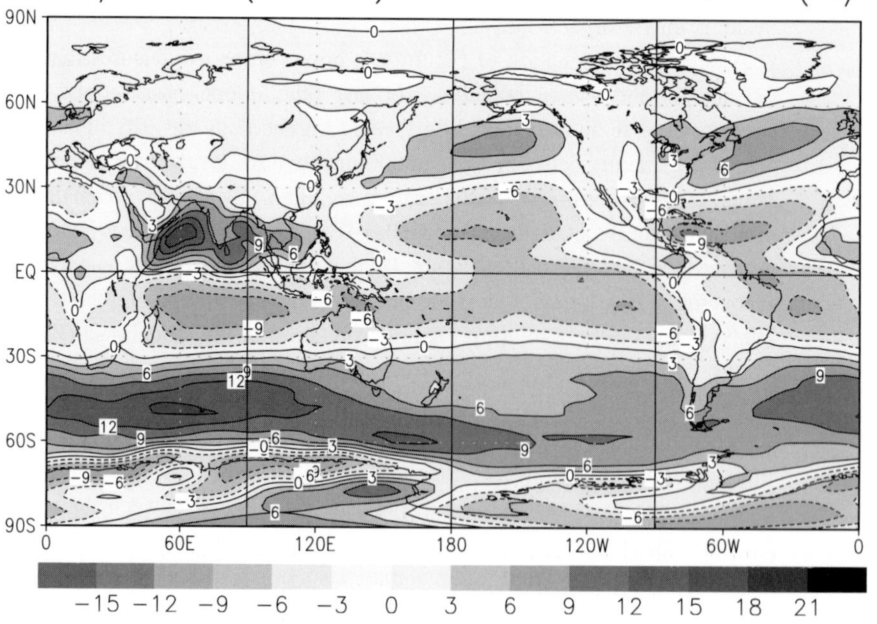

Figure 4.4 The mean zonal wind component at lower levels (925 hPa) for the 1981–2010 period based on the NCEP Reanalysis 2 (Saha et al., 2010) for DJF (top), and JJA (bottom). Yellow and red shading denotes westerly winds, blue shading denotes easterlies, (m/sec). Note the change in scale compared to the previous figure. A black-and-white version of this figure appears in some formats. For the color version, please refer to the plate section.

Figure courtesy of M. Chelliah, CPC.

of the local rotation within a fluid. In meteorology, the movement of cyclonic vorticity from one place to another is typically associated with increased upward vertical motion in the wind fields and increased probability of precipitation. The early numerical models for weather forecasting were capable only of transporting, or advection, of patches of vorticity that existed in a model's initial state of the atmosphere and thus these models could not develop new storms. Because of this, early model-based forecasts were not very successful beyond 12–24 hours. Positive vorticity advection at mid-tropospheric levels (say 500 hPa) in the atmosphere is seen in conjunction with low-pressure areas in weather systems. Negative vorticity advection is associated with high-pressure areas and little or no storm activity. Since the vorticity is so closely related to the evolution of weather events it emerged as a powerful diagnostic tool in weather forecasting. The divergence is a measure of how the winds in a weather system are either coming toward or blowing away from an area, at a given pressure level. Computation of the divergence is extremely sensitive to small fluctuations in the winds. Values of divergence based on individual weather observations over small areas (10s of kilometers) tend to have a lot of variability, some of which is noise due to small errors in the observations. Averaging wind fields over time reduces the small-scale variations and makes the calculation of divergence more meaningful in climate studies. The winds associated with storms generally converge near the surface and diverge at upper levels in the atmosphere. This means that somewhere, at intermediate levels in the atmosphere, the value of the divergence must go to zero. This level of non-divergence, historically taken to be midway through the depth of the atmosphere, around 500 hPa, is also the level where the vertical component of the wind in a storm system is generally the largest. This property of an ideal atmosphere is one reason that meteorologists have historically tended to pay attention to the winds at the 500 hPa level (even though the actual level of non-divergence is a bit different from 500 hPa from storm to storm). In climate analysis, however, many studies have focused on the mean circulation of the atmosphere at 925 hPa, near the Earth's surface, and at 200 hPa or 250 hPa, near the top of the midlatitude troposphere.[4]

Stream Function and Velocity Potential

One of the primary aims of weather analysis and forecasting is to provide guidance on transient phenomena, that is, the local weather, and thus the emphasis in weather analysis has traditionally been on vorticity and divergence. In climate studies the emphasis is on the analysis, understanding, and in some cases, prediction of the time-integrated influence of weather systems on relatively large spatial domains. Thus, in climate studies the rotational and divergent components of the wind fields are often discussed in terms of the mean stream function and velocity potential that reflect motion on larger spatial scales and can be computed directly from the divergence and vorticity. A brief mathematical description of these quantities is given in Appendix B, for those

[4] The troposphere is "That portion of the atmosphere from the earth's surface to the tropopause; that is, the lowest 10–20 km (6–12 mi) of the atmosphere; the portion of the atmosphere where most of the weather occurs." Source AMS Glossary of Meteorology (2012).

familiar with vector calculus and interested in the details. If the wind is divided into a part that has no divergence and a part that has no rotation, the contours of stream function may be thought of as the path that the non-divergent part of the wind would flow parallel to (assuming the flow isn't changing in time) and the negative and positive centers of the velocity potential field may be thought of as the areas that the nonrotating part of the wind flows from and to, respectively. The use of seasonal mean stream function and velocity potential in climate analysis is illustrated next. The 200 hPa stream function (Figure 4.5) for winter and summer show overall patterns similar to those for the zonal wind component discussed in this section.[5]

The stream function for JJA provides some indications of the mid-Pacific and Mid-Atlantic troughs,[6] features that do not appear in the seasonal mean stream function for DJF. The stream function has its greatest magnitudes at mid and high latitudes. The seasonal mean stream function is roughly symmetric about the equator for both seasons. The stream function for Southern Hemisphere summer (DJF) clearly reflects the position of an area of higher pressure, the monsoon ridge extending from Africa, eastward over Australia through the western Pacific. In Northern Hemisphere summer (JJA), the stream function reflects the position of the Asian Monsoon high that dominates the global circulation from eastern North Africa throughout southern Asia in that season. The monsoon systems over the Americas are only weakly reflected in the seasonal changes in the stream function.

The mean 200 hPa velocity potential (Figure 4.6) is a valuable diagnostic tool for understanding the annual cycle and monitoring interannual and intraseasonal climate variability, particularly in low latitudes. The magnitudes of the divergent winds, reflected in the velocity potential gradients, are often as much as a factor of ten smaller than the rotational component of the wind associated with the stream function. The seasonal mean divergent winds associated with the velocity potential at 200 hPa (Figure 4.6) reflect the global monsoon systems. The divergent winds are strongest where the gradients of velocity potential are the greatest.

The centers of negative velocity potential at 200 hPa are associated with areas of outflow or divergence that are mirrored in the lower levels by the (925 hPa) convergence centers (not shown). These areas of high-level outflow and low-level convergence, reminiscent of the large-scale overturning circulations envisioned by Hadley and others, are associated with enhanced precipitation.[7] The mean negative velocity potential centers have the largest magnitude in the summer hemisphere, that is, just north of Australia and over the Amazon Basin in the December to February season (DJF) and near the Philippines and over the Gulf of Mexico during northern summer (JJA).

[5] Since the rotational component of the wind is dominant at the spatial scales shown in the figures, stream function and height contours (on pressure levels) are nearly identical except near the equator or near the surface.

[6] A trough is an elongated area of relatively low pressure; the opposite of a ridge. Source – The AMS Glossary of Meteorology (2012).

[7] The negative velocity potential centers high in the troposphere indicate areas of upward motion. As the air rises, it cools and its capacity to hold moisture decreases. The water vapor in the air condenses to form liquid water drops or ice crystals, which then precipitate.

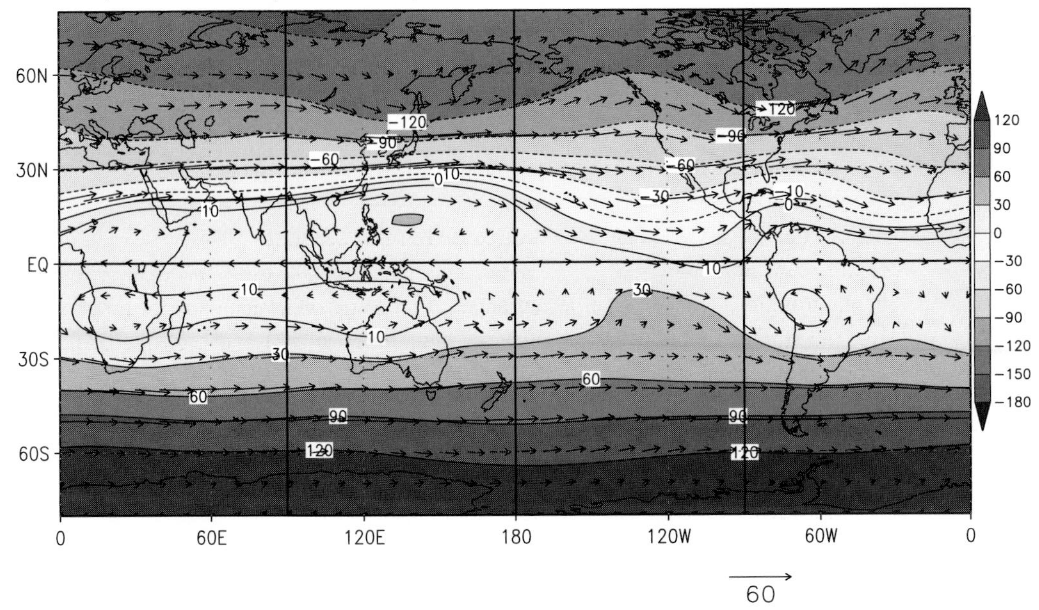

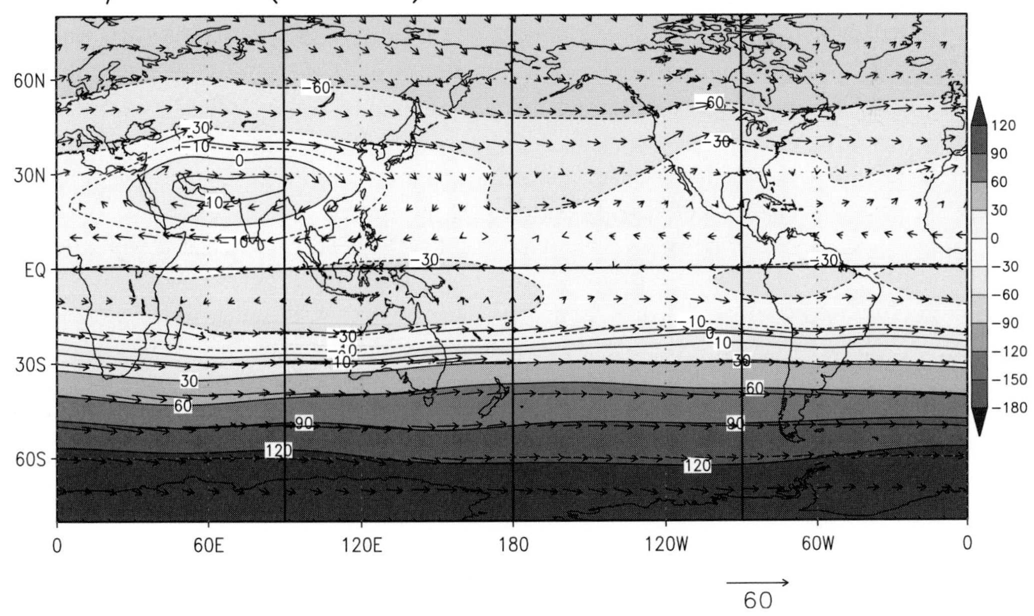

Figure 4.5 The mean 200 hPa stream function (10^6 m^2/sec), contour interval 10 units, and rotational component of the wind (m/sec) for the 1981–2010 period based on the NCEP Reanalysis 2 (Saha et al., 2010) for DJF (top), and JJA (bottom). Red shading denotes positive values, blue shading negative values. A black-and-white version of this figure appears in some formats. For the color version, please refer to the plate section.
Figure courtesy of M. Chelliah, CPC.

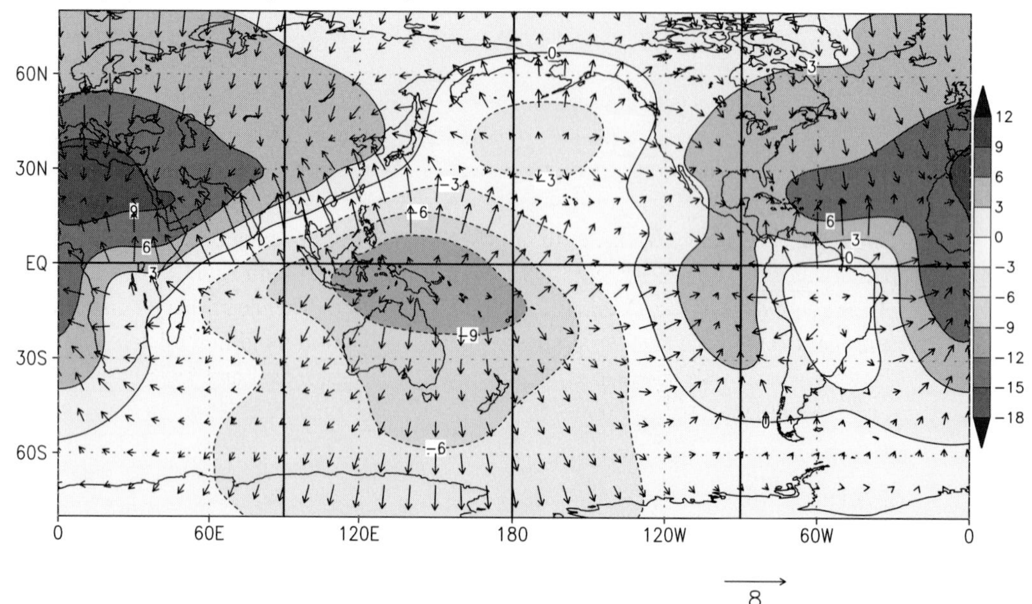

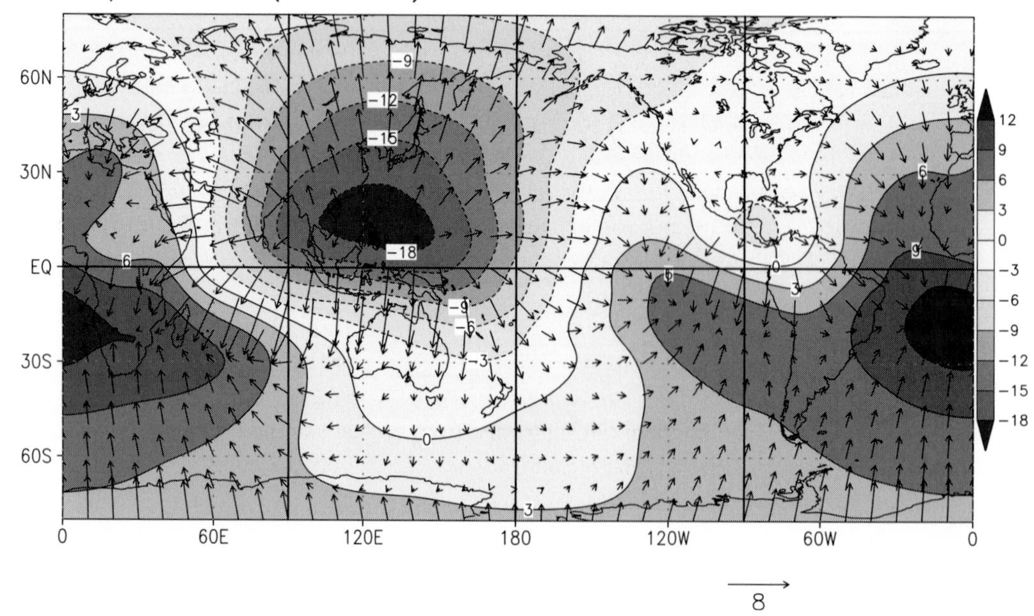

Figure 4.6 The mean 200 hPa velocity potential (10^6 m^2/sec), contour interval 1 unit, and divergent component of the wind (m/s) for the 1981 to 2010 period based on the NCEP Reanalysis 2 (Saha et al., 2010) for DJF (top), and JJA (bottom). Red shading denotes positive values, blue shading negative values. A black-and-white version of this figure appears in some formats. For the color version, please refer to the plate section.

Figure courtesy of M. Chelliah, CPC.

Conversely, the centers of positive velocity potential are associated with areas of convergence that are mirrored by low-level (925 hPa) divergence (not shown).

The areas of upper level convergence are associated with sinking motion and generally dry conditions. The mean positive velocity potential centers have their largest magnitude in the winter hemisphere, west of the South American and African continents in July through August (JJA) and over northern Africa in December through February (DJF). The seasonal hemispheric symmetry is, however, not perfect in that the African continent and the equatorial Pacific west of South America tend to show upper-level convergence in both seasons. Furthermore, the velocity potential changes sign between winter and summer in Asia, North America, and much of South America but it does not change sign over Africa, Australia, and much of Europe. The mean characteristics in the seasonal cycle of the velocity potential need to be kept in mind in the interpretation of the velocity potential anomalies used to monitor intraseasonal and interannual climate variability in equatorial regions.

4.2.2 Sea Level Pressure

The winds at any given level in the atmosphere are closely related to the pressure gradient at that level. Maps of the atmospheric pressure at sea level were of particular interest in the days of sailing ships, thus analysis of the mean seasonal cycle of sea level pressure has a long history. Since atmospheric pressure decreases with height, the sea level pressure and surface pressure differ at locations that are not at sea level. Surface pressure observations at weather stations are routinely converted, or "reduced," to the equivalent sea level value for weather and climate analysis, since the surface pressure differences between stations include the differences in pressure associated with differences in station elevations. The computation of sea level pressure from station pressure is based on the station elevation, air temperature at the time the station pressure was recorded, and the assumption of a standard atmospheric profile between the station and sea level. (In the standard atmosphere, the air temperature decreases with height by 6.5° C per km.)

The sea level pressure (SLP) changes significantly through the annual cycle (Figure 4.7a). The seasonal contrasts of SLP by season are emphasized in the maps of the differences in SLP for northern summer (JJA) subtracted from the SLP for northern winter (DJF), and northern fall subtracted from northern spring (Figure 4.7b). The SLP is relatively greater over the continents than over the oceans in the winter hemisphere and tends to be relatively lower over the continents than over the oceans in the summer hemisphere. The seasonal contrast in the mean pressure fields is particularly striking in the Northern Hemisphere and reflects the contrast between the cold dense air over the land areas during winter and the relatively warm, less dense, air over oceans.

The seasonal SLP contrast between the continents and oceans is not as strong in the Southern Hemisphere except for Australia and at higher latitudes. This is largely a consequence of the southern landmass configuration, with most of the continental areas equatorward of 30° of latitude. The magnitude of the contrasts in SLP between the transition seasons, spring and fall, is much smaller than the winter to summer contrasts

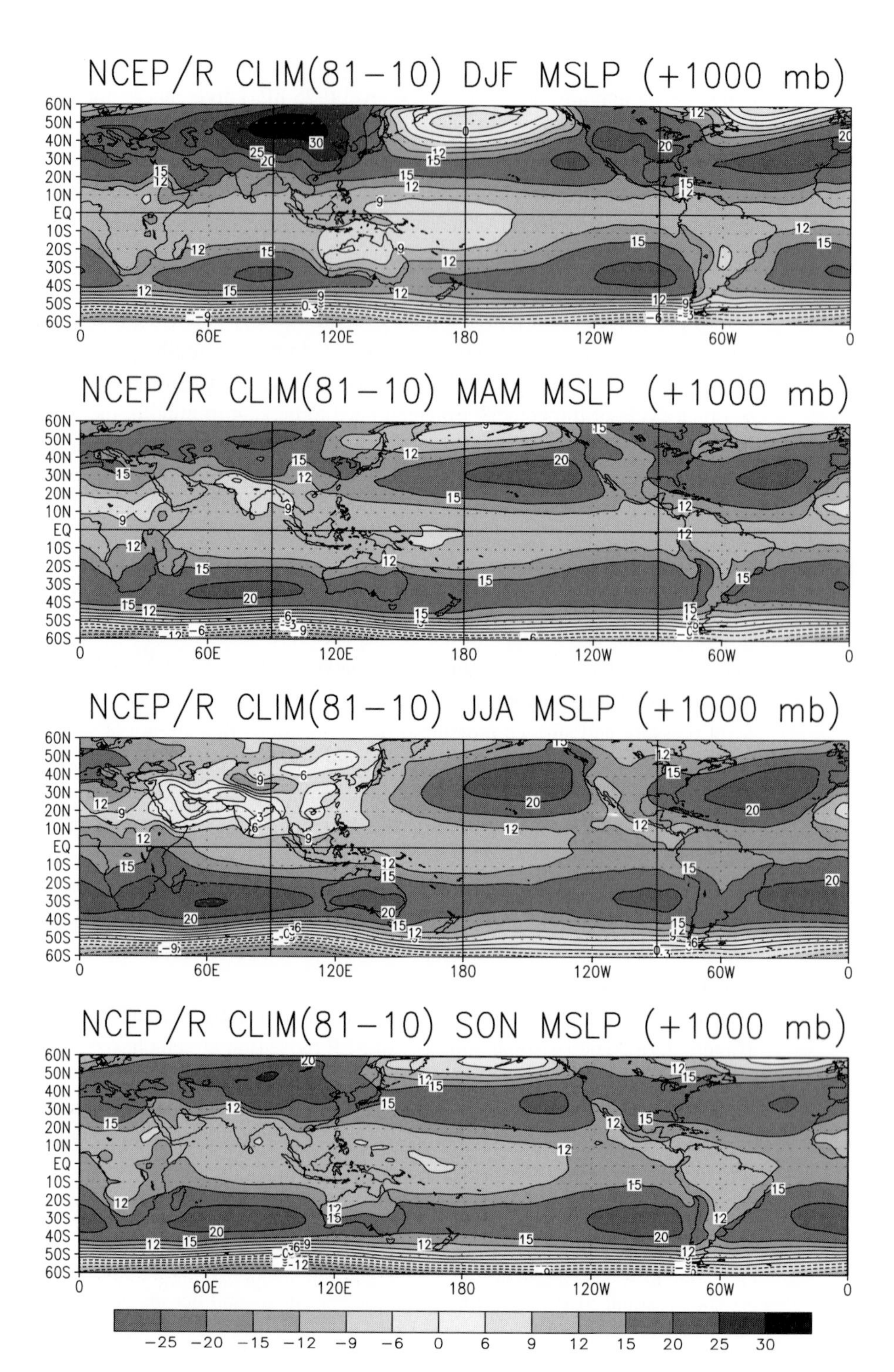

Figure 4.7a Mean sea level pressure (a) DJF, (b) MAM, (c) JJA and (d) SON for the 1981–2010 period based on the NCEP Reanalysis 2. A black-and-white version of this figure appears in some formats. For the color version, please refer to the plate section.
Figure courtesy of M. Chelliah, CPC.

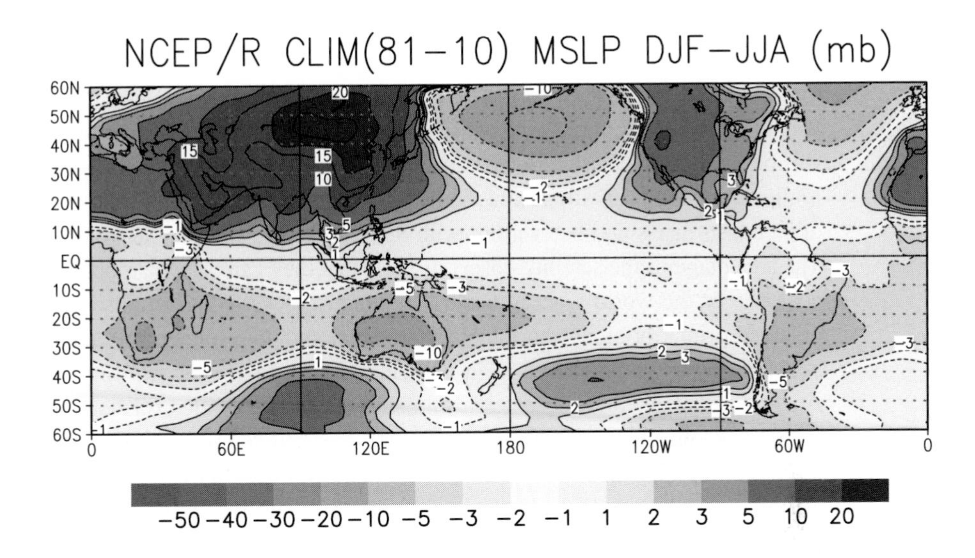

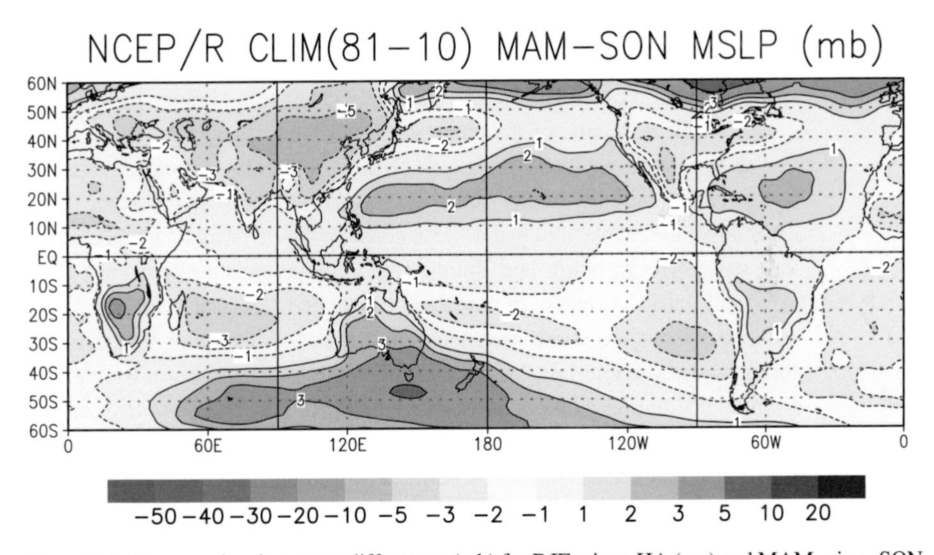

Figure 4.7b Mean sea level pressure differences (mb) for DJF minus JJA (top) and MAM minus SON for the 1981–2010 period based on the NCEP Reanalysis 2 (Saha et al., 2010). A black-and-white version of this figure appears in some formats. For the color version, please refer to the plate section. Figure courtesy of M. Chelliah, CPC.

at mid-to-high latitudes. In equatorial regions of the Atlantic and Pacific the MSLP seasonal differences are generally less than 3 hPa.

One advantage of taking seasonal differences when using the model-based analyses is that it eliminates any constant bias (although not seasonally varying biases) associated with the model, somewhat mitigating the influence of changes that may have been made to the model over the analysis period. Analysis of seasonal SLP differences also helps to minimize biases associated with the reduction to SLP in regions with very high mountains, for example in the Himalayas, where the standard atmosphere assumption may not be accurate.

4.2.3 Sea Surface Temperature (SST)

The profound importance of the SST in the climate system and in patterns of climate variability was not appreciated until the latter half of the twentieth century. In part, this was because SST over the global oceans was not adequately sampled until the advent of routine satellite observations and the development of analysis tools that would combine, or blend, surface ship observations with SST estimates from satellites (Reynolds, 1988). The advantage of the satellite observations is that they can provide global or near global coverage of the world's ocean basins, in contrast to surface ship observations that are obtained primarily along the major commercial shipping routes. The satellite observations, however, are generally biased with respect to the *in situ* ship observations. Anchoring the satellite estimates to the ship data has provided relatively accurate and complete global coverage since about 1980, as discussed in Chapter 8. These modern observations confirmed that the SSTs tend to be warmest in the tropical western Pacific and into the Indian Ocean as well as in the tropical and western Atlantic Ocean (Figure 4.8). In the western Pacific, the warmest waters, the so-called warm pool, straddles the equator throughout the annual cycle, but expands into the subtropics during the summer season in each hemisphere. The mean amplitude of the annual cycle in Western Pacific tropics is generally on the order of 1°C but, as will be further discussed in the section on precipitation, this relatively small annual range has a large influence on the atmosphere in parts of the basin where the mean temperatures are near 28°C.

The waters along the west coasts of the continents (the eastern sides of the ocean basins) are generally cooler than waters further west at the same latitude. This is especially noticeable along the west coast of South America but cooler water along the coasts of both North and South America is associated with the equatorward wind flow and ocean currents. In a mechanism called "Ekman transport,"[8] ocean currents induced by surface winds result in a net transport of water to the right (left) of the forcing wind direction in the Northern (Southern) Hemisphere. Ekman transport arises from an interaction between the surface wind forcing, the Coriolis force (See Box 4.1), and friction. The water transported away from the region of maximum forcing winds is replaced by cooler subsurface water, a process referred to as upwelling. The upwelling results in relatively cooler water along the West Coasts of the Americas. Easterly surface winds are typical along the equator in the eastern Pacific, where Ekman transport causes surface water to diverge from the equator and be replaced by relatively cool subsurface water, leading to a climatologically favored "cold tongue" (Wallace and Hobbs, 2006). This region of colder water along the equator in the eastern Pacific extends westward to the date line and is easily visible in the mean SST for both the winter and summer seasons.

The seasonal cycle of SST can also be seen in both hemispheres in the contrasts between the winter and summer as well as the contrasts between spring and fall (Figure 4.9). The seasonal SST differences are largest at mid and high latitudes, as

[8] In the late 1800s Arctic explorers noticed that icebergs tended to move at an angle to the winds forcing their motion. The phenomenon was named after V.W. Ekman (1905) who first explained it.

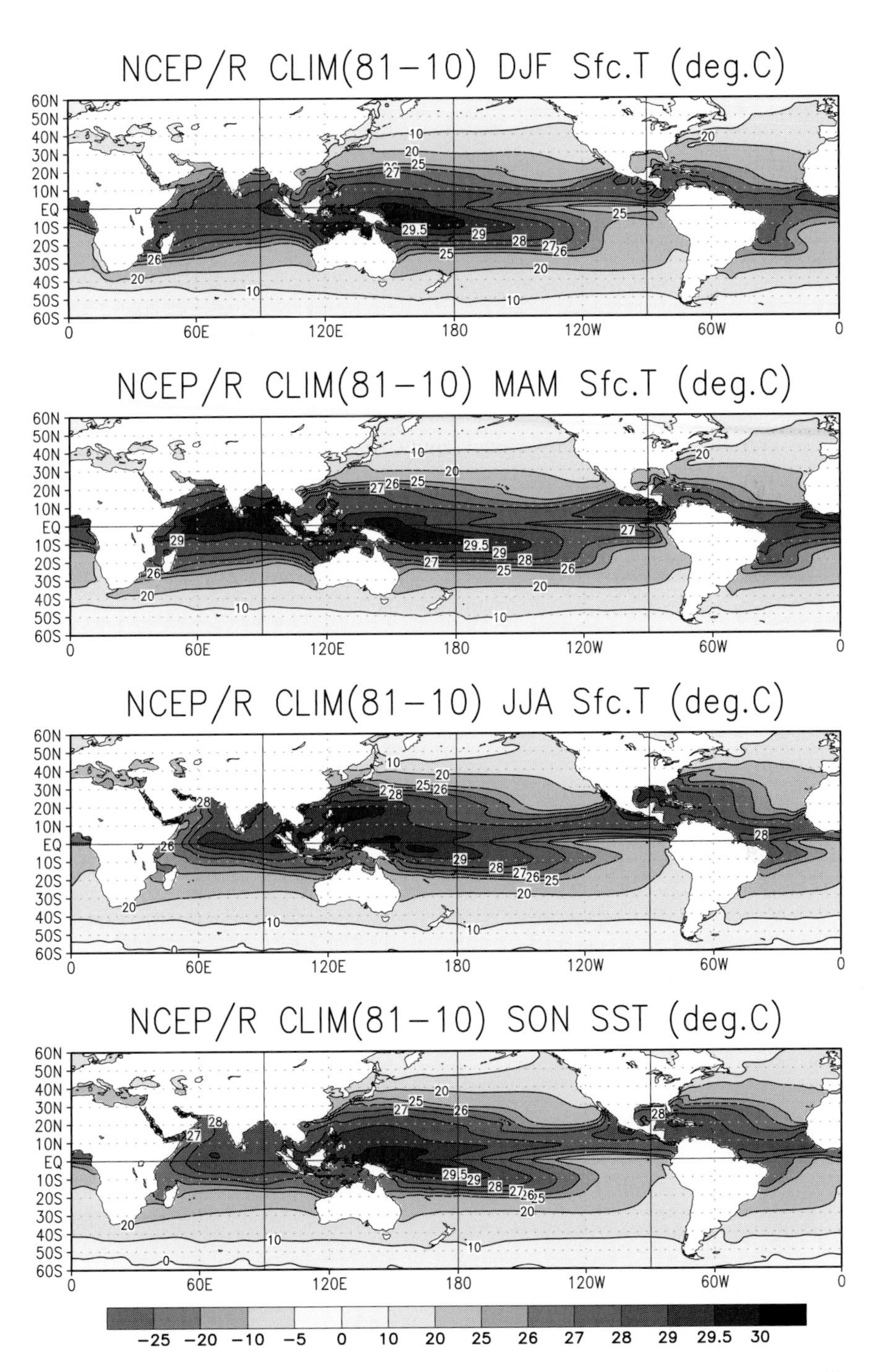

Figure 4.8 The seasonal mean (1981–2010) sea surface temperature (SST) based on satellite and *in situ* data for (a) DJF, (b) MAM, (c) JJA and (d) SON (°C). Temperatures greater than 0°C are shown in warm shades, temperatures greater than 28°C are shown in orange and red. A black-and-white version of this figure appears in some formats. For the color version, please refer to the plate section.

Figure courtesy of M. Chelliah, CPC.

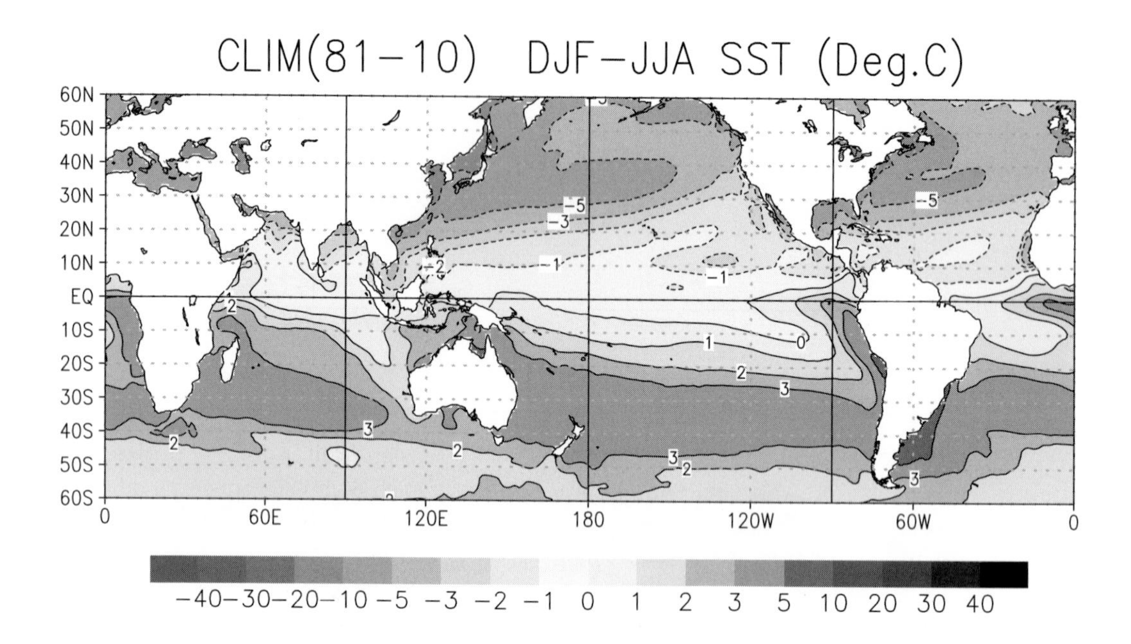

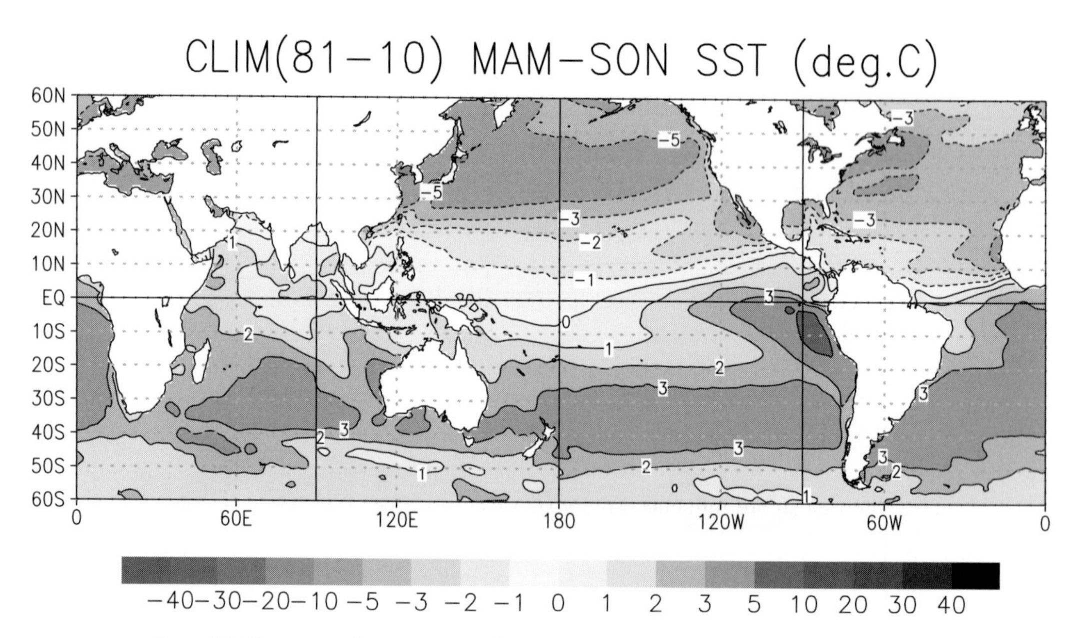

Figure 4.9 The seasonal mean sea surface temperature (SST), differences between winter and summer (DJF minus JJA, top) and between spring and autumn (MAM minus SON, bottom) in °C based on satellite and ship data for the period 1981–2010. A black-and-white version of this figure appears in some formats. For the color version, please refer to the plate section.
Figure courtesy of M. Chelliah, CPC.

would be expected, and generally less than 1°C along the equator except in the cold tongue regions and in the Indian Ocean. The cold tongue is weaker, that is, less cold, from northern winter through spring, thus, the eastern equatorial Pacific SST is relatively warmer during these seasons. This seasonal warming along the west coast of South America is a regular feature of the mean annual cycle and the onset of the annual warming was historically referred to as El Niño in Ecuador and Peru. Years when the relative warming exceeds its normal mean amplitude have come to be called "El Niño," "El Niño /Southern Oscillation," or "warm ENSO" years by much of the climate community and will be discussed in more detail later in this chapter.

4.2.4 Air Temperature Over Land

Air temperature over land is discussed in much greater detail in Chapter 6. Here, for completeness in the seasonal cycle discussion presented in this section, we present the mean seasonal cycle of air temperature over land. The analysis is based on *in situ* temperature observations in the latitudinal domain between 70°S and 70°N based on data described in Fan and van den Dool (2008). Considerably more of Earth's land mass lies in the Northern Hemisphere than the Southern Hemisphere. Thus, the seasonal mean air temperature and seasonal temperature anomalies over land tend to be dominated by Northern Hemisphere observations. The lowest seasonal mean temperatures in our analysis domain occur in the Northern Hemisphere during winter (Figure 4.10a). The highest mean seasonal temperatures over land for every season occur in Africa north of the equator and equatorial South America except during Boreal winter (DJF), when they also occur over interior regions of Australia.

The largest amplitudes in the range of the seasonal mean temperatures over land also occur in the Northern Hemisphere (Figure 4.10b). The mean seasonal DJF minus JJA differences in air temperature over land are greater than 35°C in the high Northern Hemisphere latitudes in contrast to mean seasonal differences of less than 20°C in high Southern Hemisphere land areas. The mean seasonal temperature differences for the transition seasons, MAM minus SON, are generally less than 5°C, with the SON season slightly warmer than the MAM in both the hemispheres except in central India. Global and regional land surface temperature anomalies are influenced by ENSO and the MJO (discussed later). Regional anomalies in eastern North America, Europe, and the eastern Mediterranean are influenced by NAO (also discussed later.)

4.2.5 Precipitation

The mean annual cycle in precipitation is shown in Figure 4.11 in the form of maps of the mean for each calendar season. Prominent features include continental maxima that migrate between the summer hemispheres and oceanic bands that exhibit relatively less movement through the year. In general, the strongest precipitation maxima are found in the tropics, with secondary maxima in the midlatitude oceans. Averaging over the Northern and Southern Hemispheres shows that precipitation is maximized during the Boreal summer, with more precipitation over land than ocean. Time series of globally

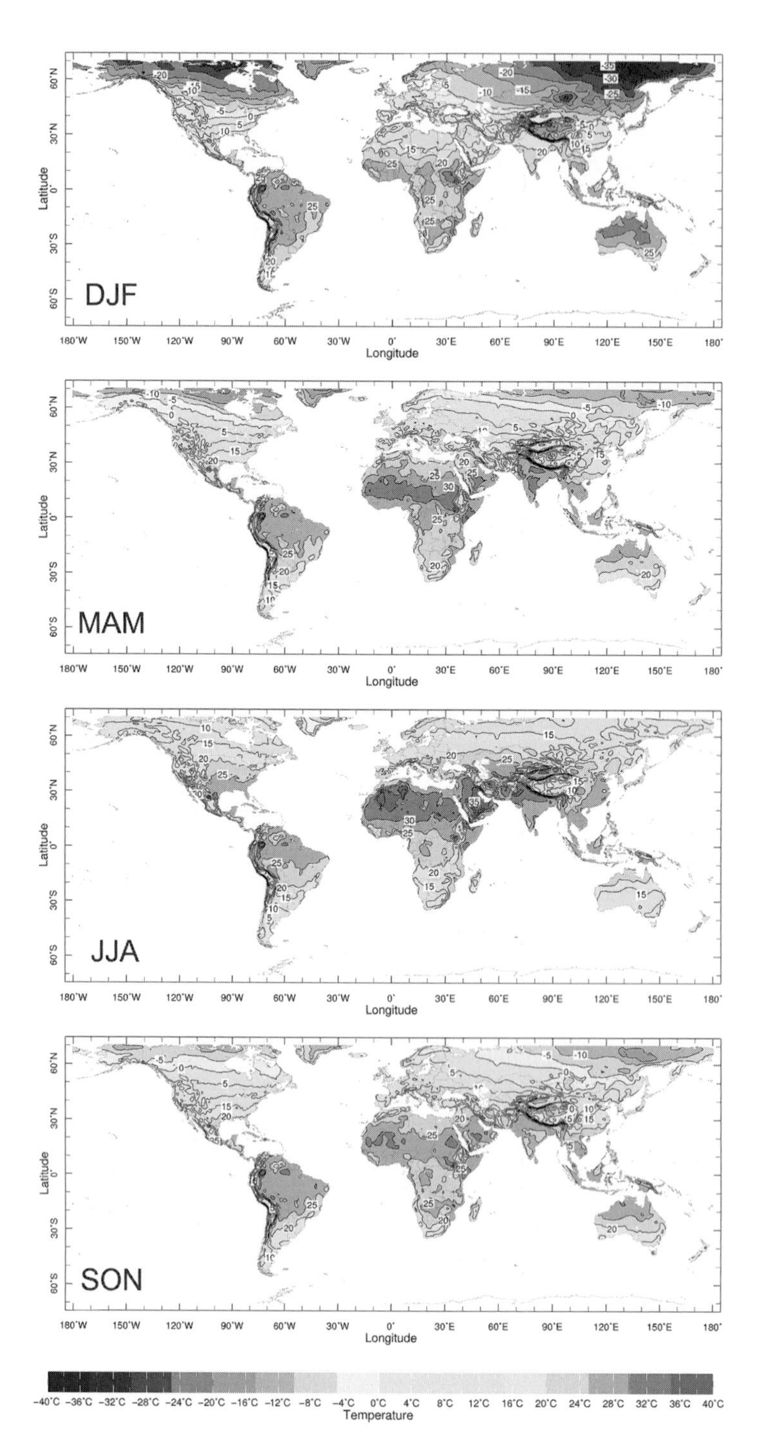

Figure 4.10a The seasonal mean surface air temperatures over land based on the GHCN-CAMS gridded monthly temperature database (Fan and van den Dool, 2008) for the 1981–2010 base period. A black-and-white version of this figure appears in some formats. For the color version, please refer to the plate section.

Figure provided by M. Bell, IRI/LDEO Data Library.

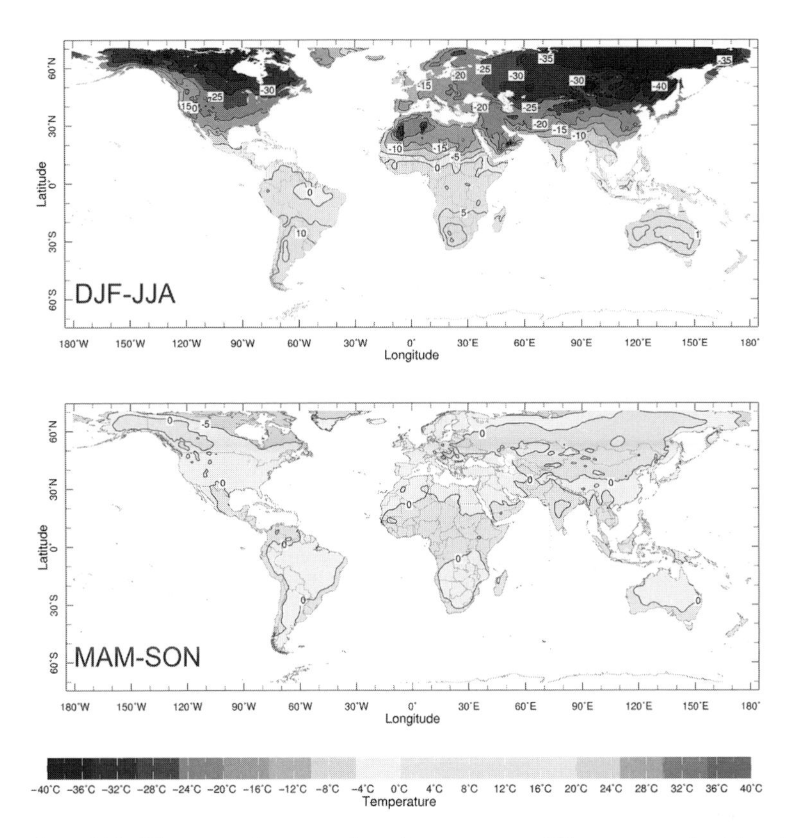

Figure 4.10b The seasonal mean surface air temperature differences over land based on the GHCN-CAMS gridded monthly temperature database (Fan and van den Dool, 2008). Northern Hemisphere winter minus summer (DJF-JJA), top, and spring and autumn (MAM-SON), bottom. A black-and-white version of this figure appears in some formats. For the color version, please refer to the plate section.
Figure provided by M. Bell, IRI/LDEO Data Library.

averaged precipitation from satellite estimates and rain gauge observations do not exhibit a clear-cut annual cycle, while those based on models that assimilate surface and satellite observations have a maximum in Boreal summer and amplitude of about 0.2 mm/day (just over 5% of the overall mean).

Prominent precipitation maxima over Africa and South/Central America progress northwestward and southeastward through the course of the year, with greatest distance from the equator occurring during the month following the solstice in each hemisphere. The reason for departure from a north–south path seems clear in the Americas, where the maximum precipitation is over the land mass north of the equator, but not so in Africa, where other mechanisms must contribute.

A similar northwestward/southeastward migration is observed in Australasia, between northern Australia and Southeast Asia. One interesting difference, compared to Africa and the Americas, is that much of the precipitation associated with this feature occurs over the Indonesian archipelago, which is at least as much ocean as land. Despite this, the annual variations in this area more closely resemble the other tropical land areas than they do oceanic structures.

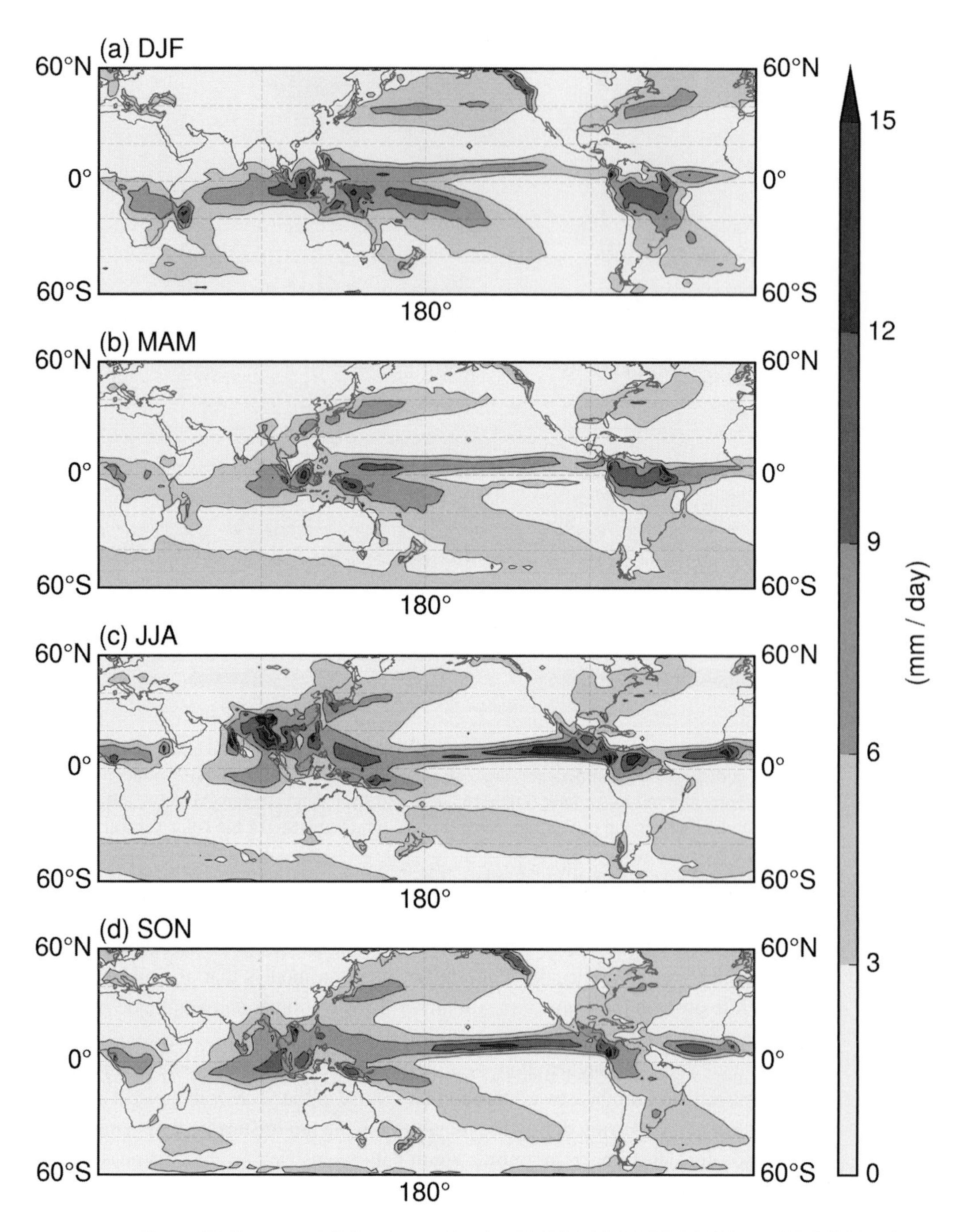

Figure 4.11 Mean precipitation in mm/day for (a) DJF, (b) MAM, (c) JJA and (d) SON for the period 1981–2010 based on satellite and surface station (GPCP) data. A black-and-white version of this figure appears in some formats. For the color version, please refer to the plate section. Figure courtesy of G. Huffman and J. Tan, NASA.

The precipitation in this region is a major component of the Australasian Monsoon system, discussed earlier in conjunction with the global wind fields. Even though many climate studies tend to describe the monsoon systems in terms of the wind fields, precipitation is often the most widely recognized aspect of the phenomenon. The connection between the evolution in the wind fields and precipitation in the Australasian Monsoon has led some to suggest that similar relationships might hold in the Americas and Africa. The situation, however, is not clear: The migration of precipitation from Central America and Mexico to the southwestern United States is connected to the North American Monsoon winds. The South American and African counterparts, however, have well-defined rainy seasons but they are not so closely tied to reversals in the winds fields as the other monsoon areas. Oceanic precipitation in both the tropics and midlatitudes tends to be organized into zonally oriented bands with greater east–west than north–south extent. The best known of these is the Inter-Tropical Convergence Zone (ITCZ), which is present, in the monthly and seasonal mean, throughout the year in the Atlantic and Pacific Oceans and for much of the year in the Indian Ocean. The ITCZ tends to migrate northward and southward, but with less latitudinal range than the continental maxima. While each ocean basin exhibits an ITCZ, considerable basin-to-basin variation is seen, and the overall patterns can be quite complex.

Two other prominent types of oceanic precipitation bands are observed, one tropical and subtropical, and the other in the midlatitudes. The South Pacific, South Atlantic, and Indian Ocean Convergence Zones (SPCZ, SACZ, and IOCZ, respectively) originate on the southeastward periphery of the continental/monsoon maxima over Australasia, South America, and Africa, respectively. The SACZ and IOCZ are summer phenomena that extend southeastward from the southernmost continental maxima into the adjacent ocean basin. During Austral winter, these features are not found, and no equivalent Northern Hemisphere features are observed. The SPCZ is more persistent in the mean, extending from near the equator southeastward during each season of the year.

The other precipitation bands over the oceans are associated with midlatitude storm tracks. In the Northern Hemisphere, the Atlantic and Pacific precipitation bands are very clearly distinct from each other, but exhibit similar seasonal cycles. During the Boreal winter, each band extends generally eastward and slightly northward from the eastern coast of the bordering continent, with more of a northward component in the eastern part of the ocean. During Boreal summer, the western end of each band exhibits a connection to the adjacent continental maximum. In the Southern Hemisphere, the corresponding storm track-associated band is significantly weaker and largely confined to the latitudes between 40°S and 60°S. Each of these features exhibits a great deal of complex behavior, far more than we describe here.

4.3 Patterns of Climate Variability

4.3.1 What Do We Mean by Climate Variability?

Weather variations are atmospheric fluctuations that are strongly dependent on, and predictable from, the initial state of the atmosphere. They have time scales ranging from minutes through 10 days or so. We focus here on climate variations that can be identified from statistics of atmospheric variations and have time scales ranging from about 10 days through decades. Analysis of climate variability aims to identify, describe, and understand recurring climate patterns, with the goal of developing methods to identify their onset and forecast their evolution and recurrence if possible. In our discussion of the analysis of climate variability we assume that very slow changes in climate, including trends, can be identified and filtered out of the datasets, allowing climate variability to be defined with respect to the mean climate of the detrended data over the period of record. While we assume here that long-term trends do not change the characteristics, such as the duration, strength, and frequency, of climate variability on shorter time scales, we note that this assumption has not been verified and may be incorrect.

Climate analysis has revealed a number of patterns of climate variability with time scales from intraseasonal (less than a season) through multidecadal. These patterns, often referred to as "modes," will be defined and discussed in this section. The ability to analyze these modes in any region is limited by the completeness, frequency, and duration of the data. Although many of the patterns are called oscillations because extremes of these patterns tend to alternate, none of them recur at fixed regular intervals, like a clock pendulum. In addition to the time scale associated with the change of a mode from the midpoint to an extreme and back to the midpoint, there may be a time scale associated with the intervals between successive occurrences of the patterns. For example, an ENSO episode (discussed in more detail later in this chapter) typically has a time scale of a year or longer (time scale of the phenomenon), but successive episodes most often occur at intervals of from two years to seven years. We will discuss in this section the general characteristics of several climate anomaly[9] patterns and the terminology and data typically used to describe, study, and monitor them. Most of the following discussion is based on atmospheric circulation phenomena, since data from this part of the climate system have some of the longest and most complete records in the instrumental era. However, one of the goals of climate analysis is to discover how these patterns interact with and manifest in other components of the climate system.

4.3.2 Submonthly to Intraseasonal Climate Variability

Submonthly

Submonthly and intraseasonal climate variations include phenomena that occur on time scales ranging from several days through several weeks. The analysis of climate patterns at these temporal scales is often performed with five-day or weekly means, but the study of some patterns may require daily data. Climate analysis on the submonthly scale may include the statistics of, for example, dry and wet spells, heat waves, cold outbreaks, and atmospheric circulation patterns. Case studies may be performed to

[9] An anomaly – in climate studies refers to the deviation of a parameter, often temperature or precipitation, in a given region over a specified period from the long-term average value for the same region. (After the Glossary of Meteorology, 2012).

diagnose individual events, but in the climate context, the statistics of several, preferably many, similar events are needed. Composite analysis, also referred to as superposed epoch analysis, is another tool used to isolate common features of a phenomenon. In composite analysis, averages of relevant parameters over several occurrences based on some index or chronology are calculated (see, for example, Reed and Recker, 1971).

Historically, meteorologists have studied anomalous circulation features in the atmosphere that persist for several days to a week or longer, sometimes called blocking[10] events, by analysis of the strength of the midlatitude upper-level westerly winds. These events were often identified and their strength measured by the zonal index.[11] Variations in the zonal index were referred to as index cycles. In the recent literature, references to the zonal index and index cycles have often been supplanted by discussions of one or more of the patterns of climate variability discussed in this chapter. Variations in the zonal index and other short-lived phenomena at the boundary between weather and climate are still not well understood. Long-lived weather patterns associated with these persistent circulation configurations, nonetheless, can have a strong influence on the mean climate over a month or season. Climate models often do not adequately replicate the magnitudes and frequency of the observed submonthly variability. The existence of such events, and others that persist for several days, illustrates the need for daily observational data in the analysis of climate.

Intraseasonal

The time scale of intraseasonal climate variability is generally taken to be less than a three-month season for the full duration of a single event; for example, the Madden-Julian Oscillation (MJO) generally occurs with periods of 20–70 days. The MJO, one of the most actively studied intraseasonal phenomena, is characterized by a propagating circulation and precipitation feature occurring in a vertical plane through the troposphere along the equator and with horizontal scale of 10's of degrees of longitude. The MJO was first noted in an observational study of atmospheric circulation in the deep tropics (Madden and Julian, 1971, 1972), and became an active research topic in the following decade (see the summary in Madden and Julian, 1994). The early studies were based on spectral analysis of 10 years of radiosonde data from island locations across the equatorial Indian and Pacific Oceans. Real-time monitoring of the MJO is now supplemented by model-based analyses of the atmospheric circulation and satellite-derived observations of Outgoing Long-wave Radiation (OLR) and precipitation. The MJO appears to originate in equatorial regions of the western Indian Ocean and propagates eastward across the Pacific before typically losing identity over South America and the equatorial Atlantic (Figure 4.12). Analyses have found modest links between the MJO and variations of rainfall in North America and other off-equatorial

[10] In midlatitudes, the progression of low and high pressure systems from west to east at the surface is associated with winds in the free atmosphere. Blocking refers to a slowing or temporary cessation of this normal progression. (Glossary of Meteorology, 2012).

[11] "A measure of strength of the middle-latitude westerlies, usually expressed as the horizontal pressure difference between 35° and 55°N latitude." (Glossary of Meteorology, 2012).

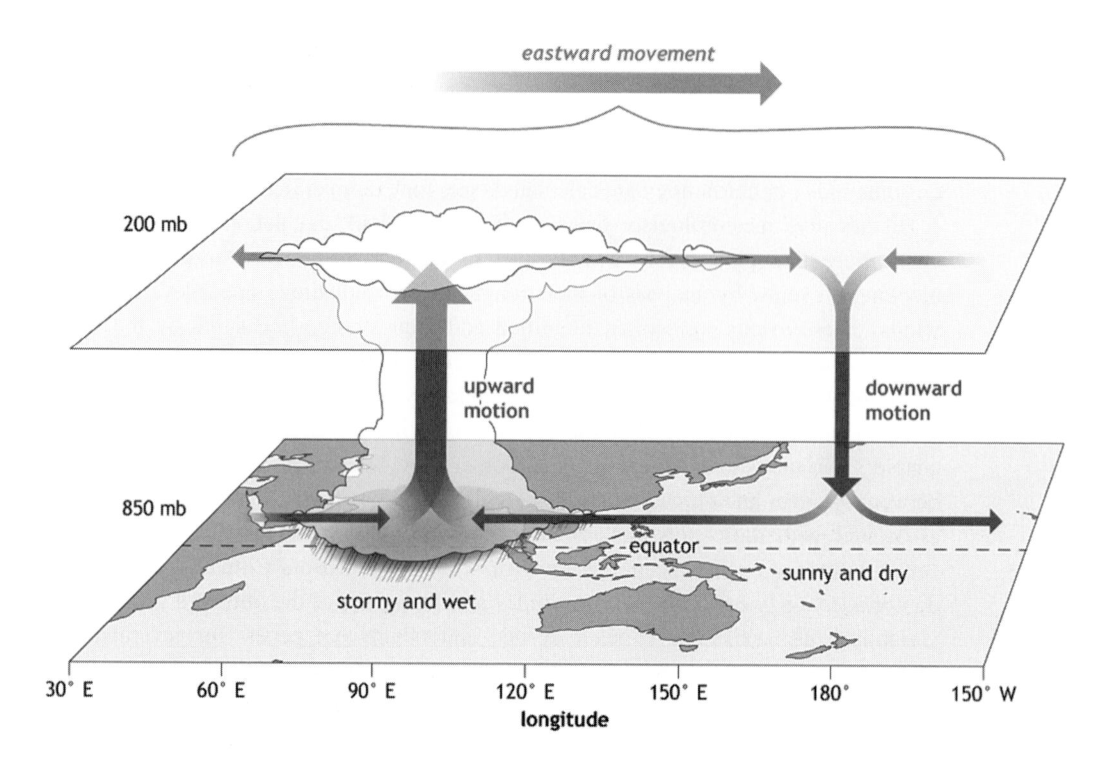

Figure 4.12 The atmospheric structure of the Madden-Julian Oscillation (MJO) for a period when the convective rainfall (indicated by the storm cloud) is centered in the Indian Ocean and less than average convective rainfall occurs in the western to central Pacific. The horizontal arrows pointing left represent wind departures from average that are from the east, and arrows pointing right represent wind departures from average that are from the west. The entire pattern moves eastward, until dissipating over South America or the Atlantic Ocean.
Source: NOAA/Climate.gov, drawing by Fiona Martin.

locations (Becker et al., 2011) as well as between the MJO and tropical storm activity (Barrett and Leslie, 2009; Klotzbach, 2014).

4.3.3 Monthly and Seasonal Phenomena

Climate studies have traditionally focused on monthly and seasonal patterns. Even though averaging over one month, or three months, may mask some climate patterns, many monthly and seasonal recurring atmospheric circulation patterns have been identified in the historical record. Before the advent of routine upper air observations, several monthly and seasonal patterns were identified in the analysis of sea level pressure, most notably in the correlation, or teleconnection,[12] studies of Walker (1924, 1928) and Walker and Bliss (1932, 1937). Later in the century, a number of

[12] Teleconnection is a statistical linkage between climate variations occurring at widely separated regions. The term was first used by Sir Gilbert Walker in his seminal studies of global pressure patterns.

modes were identified in the data from upper level atmospheric analyses (Barnston and Livezey, 1987). Most prominent among these patterns in the Northern Hemisphere are the Pacific North American (PNA), first introduced into climate studies in the 1980s by Wallace and Gutzler (1981), the North Atlantic Oscillation (NAO), first identified by Walker and Bliss (1932), and the Arctic Oscillation (AO – sometimes referred to as the Northern Annular Node, or NAM), Thompson and Wallace (2000). Figure 4.13 illustrates these phenomena. The distinction between the NAO and the AO/NAM is that the NAO refers to climate variability over the North Atlantic basin and surrounding land areas, while the domain of AO/NAM includes climate variability over the entire Northern Hemisphere. The most commonly monitored circulation pattern found in the Southern Hemisphere is the Antarctic Oscillation (AAO – Figure 4.13), also referred to at the Southern Annular (SAM) mode.

The patterns described earlier in this section, as well as others, are often linked to anomalous precipitation and temperature patterns and thus are routinely monitored using indices derived from atmospheric circulation statistics (see the Climate Diagnostics Bulletin on the Climate Prediction Center website listed in the references). The PNA and NAO, in particular, have a large influence on seasonal climate variability over North America and Europe, particularly in the cold half of the year.

The PNA is named for the locations of the pressure centers that define the pattern, which are a low pressure center in the North **P**acific coupled with two pressure centers over **N**orth **A**merica, a high pressure center usually in western Canada paired with a low pressure center usually in the southeastern United States or Gulf of Mexico (Figure 4.13). The PNA is one of the few patterns discussed here that has not been dubbed an "oscillation." The PNA has often been linked to ENSO (discussed in this chapter) in the positive, or warm phase, though not the cold phase, but sometimes appears independently of ENSO. The location of the low and high pressure centers associated with the PNA account for its association with precipitation and temperature anomalies over much of North America.

The NAO is one of the sea level pressure patterns identified by Walker and Bliss (1932). It has been monitored by indices based on differences in sea level pressure between observations in the North Atlantic and in Southern Europe. One such monthly index, based on sea level pressure data from Reykjavik, Iceland, and Lisbon, Portugal, has been extended back to the 1860s. Cold season precipitation and temperature anomalies over eastern North America and Western Europe and beyond can often be related to the circulation patterns associated with the NAO. Positive values of the NAO index are associated with colder and drier conditions over eastern Canada and Greenland, warmer and wetter conditions in the eastern United States and in much of northern Europe extending into Eurasia, and drier conditions in southern Europe into the eastern Mediterranean. Anomalies of opposite sign are associated with negative values of NAO index. A more thorough discussion of the NAO and its impacts can be found in Hurrell et al. (2003). While the influences of the NAO on climate variability over North America, Europe, and parts of Eurasia are large, model-based and empirical forecasts of the NAO magnitude and phase have not met with significant success for time scales longer than a week or two.

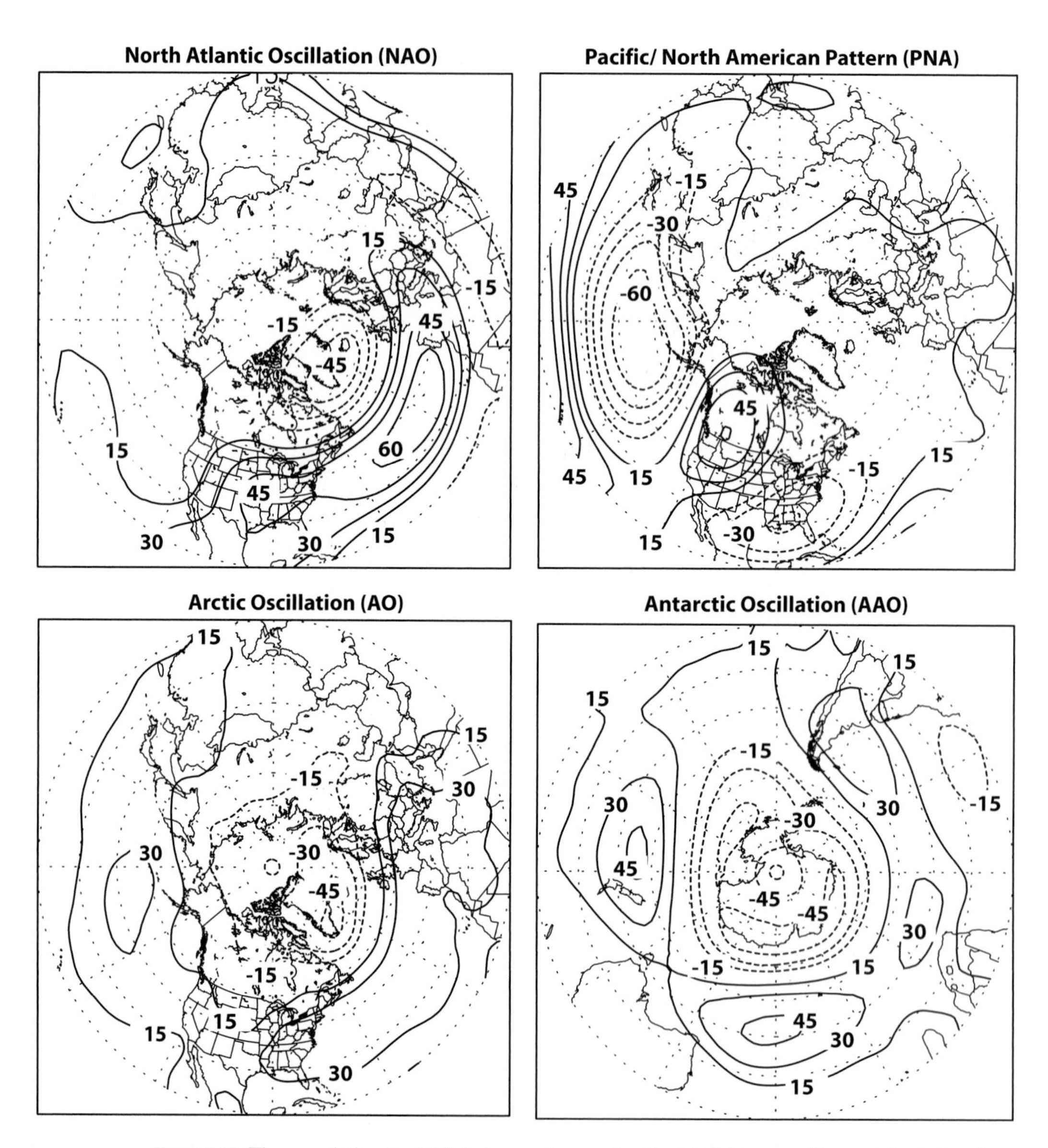

Figure 4.13 The correlation (× 100) between the standardized height anomalies and the index time series. For the NAO and PNA the correlation is based on DJF 500-hPa height data and NAO and PNA index values for 1950–2009. For the AO and AAO, the correlations are based on all of the monthly anomalies (and indices) at 1000 and 700-hPa, respectively. All months in the year go into calculating the AO and AAO correlation maps for the 1979–2009 period.
Figure courtesy of G. Bell CPC/NOAA.

The Northern Annular Mode (NAM), or Arctic Oscillation (AO), shares many features with the NAO but expands the domain of influence to include the remainder of the globe at high northern latitudes. The Southern Annular Mode (SAM)/Antarctic Oscillation (AAO), likewise, influences the weather and climate at high latitudes in the

Southern Hemisphere (Thompson and Wallace, 2000). These patterns and the others discussed in this section, apart from the MJO, are primarily or completely internal modes of the atmosphere, lacking strong coupling with the land surface or ocean. One apparent consequence of that lack of coupling is that none of these phenomena has a strongly preferred time scale.

4.3.4 Interannual Variability

In this section, we describe climate patterns that typically have durations of a year and longer. We start the discussion with the El Niño/Southern Oscillation (ENSO) and then move on to phenomena with longer time scales.

The El Niño – Southern Oscillation (ENSO)

ENSO is characterized by warm and cold "episodes" that have typical lifetimes of about a year and recur at intervals of between two and seven years. Warm and cold episodes often alternate, sometimes without a neutral gap between them, but other patterns of occurrence have been observed. ENSO accounts for the largest amount of variance in the climate system on the interannual time scale. The variance associated with ENSO is second only to that of the annual cycle itself. ENSO is a global phenomenon, with origins in the tropical Pacific Ocean basin, and represents the clearest example of coupling between the ocean and atmosphere in the climate system on interannual time scales.

Recognition of ENSO's role in interannual climate variability is the result of efforts by many scientists whose contributions span much of the twentieth century. Aspects of ENSO were recognized in the late nineteenth century (Hildenbrandsson, 1897) and early in the twentieth century primarily in a series of statistical studies. Most notable of these early studies are the seminal papers of Sir Gilbert Walker (Walker 1924, 1928; Walker and Bliss, 1932, 1937) early in the century and the groundbreaking synthesis several decades later by J. Bjerknes (1969) as well as pioneering work on the ocean component of ENSO by K. Wyrtki (1975).

One of the classic problems in seasonal climate prediction has been, and continues to be, the successful, routine, prediction of monsoon rainfall in the Indian subcontinent. In the early nineteenth century India was part of the British Empire. Following years with failed monsoon rains and extensive famine in the late nineteenth century, the British government in 1904 recruited Walker to a posting to India and asked him to devise a method to successfully predict the character of the monsoon rains over the subcontinent. His formal background being in mathematics, he attacked the problem by harnessing the computational power represented by the large number of civil servants employed by the colonial government to gather global weather data and use the data to calculate correlations among weather variables. These correlations between climate variables over large distances have come to be called "teleconnections."

In the series of studies referenced in this section, Walker identified several teleconnections in the sea level pressure fields around the globe. Among these patterns was the Southern Oscillation (SO). The SO is characterized by the tendency for sea level pressure anomalies over the western and eastern Pacific to be anticorrelated, with the

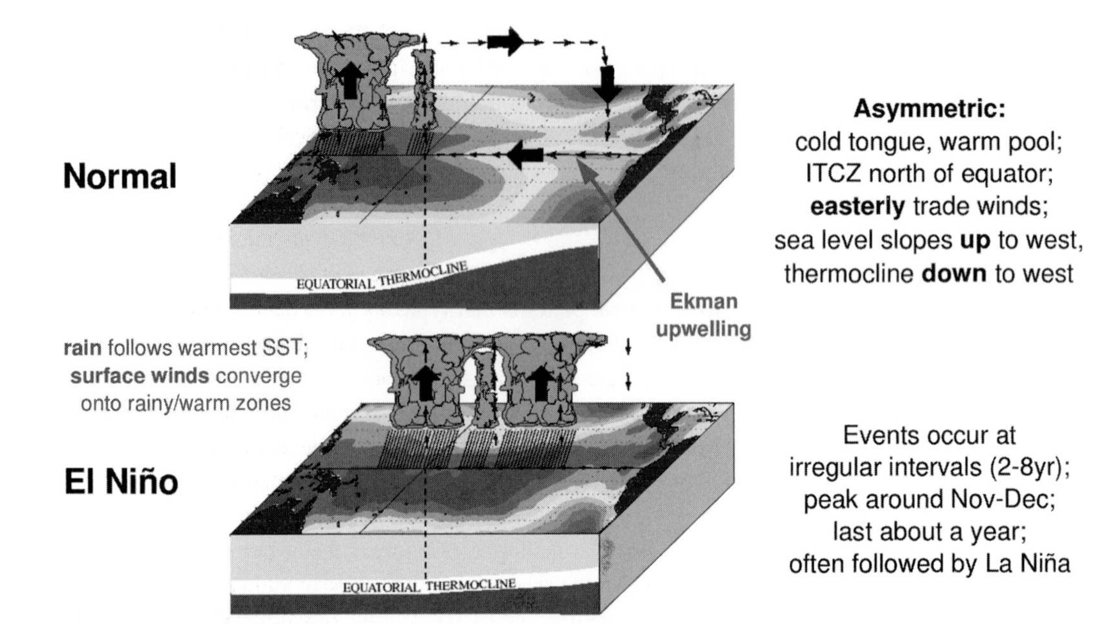
Normal
Asymmetric:
cold tongue, warm pool;
ITCZ north of equator;
easterly trade winds;
sea level slopes up to west,
thermocline down to west
EQUATORIAL THERMOCLINE
Ekman
upwelling
rain follows warmest SST;
surface winds converge
onto rainy/warm zones
El Niño
Events occur at
irregular intervals (2-8yr);
peak around Nov-Dec;
last about a year;
often followed by La Niña
EQUATORIAL THERMOCLINE

Figure 4.14 Schematic of normal versus ENSO conditions in the equatorial Pacific during northern hemisphere winter. A black-and-white version of this figure appears in some formats. For the color version, please refer to the plate section.
Source: NOAA/CPC.

largest (negative) correlations in the Southern Hemisphere. This relationship has come to be characterized and monitored through the Southern Oscillation Index (SOI). The SOI is computed from the difference between the sea level pressure anomalies in Darwin, Australia, and those at the island of Tahiti in the eastern Equatorial Pacific. Berlage (1966) was one of the first to note that the year-to-year variations in the magnitude of the yearly warming of the waters along the west coast of South America and in the eastern equatorial Pacific were correlated with the Southern Oscillation, and concluded that this correlation must be the result of interactions, or a coupling, between the atmosphere and ocean in a phenomenon that has come to be called ENSO.

In addition to the ENSO warm episodes discussed in this section the climate system sometimes evolves into what has come to be called La Niña,[13] a state characterized by anomalies generally opposite to those associated with the warm episode (Figure 4.14). Both the cold, La Niña, and warm, El Niño, ENSO episodes may occur in isolation, but the historical record shows several instances when an El Niño is followed by La Niña or visa-versa. The terminology "warm" and "cold" generally reflects the sign of the SST anomalies in the Eastern Equatorial Pacific.

[13] Fishermen in Ecuador and Peru, by tradition, called the annual ocean warming off the west coast of South America that begins late in the calendar year El Niño, or "the Child," since it roughly coincides with Christmas. The scientific community later appropriated that term to describe the greater than average warming that occurs in ENSO warm episodes, and coined the term La Niña, or "the girl," to refer to the ENSO cold episodes.

The idea of ocean–atmosphere coupling was further developed and significantly expanded in a groundbreaking study by J. Bjerknes (1969), who suggested that the atmospheric component of this interaction could be thought of as the waning and waxing of the Hadley and Walker circulations in the Pacific Basin. Improved monitoring of the tropical Pacific Ocean (see Chapter 8) has also revealed the role of the ocean subsurface in the phenomenon we now know as ENSO. Our current understanding of atmosphere and ocean coupling involved in ENSO has four main components. These are: (1) variation in the near global surface pressure anomalies characterized by the SOI, (2) anomalous seasonal warming or cooling of the sea surface temperature (SST) in the Equatorial Pacific, (3) significant alterations in the patterns of convective precipitation in the equatorial Pacific, and (4) major alterations in the ocean's substructure reflected in changes in the depth of the thermocline, all shown in schematic form in Figure 4.14. Research since the 1980s has shown that there are substantial variations in the evolution of each individual ENSO "episode" but that, in general, the phenomenon has the following characteristics in common. (1) The onset of an ENSO episode tends to occur in the latter part of the calendar year, with peak magnitudes in first quarter of the following year, a characteristic often referred to as "phase locking to the annual cycle," (2) a change in the surface pressure fields in the Pacific Basin reflected in strongly negative SOI's with warm episodes and a significant weakening, or complete collapse of the equatorial easterly winds in the Pacific, (3) anomalous SST along the equator with the largest anomaly magnitudes in the eastern Pacific but the highest correlation with the surface pressure, the SOI, in the central Pacific (near the dateline), (4) eastward displacement of convective precipitation in warm episodes and enhanced convective precipitation in the western equatorial Pacific during cold episodes. ENSO has been found to have a significant influence on seasonal precipitation and temperature for several regions of the world (Ropelewski and Halpert, 1986, 1987, 1989; Halpert and Ropelewski, 1992; Mason and Goddard, 2001), as shown in Figure 4.15.

Temperature and precipitation anomaly patterns associated with ENSO are a consequence of significant anomalies in the equatorial Pacific Ocean that influence the atmospheric circulation in the global atmosphere, for example, shifts in the Walker and Hadley circulation and midlatitude jet streams. The patterns of precipitation anomalies shown in Figures 4.15 and 4.16 are not expected to occur in conjunction with every ENSO episode for every region of the world but anomalies in each region identified in the figure have been shown to occur in 75–80% of ENSO episodes in the historical record. The strength and consistency of the ENSO-related temperature and precipitation relationship for any location varies with the time of year.

Considerable research has been performed from the 1980s onward to further the understanding and prediction of ENSO. These efforts have been supported by significant increases in the number and types of ocean observations (described in Chapter 8) as well as the development of sophisticated numerical models that couple the ocean and atmosphere (Sarachik and Cane, 2010). These observations and models have led to the routine monitoring and prediction of ENSO (discussed in Chapter 12). Further research continues to reach a more complete understanding of the transitions among warm, cold, and neutral ENSO conditions.

El Niño and Rainfall

El Niño conditions in the tropical Pacific are known to shift rainfall patterns in many different parts of the world. Although they vary somewhat from one El Niño to the next, the strongest shifts remain fairly consistent in the regions and seasons shown on the map below.

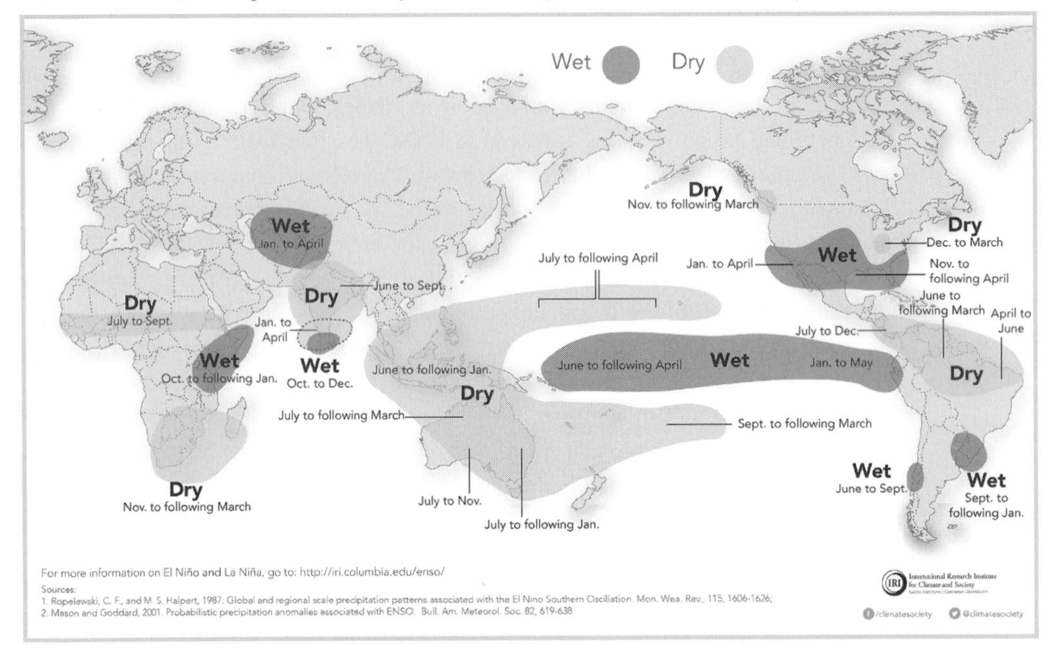

Figure 4.15 El Niño and Rainfall Schematic. A black-and-white version of this figure appears in some formats. For the color version, please refer to the plate section.
After Ropelewski and Halpert, 1987 and Mason and Goddard, 2001. Figure from the IRI Data Library, used with permission.

Quasi-Biennial Oscillations (QBO)

Analysis of tropical radiosonde observations in the early 1960s (Reed, 1961) led to the discovery of an oscillation in the strength and direction of equatorial stratospheric winds that had a period of just over two years, and was later dubbed the stratospheric QBO (Angell and Korshover, 1964). The QBO is centered on the equator in the lower stratosphere and represents interplay, or coupling, between the troposphere and lower stratosphere and continues to inspire research (e.g. Baldwin et al., 2001). In the stratosphere, the equatorial winds alternate from being primarily westerly to primarily easterly with a mean period of just over two years (hence the term "quasi-biennial"). The QBO is thought to be a natural mode of stratospheric variability that is excited, in part, by the upward propagation of atmospheric gravity waves[14] from the troposphere (Lindzen and Holton, 1968; Holton and Lindzen, 1972).

[14] The waves commonly observed on the surface of bodies of water are gravity waves. Gravity waves in the atmosphere are similar but usually not as easy to observe. Atmospheric gravity waves are most evident when they are associated with a regular pattern of "wave clouds" in the lee of mountains or when you are in an airplane being shaken by clear air turbulence.

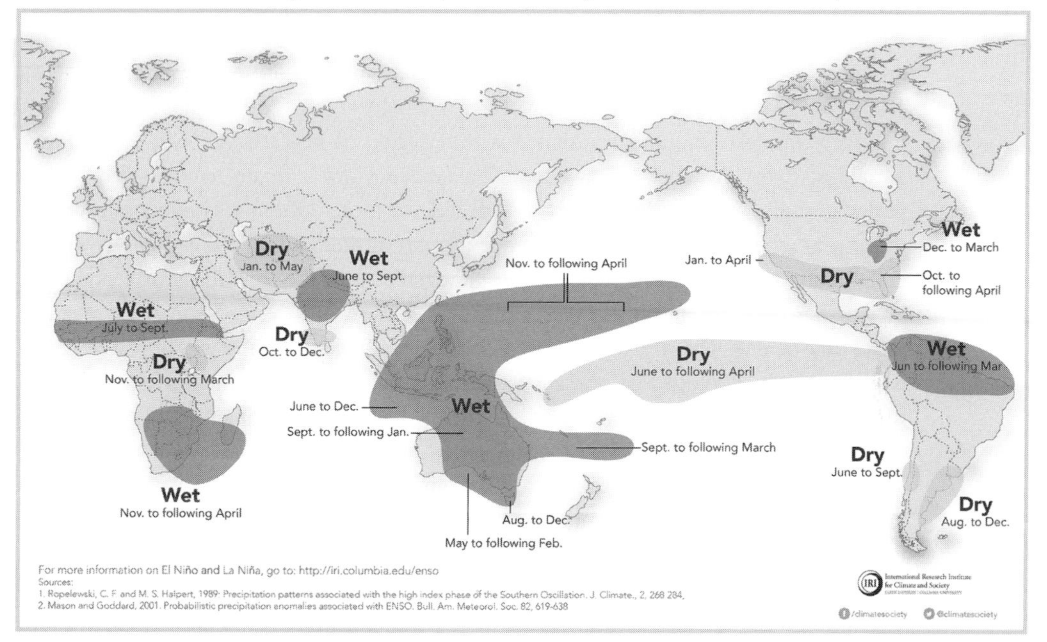

Figure 4.16 La Niña and Rainfall Schematic. A black-and-white version of this figure appears in some formats. For the color version, please refer to the plate section.
After Ropelewski and Halpert, 1989 and Mason and Goddard, 2001. Figure from the IRI Data Library, used with permission.

Early studies looking for links between tropospheric phenomena and the QBO were particularly puzzling because they uncovered statistical evidence that the troposphere has a broad tendency for quasi-biennial variability in many parameters in many places, for example in surface temperature over the United States (Rasmusson et al., 1981). Tropospheric biennial variability has been described as a "statistical will-o-the wisp" (Landsberg et al., 1963) because the historical record shows there are times when biennial variability in the troposphere appears strong and times when it is weak or nonexistent. Some research has suggested that the occasional appearance of biennial variability in some atmospheric data analysis may be a manifestation of the occurrences of paired warm and cold ENSO episodes (Ropelewski et al., 1992).

More recent studies have uncovered statistical links between the stratosphere and tropospheric weather and climate phenomena (for example, Baldwin and Dunkerton, 1999; Cohen et al., 2007; Shaw and Perlwitz, 2013). Many of the recent studies have linked variations in the stratospheric polar vortex to variations in the tropospheric AO or NAO in the Northern Hemisphere and the AAO in the Southern Hemisphere. As discussed earlier, variations in these annular modes are related to variations in in surface temperature and precipitation in Europe and North America.

Decadal and Multi-Decadal Climate Variability

Multi-year episodes of dry or wet, warm or cool, periods occur in virtually all regions of the world from time to time. A prolonged wet period in the African Sahel during the 1950s, for example, was followed by a long dry period in the 1970s and 1980s. Such droughts, including those in Australia and in the United States during the 1930s and 1950s, bring attention to slowly evolving climate states. Although there is only a relatively modest amount of variance with a period of near 10 years in most climate variables, many researchers have noted a tendency for climate to vary slowly over periods ranging from several years to a decade or more. Often, it is uncertain whether the appearance of these multi-year regimes is part of the random component of weather and climate variability, whether they are the result of weakly forced climate processes, or whether they are an artifact of an observational record that is short with regard to these time scales. This illustrates one of the major challenges in analyzing atmospheric data in search of modes of climate variability. As we have seen, a great many such patterns have been identified, but only in a few cases have clear physical mechanisms been found.

Some studies have found statistical relationships between the 11-year sunspot cycle and climate variables, but the statistical significance is generally low, analysis periods relatively short, and the nature of the statistical relationships tend to be ephemeral (also discussed in Chapter 5). The magnitude of statistical relationships, for instance, may vary greatly with the addition of one or more data points (Barnston and Livezey, 1989). The greatest difficulty in substantiating claims of sunspot–climate relationships is that, to date, there has been no widely accepted and testable hypothesis explaining the physical relationships responsible. One additional difficulty in establishing the validity of relationships between decadal climate variability and sunspots is that the instrumental climate record rarely extends much longer than a century, so that the record usually does not contain enough independent samples to establish robust statistical significance.

Decadal variability, nonetheless, appears in many climate records and can account for a modest percentage of the total variance in some cases (Greene et al., 2011). The magnitude of decadal variability may be larger at some locations than at others. For some analyses, proxy climate records, from ice cores or tree rings for example, may be useful in increasing the length of data record. One limitation of proxy records is that, very often, the location of the proxy record is not coincident with the locations of interest. Thus, the analysis must determine, or assume, how climate variability in the study region is related to climate variability at a proxy site. Considerable efforts are being made to develop a useful predictive capability on decadal time scales (Meehl et al., 2009).

Variability in the Atlantic Meridional Overturning Circulation (AMOC) is thought to influence, and be influenced by, climate variability on many time scales, including decadal and multi-decadal (Hurrell et al., 2006). Variations in the AMOC on decadal and longer time scales are determined by the physical mechanisms associated with air–sea interactions and the overturning of deep ocean water in the Atlantic Basin. The main limitation in studies involving AMOC on the longer time scales is, again, that the instrumental records are very short. AMOC remains an active area of research using coupled ocean-atmosphere climate models. Climate model simulations have been successful in replicating long period aspects of the observed global climate over the past century given observed forcing.

Multi-decadal climate variability is even more difficult to characterize, and of course more challenging to distinguish from climate change (discussed in Chapter 5). One can find evidence in climate datasets that suggests that the climate in a region has shifted after some date, for example 1976–1977 in the Pacific Basin, leaving the climate state after the shift different in some aspects than the climate earlier in the record. This late 1970s shift in the sea surface temperature and other climate variables has been identified as part of the Pacific Decadal Oscillation (PDO) (Mantua et al., 1997; Zhang et al., 1997; Mantua and Hare, 2002). The PDO spatial patterns in SST, sea level pressure, and some circulation features are similar to ENSO patterns. In the PDO, however, the magnitudes of the patterns are relatively larger at higher latitudes than in equatorial regions, in contrast to the ENSO patterns that have the largest magnitudes in equatorial regions.

Other examples of multi-decadal climate variability have been identified. Multi-decadal variability in the Atlantic Ocean with a period of 60–80 years, referred to as the Atlantic Multi-decadal Oscillation (AMO), has, for instance, been linked to variations in hurricane activity (Trenberth and Shea, 2006). The "O"s for oscillation in the AMO and PDO suggest that the climate patterns may at some future date shift back to an earlier state. An alternative interpretation of the observed multi-decadal variations in climate is that such shifts in the data may result from a superposition of climate modes on shorter and longer time scales. For example, slight systematic changes in ENSO may have interacted with decadal variability or some other mode or patterns to produce the observed climate shift (Newman et al., 2016). In either case, the changes in climate state need to be considered when analyzing climate variations over an analysis period that spans the shift. If the change in the mean climate state associated with the shift is not considered, one might conclude that the character of the MJO, ENSO, or some other mode of climate variability before, for example, the late 1970s was different than in following years. Such an analysis is ambiguous because it can't distinguish whether changes in the anomalies associated with a phenomenon reflects a change in the phenomenon or a change in background climate, or both.

4.4 Interactions among Climate Variability on Various Time Scales

All the climate phenomena discussed in this chapter might be present at the same time, along with a healthy dose of chaotic (not coherent) variability, making the analysis of variability in the climate system incredibly challenging and interesting. The annual cycle accounts for the largest amount of variance in the climate system, most notably in temperature at mid-to-high latitudes and in precipitation in monsoon regions of the world. After that, ENSO has accounted for the largest percentage of variance on the interannual time scale, at least in recent centuries. Climate patterns like the MJO or NAO that operate on shorter time scales than ENSO may be present with or without

ENSO, while, at the same time, other climate patterns like AMOC or PDO that operate on longer time scales may also be influencing other variations in the climate system.

The MJO tends to be modulated by ENSO state. Other shorter-term climate patterns, like the NAO, seemingly have no relationship with ENSO nor do there appear to be strong relationships among shorter-lived climate patterns, for example between the NAO and MJO. One of the remaining difficult challenges in understanding the climate system is to determine whether the annual and interannual climate variations discussed here are robust with respect to climate variations on longer time scales. It is currently not known how the character of the annual cycle, including the monsoon systems, ENSO, and shorter-term variability might vary as a function of changes on the decadal and multi-decadal time scales. The inadequate length of the instrumental record requires that observation-based analysis of possible changes in character of the climate patterns must rely on proxy data. While such data are valuable and useful, they have significant limitations (as discussed in Chapter 9).

Climate reanalyses, discussed in detail in Chapter 3, provide another source of information useful in the study of the interaction of climate on various time scales, especially given the longer time spans covered by some modern products. Since, in reanalysis, the models assimilate daily or more frequent observed data, their patterns of climate variability are constrained to resemble those in the observations. The availability of century or longer reanalysis-based daily data provides a rich data base for investigating the statistics of day-to-day weather as modulated by the climate patterns and may provide some insight on how these statistics change over time.

4.5 Summary

In this chapter we first provided a description of the early attempts to describe and understand the atmospheric circulation, starting with the surface winds over the oceans in the sixteenth and seventeenth centuries. We then moved to the beginning of the modern era, starting in the mid-twentieth century when radiosondes and then numerical models provided us with the ability to describe the mean global atmosphere through the troposphere and into the lower stratosphere. This was followed by descriptions of the various patterns, or modes, of climate variability that were identified through careful analysis of the data. By the late twentieth century these recurrent patterns in the pressure fields and winds provided a framework for describing climate variations and their relationships to observed monthly and seasonal temperature and precipitation. Analyses of ENSO led to the discovery that at least some interannual climate variations result from coupling and feedbacks between the oceans and the atmosphere. Later analyses of the MJO revealed that interactions between the ocean and atmosphere can also operate on sub-seasonal time scales. Analysis of other patterns of climate variability, such as the NAO and AO, have not uncovered evidence of such coupling; this may account for the difficulties in understanding and predicting variations in these patterns. We concluded this chapter with discussions of the analysis of interactions among coherent interannual, seasonal, and sub-seasonal patterns with climate variations on decadal and longer time scales.

Suggested Further Reading

Blunden, J. and D. D. Arndt (eds.), 2013: The State of the Climate in 2012, *Bulletin of the American Meteorological Society,* Vol. 94, No.8, August 2013, 240 pp Appendix 2. (This reference includes an excellent list of Relevant Datasets and Sources pp S205–S211).

Chang, C.-P. (ed.), 2011: The Global Monsoon System: Research and Forecast, World Meteorological Organization, WMO/TD **1266** (TMRP Report 70). (Proceedings of a WMO workshop that contains overviews by experts of what is known about global monsoon systems.)

Huler, S. 2004: *Defining the Wind: The Beaufort Scale and How a Nineteenth-Century Admiral Turned Science into Poetry*. Crown Publishers, Random House, 290 pp. ISBN 1-4000-4884-2. (An account of the history of the Beaufort Wind Scale)

Lorenz, E. N., 1967: *The Nature and Theory of the General Circulation of the Atmosphere*. World Meteorological Organization, Geneva, Switzerland, 161 pp. (A classic monograph on the topic including the early history of attempts to explain the general circulation.)

Rasmusson, E. M., M. Chelliah, and C. F. Ropelewski, 1999: The Observed Climate of the 20th Century, chapter 1, pp. 1–138, in *Modeling the Earth's Climate and Its Variability*, Volume 67, W. R. Holland and S. Joussaume (eds.) Elsevier Science B.V. (A detailed exposition of observed climate variability)

Sarachik, E. S. and M. A. Cane, 2010: *The El Niño-Southern Oscillation Phenomenon.* Cambridge University Press, London. 384 pp. (A graduate level text on ENSO.)

Wallace, J. M., and P. V. Hobbs, 2006: *Atmospheric Science*, 2nd edition. Academic Press, 483 pp. (A comprehensive undergraduate to graduate-level textbook and reference source covering meteorology and climate.)

Questions for Discussion

4.1 How should trends in the data be taken into account when computing teleconnections? Pattern correlations? Correlation patterns? EOFs?

4.2 Discuss advantages and disadvantages of representing precipitation anomalies in arid regions as actual anomalies, anomalies expressed as a percentage, and standardized anomalies.

4.3 How would you compute and display the mean monthly winds at a particular location?

4.4 Discuss the difficulties in defining the annual mean wind at a monsoon location where the winds blow from the southwest for four months, are light and variable for four months, and blow from the northeast for four months.

4.5 Discuss the difference between the lifetime of a coherent climate phenomenon and the interval between occurrences of the phenomenon. How would you distinguish between these two time scales?

5 Climate Change

5.1 Introduction

We have seen in Chapter 1 some of the challenges involved in defining climate and distinguishing it from weather. One key characteristic of climate is that it varies more slowly than weather, while another is that it involves all the components of the Earth system interactively, while weather is much more an atmospheric phenomenon. Aside from the change of seasons in temperate latitudes, variations in climate can be subtle, but they exist and are important. Phenomena like ENSO can be powerful influences on weather and have life cycles on a scale that make their impacts quite evident – there are many parts of the world in which El Niño and La Niña seasons differ noticeably from one another.

As discussed in Chapter 4, climate variations like ENSO can only be described through reference to some background "mean" climate, or climatology that includes the mean annual cycle. The subject of this chapter is the behavior of that background climate and ways in which it changes over time, thus the title "Climate Change." As we will see, climate, however defined, appears to be always changing and so the notion of a stationary background climate against which all observations, analyses, datasets, predictions, and applications occur is never fully accurate. Therefore, some understanding of and measurements, or more correctly estimates, of changes in climate are essential to all aspects of climate analysis.

Climate change is a complex topic that requires us to explain our understanding of several concepts. In this chapter, we begin by discussing how we distinguish climate change from climate variability and characterize change as free or forced, and oscillatory or trend. The overwhelming strength of the annual cycle (see Chapter 1) leads us to use the mean of that cycle, generally expressed as the time series of monthly means over some averaging period, as the representation of the mean climate. The temporal perspective is a crucial aspect of these concepts: climate defined as a 30-year mean, which is the conventional definition discussed in Chapter 3, is generally very different from the climate based on 100-year or 1000-year means. We then consider the information sources available to any investigation of climate change, however defined. Since we must consider long time scales, our observations from the modern suite of systems are of limited value – they simply do not cover a long enough time span to fully characterize the changes that Earth's climate can undergo. We will discuss the various ways in which this limitation is addressed, including proxies, reconstructions, and model simulations based on theory constrained by observations. We use this information to briefly describe observed changes on various time scales, and illustrate the differences between variability and change. Recent information indicates that humans have modified the climate, and the various methods through which this might have occurred over the past several thousand years will be discussed.

5.2 Discussion of Definitions

5.2.1 Change vs. Variability

Just as the boundaries separating weather from climate are fuzzy, so too is the distinction between climate variations, such as those discussed in Chapter 4, and climate changes. In Chapter 1 we use time and space scales to distinguish weather variations from climate variations; the temporal boundary we use is about two weeks. This time span is based on the length of time that atmospheric variations can be predicted from the initial state of the atmosphere; predictions of longer time scale phenomena depend on the impact of boundary conditions. A somewhat analogous approach can be applied to climate: Climate changes result from variations in external forcing, such as solar insolation, while climate variations are the consequence of internal interactions among the components of the climate system. Both variations and changes depend critically on internal feedbacks within the climate system, which can make it difficult to classify exactly what's going on.

5.2.2 Forced vs. Free/Internal

Earth's climate system is complex, consisting of interactions between two dynamic fluids, the atmosphere and oceans, as well as interactions with the land surface and external forcing. The system is driven by the radiation arriving from the Sun, and our climate is, to a large extent, the result of the balance between the incoming solar radiation and outgoing radiation from Earth. Even if there were absolutely no variations in the solar radiation, or changes in the constituent gases of the atmosphere, or in the configuration of the continents, there would still be internal variations in the climate. Free, or internal, climate variations, for example in the surface temperature at any location on Earth, would arise from differences in the year-to-year integrated effects of the day-to-day weather or from internal interactions among elements of the climate system. We expect that these local climate variations would be limited[1] and that the mean climate of Earth, given a sufficiently long averaging time, would be essentially constant. The climate system, however, may be altered by variations in the external forcing elements that affect the climate system. These could include changes in the incoming solar radiation, impacts of meteors or other celestial bodies, changes in the composition of the atmosphere and oceans, volcanic eruptions, changes in the locations and shape of continents, as well as changes in the orbital characteristics of the Earth/Sun system.

[1] By limited we mean that the statistics of climate parameters like temperature or precipitation at any particular location or for the planet as a whole would be restricted to a neighborhood of the statistics of the mean climate; i.e. anomalies would be limited to some magnitude and would not exhibit a sustained trend.

5.2.3 Trends vs. Oscillations

Climate variations and changes can be characterized as trends or oscillations but are nearly always some sort of combination. For example, in midlatitudes the annual temperature cycle is strongly oscillatory, with a peak sometime during summer and a minimum during winter. However, during a particular fall and spring any analysis of temperature will reveal a pronounced trend. Modern datasets include many complete annual cycles, making it easy to understand that the annual cycle is an oscillation. However, on a much longer time scale, temperatures over the past few million years have fluctuated by 6–7°C between interglacial periods like the present to the coldest portions of ice ages. The decline from interglacial to glacial, and the subsequent warming at the end of the glacial period, would appear as trends for quite a long time before enough data were available to define the oscillation. Practically speaking, oscillations can be distinguished from trends if long enough time series of observations are available. Whether the climate is viewed as varying between different states, oscillating, or experiencing a trend depends to a large extent on the time span being considered.

Most climate datasets will show trends over some portion of their historical record. Estimates of the magnitude of a trend are sensitive to the duration of the dataset being analyzed and the number of spatially and temporally independent data points available over the entire analysis period, or alternatively the duration of the data set and frequency of independent observations. Climate scientists generally consider all variations to be oscillatory unless, importantly, some external forcing that exhibits a clear trend can be identified.

5.2.4 Relevant Time Scales

Informally, people consider the mean climate to be what they have experienced over their lifetime, typically a time span of several decades. The climate, however, has been changing throughout Earth's 4.5 billion-year history. During that time the composition of the atmosphere has changed, the continents have formed and moved about the planet, coming together and apart, the output from the Sun has changed, and ice ages have come and gone. Understanding how the climate has varied in response to these changes is the proper study of climate as a scientific discipline. A central question is, however, what time scales of climate variability and change are relevant to human activities, especially over the coming few generations? Of considerable concern are questions about climate changes that could threaten the planet's ability to support the population. In addition, just as day-to-day activities are influenced by the weather, there is great interest and value in understanding, and predicting, variations in climate on time scales of weeks, months, years, and decades.

5.2.5 Manifestations of Change in the Climate System

Since we define the background climate as the mean annual cycle over some extended period, we can expect to observe variations and changes most readily in parameters for which the annual cycle is clear and pronounced. Surface air temperature, both over land and ocean, is one such parameter, and a reasonably clear decadal trend in global mean temperature is evident in temperature datasets. In most parameters, however, climate variations and changes are challenging to detect simply because the changes are modest in magnitude and adequate datasets are difficult to obtain. For example, a systematic warming of the planetary average temperature is now clear in temperature data, and a strong trend is clear in the time series of atmospheric concentrations of carbon dioxide. Well-established theory tells us that increases in CO_2 should lead to an increase in near surface temperature due to its effect on atmospheric radiative transfer, and so we are comfortable with the conclusion that increased fossil fuel burning since the beginning of the industrial revolution has led to the observed temperature increases. However, it is much harder to identify accompanying climate changes in observations of other parameters, even those for which the theory is straightforward. Sea level is a good example: warming temperatures melt more ice and snow, and the resulting liquid water should find its way to the oceans and be seen as an increasing trend in sea level. In addition, warmer ocean temperatures should expand the volume of the oceans, because water volume expands as temperatures increase (above a temperature of 4°C), leading to further increases in sea level. However, it turns out (as discussed in Chapter 8) that observing sea level is more difficult than observing temperature or CO_2, and we are only now managing to arrive at confident estimates of sea level change based on observations. Most climate parameters present similar challenges, and clear observations of climate changes are not common.

5.3 Information Sources

People develop an intuitive notion of climate through their own experiences. Since human memory is optimized to recognize patterns on time scales relevant to survival, few people can recognize changes even from a single year to the next without written records and quantitative observations. Identifying climate variations and changes is impossible without accurate and stable observations accompanied by well-founded theory and realistic models. Our current capabilities have improved significantly in the past several decades, but much remains to be accomplished.

5.3.1 Observations

What do we know about past climate and how do we know it? The earliest climate datasets were probably written records of the dates of annually repeating phenomena. Many of these are purely astronomical, such as the solstices, but others, like the Nile flood, are features of the climate. Meteorological observations such as rainfall amounts or daily temperatures began to be made and recorded within the past few hundred years, and compilations of those data constitute the longest duration climate datasets. In recent years, both the number and kinds of observations available have expanded enormously, particularly in the form of satellite observations. Fully characterizing global climate change and variability using these observations alone is not possible. Sufficiently long datasets are available for a few parameters like precipitation (Chapter 7) for a few thousand locations, but these global datasets rely on statistical methods to infer values for unobserved locations. Satellite estimates have been used to fill these gaps in recent decades. The use of satellite estimates of other parameters in climate change studies is an active area of research.

5.3.2 Proxy Records

Actual observations of fundamental climate parameters are only available for limited areas and periods and must generally be combined with other information (see derived datasets in this section) to be useful in studies of large-scale climate variations and change. Proxy datasets, ice cores, for example (see Chapter 9), provide useful supplemental information by integrating climatic conditions over much broader areas and over longer duration time spans. However, the connection between proxy data and the physical parameters of the climate system is often obscure and poorly understood, limiting their value in climate studies. Proxy records are especially valuable where they overlap with instrumental records from the areas that they represent, because the information from both can be combined into a more complete picture of climatic behavior. Proxy data provide the only observational tool that allows us to delve back into the earliest periods of climate's evolution.

5.3.3 Theory

Scientists do not have a complete theory that explains the climate system, but the dominant roles played by solar forcing, the physics of radiative transfer within the atmosphere, the fluid dynamics of the atmosphere and ocean, and the interactions between them mean that we have many of the elements of such a theory. We use these elements to understand the behaviors we can observe and to create numerical models of the components of the climate system that we use to simulate past and future behaviors.

5.3.4 Numerical Models

Numerical models of the climate system are subject to many limitations, and they have only a limited ability to predict unforced climate variations. However, the current generation of climate models, when driven by observed external forcings, is capable of simulating important aspects of the seasonal cycle, such as extratropical atmospheric temperatures, as well as the observed trend in global mean surface temperatures. Numerical model simulations are being used to provide projections of the range of potential climate changes associated with increases in CO_2 and other greenhouse gases, and comparisons between their results and observations will improve our understanding of climate variability and change and our ability to predict future events.

5.3.5 Derived Datasets

Derived[2] datasets are central to the study of past climate, including the background climate as well as variations and changes on various time scales. Most studies of global climatic behavior are based on datasets derived in various ways from combinations of observations and model results. The datasets most frequently used for such studies are reanalyses (see Chapters 3 and 11), in which all available observations are combined with a numerical model of the atmosphere or ocean to obtain the most accurate possible depiction of the total state of the system at a specific time. Even in cases where numerical models of the atmosphere and ocean are not used, statistical models are required to convert scattered observations into complete gridded analyses over large areas. Probably the most familiar of these datasets are the various global surface temperature products (discussed in more detail in Chapter 6), which are most often reported as a single number representing the global temperature anomaly for a single month, season, or year, but which are actually gridded analyses covering most of the globe with a spatial resolution of 100 km or less. These analyses are based on observations from thermometers on land and a variety of systems over the oceans, including satellites, drifting and moored buoys, and ships, that are combined using statistical models that compensate for known errors of the different observations. Derived datasets are our only useful descriptors of global climatic behavior for such parameters as global precipitation and atmospheric and oceanic circulation.

5.4 Natural Climate Change on Long Time Scales

The climate of our planet has been changing, almost continuously, since its formation, an estimated 4.5 billion years ago. The age of Earth is based on measurements of the radioactive decay of uranium and other materials that make up the planet as well as in meteorites (Dalrymple, 1991). The changes in Earth's climate on very long time scales can be ascribed to changes in the composition and relative concentration of gases that form our atmosphere, the movement and subsequent changing configuration of the continents and oceans in response to plate tectonics, and variations in Earth's orbit and orientation in relation to the Sun as well as to variations in the amount of solar radiation.

The first 4.0 billion years of Earth's climate, referred to as the Precambrian, can only be inferred from the geologic records. The most recent 542 million years is the Phanerozoic eon, consisting of three eras: Paleozoic, Mesozoic, and Cenozoic (Plummer et al., 2003). Proxy surface temperature and CO_2 estimates for the past 500 million years (see Barry and Gan, 2011) have been made from delta-O_{18} measurements, as discussed in Chapter 9. These estimates have suggested that periods in which the surface temperatures were relatively high/low corresponded to periods in which there were higher/lower concentrations of CO_2. These studies suggest that surface temperature and atmospheric

[2] Derived datasets are those that have been constructed from observations through the use of theoretical, statistical, and/or empirical transformations or algorithms. After the definition in data.gov.uk glossary (2015).

CO_2 co-vary in a manner that serves to amplify changes in either one in response to the other (Royer, 2006). The Phanerozoic eon has been punctuated by three major periods of glaciation. The climate changes documented in the geological records occurred on time scales longer than human recorded existence. The most recent glacial maximum is estimated to have occurred some 18,000 years ago. The most recent geologic epoch, the Holocene,[3] is generally considered to date from 11,000 years ago (Barry and Gan, 2011) and is characterized as being a relatively warm interglacial period, similar to many that have appeared during the past few million years between glacial maxima. Proxy climate records based on the analysis of ice cores for the pre-instrumental part of the Holocene are discussed in Chapter 9. Estimates of climate variations during the Holocene have also been made through analysis of corals, tree rings, and lake and ocean sediments. In a very general sense, any climate variations larger in magnitude and/or occurring much more rapidly than those previously observed in the Holocene are taken to be possible harbingers, or early indicators, of contemporary human-induced climate change.

5.4.1 Solar Output

The amount of energy produced by the Sun that arrives at the top of Earth's atmosphere is the main influence on global climate, and since the distance between the Sun and Earth has not changed significantly over time, long-term changes in solar output, or luminosity, are a potential source of climate changes over very long time scales. According to the Standard Solar Model, the luminosity of our Sun about 4 billion years ago was about 70% of its current value and increased fairly steadily to its present value (Sagan and Mullen, 1972; National Academy Press, 1982). This rate of change would produce an increase in total solar irradiance (TSI) of 0.1%, about equal to the difference between the minimum and maximum TSI during an 11-year sunspot cycle, in about 130 million years, certainly too small to detect during the period for which we have useful global temperature proxies. This long-term scenario does pose a problem, however, because calculations show that an atmosphere with greenhouse characteristics comparable to the current state combined with solar luminosity at 70% of the present value would not permit liquid water at the surface, while geologic evidence indicates the presence of sediments deposited from liquid water as long ago as 3.8 billion years. The current explanation relies on an atmospheric composition with much higher concentrations of greenhouse gases than at present, but research continues.

5.4.2 Land/Ocean Configuration

The distribution of continents and oceans on Earth's surface evolves as the crustal plates that make up the surface move and interact, and the changing distribution leads to variations in climate on time scales of many millions of years (see, for example, Plummer et al., 2003, pp. 463–465). Representations of Earth's surface at various times

[3] It has been suggested that the terminology for the current era, say starting with the mid-nineteenth century, be changed to the "Anthropocene," indicating an epoch when human influence started to have a significant impact on climate. This terminology has not been formally adopted.

in the past can be combined with evidence from fossils and geological data to depict the presence and absence of ice caps and permanent sea ice. Global atmospheric and oceanic temperatures can be inferred from a number of proxies, such as the nature of fossils in ocean sediments and the ratio of isotopes of oxygen, while the concentration of atmospheric CO_2 can be determined by analysis of air trapped in bubbles in ice cores. These observations make it clear that the global climate has varied dramatically over the past 500 million years, with global average temperatures ranging from a subtropical 25°C during the Cambrian period, the first 54 million years of the Phanerozoic (Plummer et al., 2003), and other epochs to 10°C during ice age periods of the past several million years. Generally speaking, the processes that drove these changes are the same as those at work in today's climate: continents in high latitudes and high elevation terrain everywhere permit the accumulation of snow and ice, and ice-albedo feedback leads to further cooling, while continents and archipelagoes modify ocean currents in ways that enhance or suppress high latitude warming. For example, today's climate is strongly influenced by the twin cooling engines of Antarctica, whose location surrounding the South Pole greatly facilitates a permanent ice cap, and the Arctic Ocean, whose isolation from warming tropical ocean currents permits the establishment of semi-permanent sea ice cover. Oscillations in the timing of high latitude solar heating resulting from fluctuations in the orbit and rotational axis of Earth[4] are thought to have led to variations between full ice age conditions and the type of warmer interglacial conditions of the present, but the overall configuration of continents and oceans has led to the generally cold conditions of the past several million years.

5.5 Shorter Time Scale Climate Variations

In the context of this chapter the phrase "shorter term" refers to time scales from several years through several decades. In some cases, these variations may be associated with some known physical forcing, or from some internal oscillation (as in ENSO) or feedback loop (as in the positive feedback relationship between planetary albedo and high latitude surface ice and snow). In many cases, short-term variations in the circulation and other characteristics of the climate system have been observed, but their root cause is not well understood.

5.5.1 Forced Variations

Sunspots

The radiation reaching Earth from the Sun during the current era is very nearly constant but subject to small variations associated with sunspots and the annual cycle in the distance of Earth from the Sun. The number of sunspots, or transient small dark areas on

[4] During the first half of the twentieth century, the Serbian mathematician M. Milankovitch found that variations in the precession of the equinoxes and solstices, the varying tilt of Earth's rotational axis, and varying eccentricity of Earth's orbit are responsible for the sequence of ice ages during the Pleistocene era. Quoted from the AMS Glossary of Meteorology (2012).

the Sun's surface, varies with a period of approximately 11 years. Historical records over 400 years also show some modulation of the 11-year cycle over longer-term periods. Sunspots are associated with variations in the solar constant[5] of around 1 watt/m^2 or 0.1% (IPCC 3rd Assessment Report 2001). Some studies have found modest statistical links between variations in the number of sunspots and climate variations, but none of these studies has been able to demonstrate a physical mechanism that would explain such a relationship.

Volcanoes – The Wild Cards in the Climate System

Volcanoes have had a profound influence on climate in the past and are likely to continue to influence the climate in the future. Volcanic eruptions drastically alter the surface conditions in the local area near the eruption, often reshaping the topography and character of the land surface. Such large changes in the land surface fundamentally change the local climate. Local climate processes, as discussed in Chapter 10, are very sensitive to the type of soil, vegetation, and topography that are all altered in the vicinity of a volcanic eruption. As discussed in Chapter 1, volcanoes do not, in the current geologic era, contribute significant amounts of greenhouse gases, like carbon dioxide and water vapor, into the atmosphere, although they were likely major contributors at times in the history of Earth. Volcanic eruptions can, however, influence the global temperature. In some cases, sulfur dioxide and water vapor associated with an eruption act to form sulfate aerosols once they get into the stratosphere. In the stratosphere, these aerosols increase Earth's albedo, for periods of one to two years and sometimes longer (Robock and Mao, 1995), and thus influence Earth's radiation balance and surface temperature. During the instrumental era, globally averaged temperatures have been shown to drop by 0.5°C–0.6°C for as much as a year following large eruptions. Temperature decreases of greater magnitude are likely to have occurred before the instrumental record. For example, the 1815 eruption of Mount Tambora in Indonesia is credited as the primary cause of the 1816 "year without a summer," characterized by June snowfalls in New England and parts of Europe and a general collapse of the growing season at midlatitudes (Stommel and Stommel, 1983). Another large Indonesian volcanic eruption, Krakatoa in 1883, was responsible for the vividly colored sunsets in the Northern Hemisphere for the following year that were captured in background of Edvard Munch's well known painting "The Scream." The largest volcanic eruption on record is that of Toba, also in Indonesia, some 74,000 years ago (Zielinski, 1996). That eruption has been credited with a 3°C–5°C drop in the global temperature, resulting in several years with little or no growing seasons and the near extinction of humanity.

In addition to influencing the global temperature, the stratospheric aerosols associated with volcanoes also bias SST estimates derived from satellite observations. An example of this occurred when atmospheric aerosols from the 1982 eruption of the El Chichon volcano in southern Mexico resulted in a cold bias in satellite estimates of SST in the equatorial Pacific. This cold bias in the observations resulted in significant confusion

[5] The solar constant is defined as the amount of solar radiation received outside Earth's atmosphere on a surface normal to the incident radiation, and at Earth's mean distance from the Sun. It is estimated to be 1368 w/m^2 +/- 0.2% (from the AMS Glossary of Meteorology 2012).

and delay in identifying the inception of the great 1982/83 ENSO warm episode. Even in cases where an eruption doesn't significantly bias the observations, interpretation of the influence of a particular volcanic eruption on climate can become more difficult when the eruption occurs in conjunction with some mode of climate variability. The occurrence of volcanic eruptions should be noted and taken into account when analyzing climate data over time periods that span the occurrence of eruptions.

The geologic processes that determine volcanic activity have very long time scales in comparison to human activities, making it difficult to distinguish among extinct, dormant, and active volcanoes. Currently, there are 1500 or more volcanoes identified as active. This generally means that they have erupted sometime in the past several thousand years. Robock (2000) lists the major known volcanic eruptions that occurred in the past 250 years. Since the timing and magnitude of volcanic eruptions have no clear temporal pattern, and since models of the physical processes involved are not yet able to provide specific predictions, volcanoes can be thought of as the "wild card" in the climate system. On the other hand, once an eruption occurs, the influence of the eruption on climate may be anticipated based on past climate analysis and observations of volcanic events together with models of the climate system.

5.5.2 Free/Internal Variations

The coherent climate variations discussed in the previous chapter have been defined, primarily, by observations taken since the late nineteenth century. We do not know how stable these climate variations are over long time scales, and how they might change in the context of long-term trends in the global climate. Coupled ocean–atmosphere variations like the MJO and ENSO depend on the current configuration of the SST in the tropical Pacific Ocean and the Walker and Hadley circulations. On the global scale, ocean temperature, salinity, and the three-dimensional circulation of the ocean are influenced not only by the Walker and Hadley circulations, but also by the fresh water influx from rivers and ice melt, discussed in further detail in Chapter 8. It is not yet clear how the interactions among these internal components of the climate system will be influenced by climate change.

In the last chapter, we identified and discussed some of the free oscillations in the atmosphere, including the PNA, NAO, AO, and AAO, but many other patterns have been identified in studies of the atmospheric circulation. These patterns are often thought of as sort of resonant modes in the atmosphere, features that recur frequently due to boundary constraints of various sorts in the same way that a rope fastened at each end will exhibit oscillations of specific frequencies that depend on the distance between the end points. In a landmark study, Barnston and Livezey (1987) identified 13 patterns in the analysis of the monthly 700 hPa height field based on 35 years of data (1950 to 1985). Among the patterns they identified were those we have already discussed, along with others that have not been prominent in the years following the study. Observations of many climate variables show that a substantial amount of decadal and longer time scale variability is present in the climate system, which suggests that the occurrence of some of the internal circulation patterns is influenced by decadal to multi-decadal climate variability like the PDO or AMOC, also discussed in the previous chapter.

The only internal pattern to have been clearly shown to influence global atmospheric and oceanic temperatures is ENSO (Ropelewski and Halpert, 1991; Halpert and Ropelewski, 1992). Warm episode years tend to be associated with extremely warm years across the planet, often becoming the years with record high global temperature up to that point in the record. Cold episode years, on the other hand, tend to be relatively cooler than previous years.

5.6 Anthropogenic Climate Change

As we have seen, the climate is not stationary – it changes on nearly all time and space scales due both to external influences and internal oscillations. While we have focused on changes related to the physical components and forcings of the system, the biosphere has affected the climate in various ways since it first appeared. For example, the appearance of photosynthetic plants that absorb CO_2 and release oxygen resulted in an enormous change in atmospheric composition, with oxygen absent before and stabilizing at approximately 21% at present. Phenomena like changes in ocean color due to plankton and changes in land surface due to changing populations and behavior of plants and animals are appropriately seen as engendering climate variations and changes. Humans are extremely effective at changing climate by modifying surface characteristics through agriculture and city building and by changing atmospheric composition through fossil fuel burning. This section of the chapter summarizes observation-based evidence for climate change related to human activity.

5.6.1 Context

Before the industrial revolution it was inconceivable to most people that humans were capable of influencing global climate. The forces of nature were assumed to be immune to the activities of our species. By the mid-twentieth century, however, some scientists began to become curious, then concerned, by the possibility that human activities might be having an influence on global climate. This concern was spurred by the realization that much of the progress in the developed world was dependent on burning carbon-based fuels and releasing the products of their combustion into the open atmosphere. At first, the main concerns were the negative effects of various kinds of air pollution, such as smog in urban areas and acid rain farther from sources, but eventually global changes due to increasing CO_2 and the impact of CFCs on stratospheric ozone raised the possibility that the effects of humanity could affect the entire global climate. In addition, scientists began to detect a systematic influence of urbanization on temperatures and to understand that large-scale agriculture and deforestation should have some impact on the climate.

5.6.2 Greenhouse Gases

The primary gases that comprise our atmosphere are nitrogen (78%) and oxygen (21%). Argon accounts for the bulk of the remaining 1%. A number of trace gases account for roughly 0.07% of dry air (Smithsonian Meteorological Tables, List 1951). These trace gases mainly include carbon dioxide, methane, and a variable amount of water vapor. While their concentration is very small compared to oxygen and nitrogen, their influence on the climate system can be much more significant than their relative abundance suggests.

Summary of the Theory

As discussed in this chapter, at the present and for the recent past[6] the Sun has provided Earth with a nearly constant source of energy. The radiation from the Sun covers a broad spectrum that peaks in the visible range. This spectral peak[7] is determined by the Sun's temperature, approximately 5000°K. The incoming solar radiation heats Earth's surface, which then radiates energy back into space. If Earth did not have an atmosphere and was in thermal equilibrium with its surroundings, the mean annual temperature of the planet would be about −19°C rather than the (currently) observed mean temperature of about 14°C. This 33°C temperature difference arises because Earth's atmosphere is almost completely transparent, when free of clouds, to the incoming solar radiation but not to the outgoing thermal radiation from Earth's surface. The greenhouse gases, including carbon dioxide, methane, and water vapor, absorb and reradiate much of the outgoing radiation, raising the temperature of the surface and the lower atmosphere. Radiative transfer theory suggests that increases in the concentration of the trace gases will result in increases of surface and atmospheric temperatures.

Observations

While the fundamental concept of near surface warming due to atmospheric greenhouse gases dates to the nineteenth century (Arrhenius, 1896), accurate measurements of global atmospheric CO_2 were not available until 1958 when regular, continuous, measurements of CO_2 were initiated. The observations, initially taken at an observatory on Mauna Loa, Hawaii, were made possible by the development of techniques to obtain accurate estimates of CO_2 concentration from flask sample data (Keeling 1960). This site was chosen, in part, because the air was relatively free from most local sources of CO_2 and thus representative of the concentration of CO_2 in the free atmosphere. Regular CO_2 measurements have since begun at several other locations, including at the South Pole, and have been used to estimate global CO_2 as well as regional variations. In addition to a relatively steady trend of about 1.4 ppm/year, these observations show a regular annual cycle in CO_2 concentrations (Figure 5.1).

Peak CO_2 values occur in April/May and annual minima in September, with an amplitude of about 5–6 ppm. This annual cycle follows the annual growing cycle of land-based vegetation in the midlatitudes of the Northern Hemisphere,[8] which

[6] In this case, "recent past" means tens to hundreds of million years of relative stability.

[7] One of the landmark discoveries of nineteenth-century physics was the discovery that all bodies will emit electromagnetic radiation at a wavelength determined by their temperature. The reasons for this were not fully explained until the development of quantum mechanics in the early part of the 20th century.

[8] While the same phenomenon occurs in the Southern Hemisphere, the areal extent of land there is far less, and so the Northern Hemisphere dominates the global annual cycle.

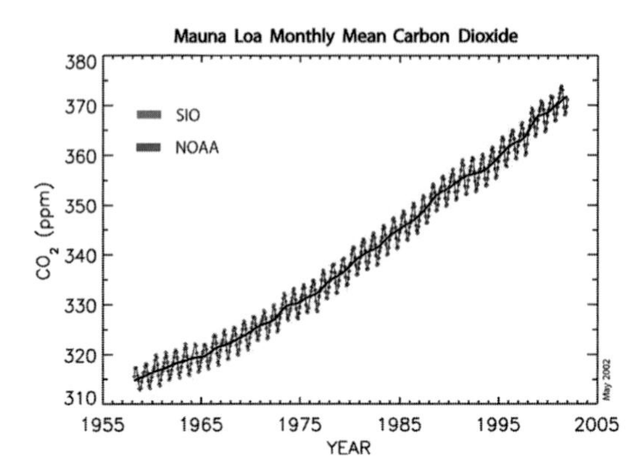

Figure 5.1 Time series of CO_2 concentration (ppm) from the Mauna Loa Observatory, 1957–2005. Early data (blue) were taken by the Scripps Institute of Oceanography. The CO_2 concentration had risen to over 400 ppm in 2015 and continues to rise. A black-and-white version of this figure appears in some formats. For the color version, please refer to the plate section.
Source: NOAA Earth System Research Laboratory, Global Monitoring Division.

absorbs atmospheric CO_2 during the May to September growing season and releases it at the end of the season.

Studies of the variations in trace gases and their role in climate change always cite these changes in CO_2 for several reasons. CO_2 is nearly chemically inert in the atmosphere, and so releases from fossil fuel burning remain in the atmosphere until transferred to some other component of the climate system. While the residence time of CO_2 is only a few years, the lifetime of the perturbation that humanity is currently imposing is a more complex question. Current research suggests that, were humanity to stop emitting CO_2 entirely now, the excess (above the pre-industrial value) CO_2 would take thousands of years to disappear from the atmosphere. The amount of CO_2 contained in the upper oceans tends to be in equilibrium with the atmosphere, and so atmospheric increases are reduced somewhat by oceanic absorption. Growth of new vegetation also absorbs some, although most of that is re-released to the atmosphere after the end of the growing season, leading to the observed annual cycle in CO_2 concentration. Theory suggests that increasing atmospheric CO_2 will lead to increasing temperatures at the surface and in the lower atmosphere, and a great deal of research is being devoted to exactly how that will play out.

Modern observations of atmospheric CO_2 (Figure. 5.1) exhibit a pronounced increasing trend, with annual mean CO_2 concentration rising from roughly 314 ppm in 1958 to above 400 ppm by 2016. Historical estimates of CO_2 concentrations for earlier periods obtained from analysis of air bubbles trapped in ice cores (for more details, see Chapter 9) suggest a CO_2 concentration of 280 ppm as a reasonable estimate for the preindustrial period (IPCC 1990). Calculations using a variety of models lead to the conclusion that this increase in atmospheric CO_2 has resulted from human activities, largely the burning of fossil fuels, and has led to the detected increase in global surface temperatures.

Methane and water vapor also absorb outgoing radiation from Earth. While methane occurs in smaller concentrations than CO_2, it is a more efficient greenhouse gas.

Methane is released in conjunction with fossil fuel extraction, decay of organic matter, and as a by-product of the digestive process of cattle and other livestock. Methane has an estimated nine-year residence time in the atmosphere and, unlike CO_2, is removed through chemical interactions within the atmosphere. The lifetime of perturbations in atmospheric methane due to human activity is not well known, and variations in atmospheric concentration and in anthropogenic sources are more varied and less well understood.

The concentration of water vapor, another greenhouse gas, varies spatially on short time scales and is an important part of the hydrologic cycle. As the atmosphere warms it is capable of holding more water vapor, thus increasing its concentration. The amount of water vapor in the atmosphere is very strongly controlled by global oceanic surface temperatures and has been increasing in recent decades. Local concentrations of water vapor are strongly controlled by atmospheric thermodynamics and exhibit extremely large variations on relatively small time and space scales. Unlike CO_2, monitoring total global water vapor concentrations depends on a combination of satellite observations and derived atmospheric datasets.

5.6.3 Observations That Show Evidence of Climate Change Related to Increasing Greenhouse Gas Concentrations

Observations of CO_2 show a clear, unrelenting, increase in its annual mean concentration in the instrumental record from 1958 forward. Do we see that increase in other observations of the climate system? In this section we examine other observational datasets for evidence of global climate change.

Surface Temperature

The mean global surface temperature has become the most closely monitored index of climate change. Interest in climate change was stirred by a series of papers analyzing Northern Hemisphere surface temperatures that appeared in the late 1970s into the early 1980s (Vinnikov, 1977; Barnett, 1978; Jones et al., 1982). These were not, by any means, the first studies of the Northern Hemisphere or global surface land temperature but were among the first to stir significant interest and attention of the wider research community. Earlier published studies of global temperature (Callendar, 1938, 1961; Willett, 1950; Mitchell, 1961, 1963) gave indications of large-scale warming but did not capture the attention of the climate research community or the general public at that time. These early estimates of global temperature trends were looked upon with some skepticism, but a historical review of climate change studies (Le Treut et al., 2007) points out there is generally close agreement in the global temperature record presented in these earlier studies and contemporary studies. Le Treut et al. further note that Callendar (1938) was the first to explicitly demonstrate a relationship between rising surface land temperatures and increases in anthropogenic CO_2. Callendar was building on what was, when it first appeared, a controversial study by Arrhenius (1896) that linked measurements of "carbonic acid," which is related to concentration of CO_2 in the atmosphere, and air temperature. In the modern era, an early assessment report to the

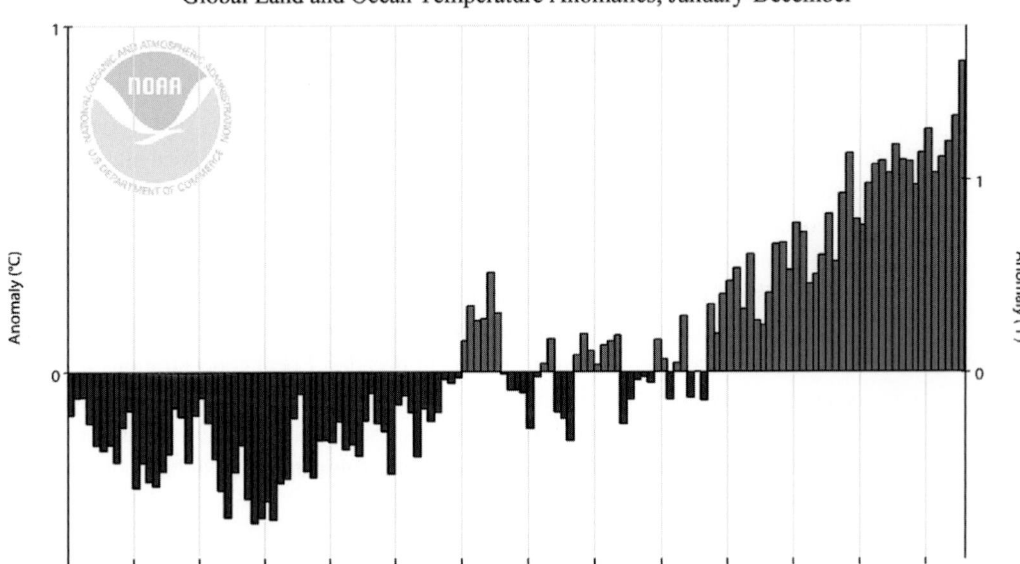

Figure 5.2 Time series of the global mean annual temperature anomaly, differences from the twentieth-century average, for 1880 through 2015.
Source: NOAA's National Centers for Environmental Information/NCDC website.

Climate Research Board of the National Research Council (Charney et al., 1979) concluded there was a clear relationship between temperature and the concentration of CO_2 in the atmosphere. Studies of Northern Hemisphere temperatures by Jones et al. (1982) were later expanded by a number of researchers to include land surface and sea surface temperatures for the entire globe. Time series of mean global temperature have been extended back to the mid-nineteenth century and are now routinely updated on a monthly basis and serve as a default index for monitoring climate change. The climate system, however, is not so simple. The observed temperature index time series, while showing the general increase in surface air temperature for the past century, also shows that this increase has not been monotonic (Figure 5.2).

Each successive year is not warmer than the previous, at least partially because of the oscillations and variability in the climate system discussed in Chapter 4. We also note that even though annual mean global temperatures in recent decades have all been higher than the twentieth-century average, temperatures have not been uniformly above average for all locations on Earth. The observed magnitude of temperature anomalies at various locations on Earth, for any season, may have either higher or lower than average temperature for that location, for example as shown in Figure 5.3a and b.

Sea Level

Sea level as a global quantity depends on the volume of liquid water on the surface of Earth, the average temperature of that water, and the volume of the basins in which it

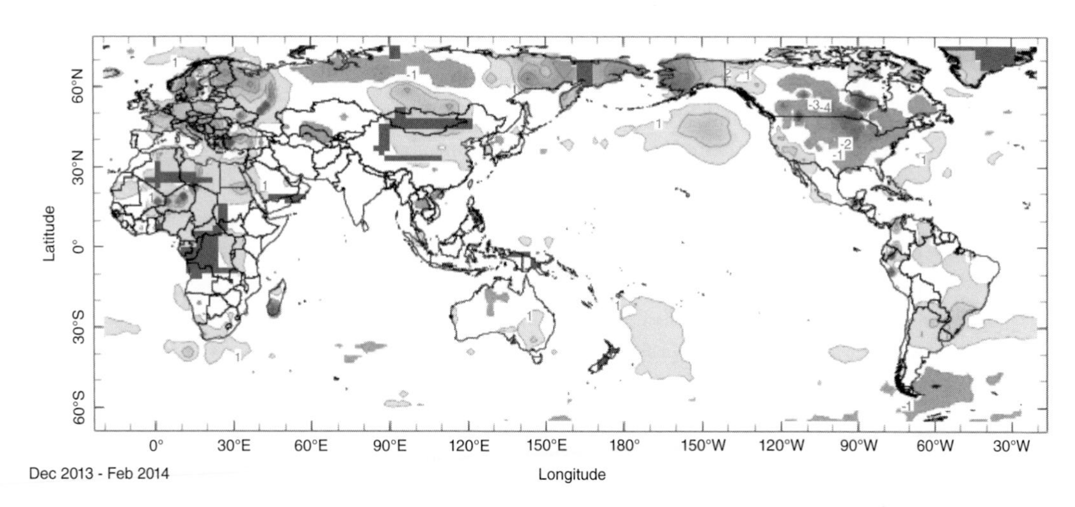

Figure 5.3a Map showing the pattern of surface temperature anomalies for a boreal winter (December 2013–February 2014). A black-and-white version of this figure appears in some formats. For the color version, please refer to the plate section.
Source: Climate Prediction Center NOAA, figure plotted using the IRI Data Library.

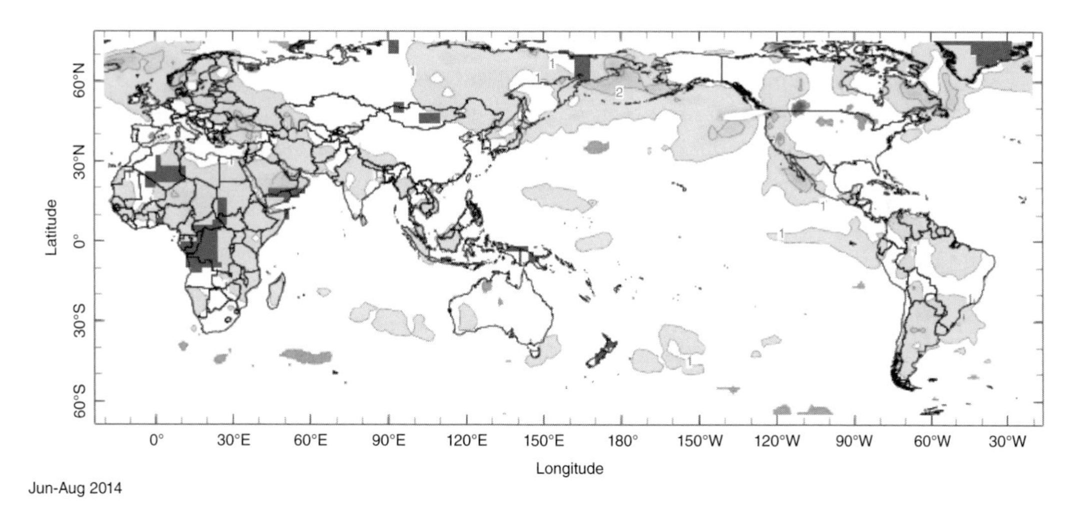

Figure 5.3b Map showing the pattern of surface temperature anomalies for a boreal summer (June–August 2014). A black-and-white version of this figure appears in some formats. For the color version, please refer to the plate section.
Source: Climate Prediction Center NOAA, figure plotted using the IRI Data Library.

resides. Increases in global mean temperature will increase the total volume and therefore sea level through thermal expansion, and will increase the rate at which glaciers and ice caps melt, thus increasing the total volume of liquid water. Sea level is measured at coastal sites by tide gauges, and many locations have time series of more than a century. A number of factors strongly affect coastal sea level in addition to global temperature changes: Lunar and, to a lesser extent, solar tides change sea level on a daily time scale, while subsidence and uplift in coastal urban areas related to human activities or tectonic processes, and ongoing elevation changes resulting from glacial rebound lead to trend-like sea level changes of either sign. Many analyses of tide gauge data conclude that, after correcting for other influences, an increasing trend in sea level is observed (see the Sea Level URL in the references). In the past 25 years, radar altimeters aboard satellites that measure the distance to the surface of the ocean have provided estimates of changes in sea level over the global ocean. Open ocean sea level is influenced by a variety of regional factors in addition to global mean temperature, including ocean circulation and seasonal to interannual climate variations like ENSO, but a long-term global sea level rise signal in the satellite data is clearly seen (see the sea level URL in the references – Watson et al., 2015). Disentangling the effects of all the influences on sea level rise is challenging, but scientific consensus on a global rise of 0.15–0.3 cm/year over the past century is strong.

Sea Ice

Regular observations of Arctic and Antarctic sea ice based on satellite data became available starting in 1979 (Chapter 9). No consistent changes in the minimum (September) sea ice extent were observed in the Arctic until about 1990. From that time forward, there has been a notable decrease in the September minimum that has been attributed to rising temperatures (Figure 9.3). The decrease in sea ice extent has been large enough to allow for a limited amount of commercial shipping to take place during the late summer in recent years. There has been no observed decrease in the maximum Arctic sea ice extent but sea ice thickness has been decreasing at some locations. The behavior of Antarctic sea ice extent has been more complex with both increases and decreases noted in recent years (Chapter 9).

Ocean Acidification

As CO_2 in the atmosphere increases, the oceans absorb a fraction of it. Scientists estimate that 25–40% of the CO_2 added to the atmosphere dissolves in the oceans, leading to changes that make seawater more acidic (although still slightly basic) and decrease the availability of materials used by many marine organisms to build their skeletons and shells. Since the beginning of the Industrial Revolution, the acidity of surface ocean waters has increased by about 30%, and projections indicate that the oceans will continue to absorb carbon dioxide and become even more acidic as atmospheric CO_2 increases. Estimates of future carbon dioxide levels, based on business-as-usual emission scenarios, indicate that by the end of this century the surface waters of the ocean could reach a pH the oceans haven't experienced for more than 20 million years, with potentially significant impacts on ocean life (see the ocean acidification URL in the references).

Mountain Glaciers and Ice Sheets

Most mountain glaciers have been retreating, decreasing in area and thickness, since the end of the last ice age, some 18,000 years ago. In recent years, several glaciers appear to be retreating at an accelerated pace, leading to the hypothesis that increasing temperatures are responsible.

Glaciers are generally contained in high mountain valleys. In contrast, ice sheets are much larger, overtopping all of the topography, including the highest mountain ranges. Greenland and Antarctica are covered by ice sheets. Evidence of increased melting of both ice sheets has been observed in recent years. The addition of melt water to the oceans increases the amount of water contributing to sea level rise. The ice sheets terminate in calving glaciers that also add cold and fresh water to the oceans. There has been some concern that the increased addition of cold and fresh water at high latitudes may significantly influence the deep ocean circulation, but sufficiently long-period observations are not yet available.

Derived Datasets and Model-Derived Information

The anomalies of global surface temperature compared to some period average have become the most widely used index of global warming. Climate models that are used for prediction of long-term climate change must first demonstrate they can faithfully replicate the observed past global temperature index behavior when given observed retrospective data that includes changes in greenhouse gas concentration. As with all such indices, a single number is useful to characterize the gross behavior of a quantity, but a single annual temperature index for the globe does not give any indication of the spatial variability nor of the monthly or seasonal variations observed in this chapter. The observed temperature data over the globe are now available on daily time scales and on spatial scales of a few degrees of longitude and latitude. Numerical modeling centers continuously strive to improve their models' abilities to replicate the characteristics of past global climate on these smaller spatial and temporal scales along with the global temperature index. The ability of a numerical model to replicate the past temperature record is often taken as a measure of a model's potential ability to produce reliable forecasts.

In addition to replicating the observed temperature patterns, model results are also compared to the observed patterns of coherent climate variations, particularly ENSO. Coupled ocean-atmosphere models are capable of reproducing the large-scale characteristics of ENSO behavior, but, in general, the models tend to produce weaker ENSO episodes than observed. In addition, the patterns of relatively warm and cold water associated with ENSO in the Pacific Ocean basin tend to differ from observations in some significant ways.

Numerical models are increasingly being employed in conjunction with observational data to provide spatially complete estimates of climate variables that reach back to earlier periods than possible using the observational data alone. These derived datasets have been used to extend the SST records back to the mid-nineteenth century (Section 2.2.2, Chapter 8, and Chapter 11). Numerical models have also been used to create derived datasets based on proxy data in attempts to reconstruct the basic climate before the instrumental era.

Proxy-Based Datasets

Proxy data provide a basis for establishing the long-term context for current climate variations and trends in the instrumental era. Proxy data come from a variety of sources based on a number of similar but differing techniques to infer climate variables like temperature and precipitation. These proxies include thickness of annual layers in ice cores, corals, lake sediments, and tree rings, as well as other sources of such data. Ice cores have been particularly useful because the air bubbles trapped in the various layers of ice give a basis for examining the chemical composition of the atmosphere for several hundreds of thousand years into the past (Chapter 9). Proxy data are limited, however, because there are very few locations that have a long continuous record. In addition, not all records from different proxies agree with each other. This may, in part, be due to the different geographical locations of various proxy data. The coral record, for instance, comes from warm tropical and subtropical oceans, ice cores from the Antarctic or from mountain high elevations, and most tree-ring data from midlatitudes. Global climate variations and local climate variations, for instance a nearby, relatively small volcanic eruption, may confound the proxy record. Nonetheless, the proxy records all tend to agree that the recent 100 years show exceptional changes from the past as reflected in the reconstructed temperature records.

5.6.4 Other Human-Induced Climate Change

While warming due to increasing atmospheric CO_2 concentrations is the most prominent form of anthropogenic climate change, others exist as well.

Aerosols

In the context of climate change, aerosols refer to small particles that are introduced into the atmosphere either by natural processes or human activities. Natural processes include dust storms, volcanic eruptions, salt particles from ocean spray, and smoke from naturally occurring brush and forest fires. Human activities that introduce aerosols to the atmosphere include burning of fossil fuels for power generation, manufacturing and transportation, and dust particles introduced in conjunction with farming and construction.

Except for soot and black carbon on the surface, which can enhance local warming, the dominant impact of aerosols is to scatter solar radiation back into space, which serves to cool the troposphere. In addition, aerosols serve as nuclei for water vapor condensation with complex effects on clouds. The net effect of aerosols introduced by human activities is to cool Earth, partially compensating for greenhouse gas-induced warming. Aerosols may also be produced when sulfur dioxide, from volcanic eruptions or burning of fossil fuels, combines with water vapor to produce droplets of sulfuric acid. When these aerosols are introduced into the stratosphere, by strong volcanic eruptions or by mixing associated with deep convection, they further cool Earth by reflecting more radiation back into space, an impact that can be seen in global temperature observations following large volcanic eruptions.

Ozone

Ozone is a molecule composed of three oxygen atoms, designated as O_3. It is found in highest concentration in the stratosphere and, in lower concentrations, near the ground. In the stratosphere, molecular oxygen, O_2, and ultraviolet (UV) light from the Sun interact to form O_3. Ozone is destroyed by chemical interactions in the stratosphere, leading to an equilibrium depending on the rate of creation. In the winter hemisphere, where sunlight is weak or absent, concentrations drop. UV radiation can harm living creatures, including humans, and stratospheric ozone absorbs enough UV radiation to shield Earth from most of its harmful effects. In the early 1970s research scientists discovered that chlorofluorocarbons (CFCs), used in air conditioning, manufacturing, and as a propellant in spray cans, could theoretically break down into chlorine and other gases when exposed to UV radiation at low temperatures; conditions that occur in the stratosphere as sunlight returns to the Southern Hemisphere polar regions at the end of austral winter. Since CFCs had recently been detected in the stratosphere and since chlorine and ozone combine very efficiently, there was concern that the CFCs could eventually destroy the ozone layer. Observations in the early 1980s confirmed that Southern Hemispheric stratospheric ozone concentrations were decreasing, most noticeably in September/October when the stratosphere is at its coldest and UV radiation is returning to high latitudes. This ozone decrease is concentrated around the southern polar regions and was dubbed the "ozone hole." This discovery prompted the nations of the world to agree to ban CFCs through the Montreal Protocol in 1989. The original unprecedented agreement has since been modified to also ban hydrofluorocarbons (HFCs) that were originally thought to be safer replacements for CFCs. Recent observations have confirmed that ozone concentrations appear to be recovering.

Near ground level, ozone is produced through interactions with other pollutants, like oxides of nitrogen (NO_x), but not directly emitted into the atmosphere. Ozone is produced in association with manufacturing and exhausts from internal combustion engines. It is one of the active components of smog and is harmful to the human respiratory system as well to plants and animals. Ozone is a powerful greenhouse gas but is generated locally, has a short residence time in the atmosphere, occurs in low concentrations compared to CO_2, and is not a major contributor to global warming (EPA, 2014).

Agriculture and Animal Husbandry

Agricultural activities may influence climate by altering the properties of the land surface and through the use of chemical fertilizers. Forests are cleared and trees burned, reducing the amount of carbon sequestered and reducing the amount that would have been sequestered in future years. Animal husbandry, raising cows and other domestic animals for their meat or milk, introduces methane as part of the animal's digestive process. The amount of land under cultivation has generally been increasing along with the world's population for several thousand years. This has led some climate scientists to conclude that human activities have served to influence climate since the advent of agriculture and, perhaps, to delay or completely prevent the start of the next ice age (Ruddiman, 2005; Kutzbach, 2010).

5.7 Future Climate Change

The positive feedback between increasing temperature and increasing water vapor gives rise to some concern about the possibility of a runaway greenhouse effect that would render Earth uninhabitable for humans. Current theory and modeling suggest this is not likely to occur on Earth until the Sun evolves into a red-giant some 100s of million years in the future (IPCC Fifth Assessment Report, 2014), long after all of the fossil fuels are gone. Human-caused increases in greenhouse gases, however, continue to give reason for concern.

As the population of the world continues to increase and larger fractions of the population have demands for meat products, electricity, and modern modes of transportation, the amount of CO_2 and other greenhouse gases in the atmosphere is bound to increase unless societies adopt alternatives to carbon-based fossil fuels. The IPCC Assessment Reports have estimated the modeled changes in concentrations of greenhouse gases and in the global mean annual temperature based on various assumptions about the amounts of greenhouse gases that will be emitted into the atmosphere in the future. These projected changes in the concentration of CO_2 and other greenhouse gases are used to drive climate models that attempt to forecast future climate change. [See Box 5.1 on the IPCC Assessment Reports].

One of the profound consequences of the many substantial modeling efforts over recent years is the conclusion that society can no longer assume that current climate conditions will continue indefinitely into the future. This realization results from an increased understanding of past climate variability as well as model-based projections that forecast future climate changes that will be tied to changes in the concentration of greenhouse gases in our atmosphere. Both natural climate variability and changes in the proportion of greenhouse gases in the atmosphere strongly suggest that society needs to prepare for the coming changes. The nature of these changes is not completely known, and future changes in climate at particular locations on Earth are likely to remain uncertain for some time to come. In addition, a better understanding is needed of the uncertainties that result from the approximations inherent in climate models, and the need to quantify and understand model uncertainties is a major challenge to the climate modeling and prediction community. A more detailed discussion of model uncertainties is found in Chapter 11.

Climate change is often discussed in the context of the expected increases in greenhouse gases and the accompanying increases in surface air temperature. This paradigm envisions changes that occur at a slow and (mostly) steady rate but, perhaps, changes that may temporarily be confounded by shorter-term climate variations. Given the complex nature of internal interactions among the components of the climate system, we cannot rule out the possibility of abrupt and unexpected climate change. For example, multiple studies have suggested there was a major shift in the climate in the North Pacific Ocean in the late 1970s that resulted in changes in the character of ENSO in years after the shift compared to the years before. It is, however, extremely difficult to distinguish a "sudden" jump in the background climate state from large shifts

Box 5.1 IPCC Assessment Reports

In December 1988 the Intergovernmental Panel on Climate Change (IPCC) was established by the United Nations to meet the growing need to synthesize observations and research on climate change. Since its inception the IPCC has issued assessment reports at five- to seven-year intervals. In the literature, these Reports are often referred to as the FAR or First Assessment Report (1990), the SAR or Second Assessment Report (1995), TAR or Third Assessment Report (2001), AR4 or the Fourth Assessment Report (2007), and AR5 the Fifth Assessment Report (2014). Each of the Assessment Reports (ARs) is the result of extraordinary efforts by hundreds of research scientists to synthesize the relevant climate change research since the previous report. Many of the world's leading climate experts formed into working groups to address the state of knowledge, progress in the theoretical framework, the quality of climate data, variations in the past observed climate, the ability of climate models to replicate past climates, and prospects for improved model-based climate predictions. In addition, the IPCC served to guide climate modeling research, in particular, to enable climate modelers to come to agreement on various options, or scenarios, for estimates of future greenhouse gas emissions on a century time scale. The ARs helped to spur support for the development of new observation systems (both traditional surface observations and those from satellites), more complete historical datasets, improvements to numerical climate models, and to focus climate scientists on global change issues.

The ARs are an invaluable resource for information and improved understanding of climate change. While the potential for significant long-term climate changes cannot be ignored, the time scales of these changes may or may not be a major factor on shorter time scales. On time scales of one or two decades, interannual and decadal variability are most likely to significantly influence local and regional climate. A major challenge for climate analysis is to improve our understanding of climate variability on these time scales under the steady onslaught of natural and human-induced climate change.

that may result from the coincidence of extremes in several of the internal oscillations that characterize the system. The detection of a change in the background climate is greatly complicated by continuing changes in the observing system, such as the addition of large amounts of global satellite climate data beginning in the late 1970s.

While global mean quantities are useful for understanding the macro-scale climate, changes in regional and local climates will certainly have the greatest impact on society. Global changes in temperature are likely to be associated with changes in atmospheric and oceanic circulation features, the global and regional hydrologic cycles, and heat, moisture, and radiation budgets that can manifest locally in any number of different ways. For example, climate model projections suggest that we should expect more pronounced heat waves, and, with less confidence, more severe droughts and floods. While the details remain uncertain, our climate will continue to change. Society must take steps to deal with coming climate changes even while the exact nature of those changes remains unknown.

Suggested Further Reading

Climate in Earth History: Studies in Geophysics, National Academy Press, May be downloaded from www.nap.edu/catalog/11798/climate-in-earth-history-studies-in-geophysics.

Fleming, J. R., 2007: *The Callendar Effect: The Life and Work of Guy Stewart Callendar (1898–1964)*, published by the American Meteorological Society, 176. (The biography of the first to link rising global temperatures to increases in atmospheric carbon dioxide.) History of the Intergovernmental Panel on Climate Change www.ipcc.ch/organization/organization_history.shtml.

The Greenhouse Effect online http://acmg.seas.harvard.edu/people/faculty/djj/book/bookchap7.html.

Lorenz, E. 1970: Climate Change as a Mathematical Problem, *Journal of Applied Meteorology*, 9, 325–329.

Karl, T. R. (ed.) 1996: *Long-Term Climate Monitoring by the Global Climate Observing System*, Kluwer Academic Publishers, 648 pp.

Report prepared for Intergovernmental Panel on Climate Change by Working Group I J. T. Houghton, G. J. Jenkins, and J. J. Ephraums (eds.) Cambridge University Press, Cambridge, Great Britain, New York, NY, USA and Melbourne, Australia 410 pp, chapters 1 and 2. Also online at www.ipcc.ch/publications_and_data/publications_ipcc_first_assessment_1990_wg1.shtml.

Questions for Discussion

5.1 How should the mean annual cycle be defined if the climate is changing?

5.2 How is climate change distinguished from climate variability using theory or observations?

5.3 Describe the physical mechanisms that might lead to climate change and explain which of them result from human actions.

5.4 Which climate datasets can be used, and how, in defining and understanding recent climate change on centennial and millennial time scales?

5.5 What are some causes and manifestations of human-induced regional climate change?

6 Temperature
Building Climate Datasets

6.1 Introduction

In the previous chapters, we concentrated on various aspects of the climate system, climate analysis, types of climate data, and climate variability and change. In this chapter we illustrate, using air temperature over land as an example, the practical challenges encountered in observing climate and the challenges in building useful climate datasets from those observations. Our intent is to illustrate the care that is given to building the climate datasets found in the climate data centers referenced in this book.

In recent years, frequent discussion in popular media of the current monthly, seasonal, or annual temperatures compared to those experienced in past years has become quite common. This familiarity might lead to the impression that the data on which these comparisons are founded are easy to obtain and maintain. The details of how temperature datasets are derived and the care needed to sustain them, however, are quite complex. In this chapter, we focus on the measurement of surface air temperature over land and the construction of datasets from these observations as an example of how climate datasets are produced. Upper air temperatures, introduced in Chapter 3, are not included in our discussions here. Sea surface temperature and air temperature over the oceans, also crucial for the derivations of global mean temperatures, are discussed in Chapter 8. Here we describe some of the early history of land surface temperature observations, identify the various sources of temperature data, and describe considerations that must be taken into account to produce high quality climate datasets.

Understanding of the observations and the datasets derived from them provides the context needed to understand their strengths and uncertainties. This chapter provides guidance on what considerations need to be addressed in processing raw data when appropriate datasets are not available for a specific climate study. This introduction outlines the challenges associated with climate analysis and reinforces the basics discussed in Chapter 2. Topics discussed include the instrumentation, as well as the biases, errors, and trends associated with changes in observing practices, location, and character of the surroundings. We also discuss the analysis of temperature versus anomalies of temperatures, as well as point measurements versus spatial averages. The topics introduced in our discussion of temperature data here are relevant to other variables discussed in the following chapters.

6.2 Surface Air Temperature Over Land

The casual references to temperature in day-to-day conversation illustrate its importance in defining the climate. When meteorologists and public use the term "temperature," unadorned by modifiers, they are usually referring to the air temperature where they directly experience it – near the ground. By international agreement in standards maintained by the World Meteorological Organization (WMO), surface air temperature refers to the temperature of the air at a height of 2 m (WMO, 2011 Guide to Climate Practices). Included in the standard is that the thermometer be housed in a louvered box painted white. This instrument shelter is commonly referred to as a Stevenson Screen, after the Scottish engineer Thomas Stevenson[1] who designed it (Figure 6.1a).

In the United States, it is also referred to as a Cotton Region shelter, presumably because it was first commonly used in the cotton growing regions of the country. The advent of modern instrumentation and the growing use of automated weather observations have resulted in changes in the instruments themselves and in the decreasing use of the traditional instrument shelter. In the United States, the Maximum-Minimum Temperature System (MMTS) has replaced the traditional shelter at many Cooperative Observation sites (Figure 6.1b). Similarly, the U.S. National Weather Service adopted the Automated Surface Observing System (ASOS) at major airports and other locations starting in 1991. A complimentary system, the Automated Weather Observing System (AWOS), had been implemented earlier at many other airport locations by the Federal Aviation Administration. The detailed description of the ASOS temperature sensor may be found on the Internet at the ASOS URLs listed in the References. These electronic instruments provide improved overall accuracy but introduce inhomogeneities in the climate record that are routinely taken into account by data centers when forming long-term time series (Quayle et al., 1991).

The required accuracy[2] and precision[3] of the temperature observations and the completeness of the temperature dataset depend on their primary purpose. When intended to support specific activities, the data required for climate analysis of past temperature observations at the site may be unavailable. For instance, a citrus grove may have a remote temperature measurement that is set to sound a warning if the air temperature in the grove falls below some critical value. In this case, although the temperature is continuously measured (at least during the cold season) only the occurrence of the threshold temperature is likely to be recorded. On the other hand, an observing site may record temperatures hourly, or at some other regular interval, allowing computation of the observed probability distribution (see Appendix A) of temperature at that site.

If the purpose of the temperature measurement is to support real-time weather and climate analysis and the prediction of regional, national, or global temperature patterns,

[1] Father of the writer Robert Louis Stevenson.

[2] Accuracy – the extent to which an instrument measures the true value of a quantity.

[3] Precision – in common meteorological practice precision is the number of significant digits used to represent a quantity. Temperatures are generally given in whole Fahrenheit degrees and to the nearest tenth of a degree Celsius.

Figure 6.1a An example of a Stevenson Screen historically used to house temperature and humidity sensors and still in use by many Cooperative Observer Network (COOP) sites. The Stevenson Screen traditionally houses maximum/minimum thermometers that are reset once daily.
Source: NOAA Central Library Photo.

Figure 6.1b An example of a Maximum, Minimum Temperature Sensor (MMTS) thermometer that uses shielded thermocouple for temperature observations. The U.S. National Weather Service has adopted MMTS as part of the Automated Surface Observation System (ASOS). Many Cooperative Observer Network sites are also adopting MMTS instruments.
Source: NOAA.

a great deal of effort is needed to ensure that temperature observations at different locations are consistent. This is the most fundamental challenge to collecting data useful for large-scale analysis. Simple things, such as using the same observing times in different regions, must be maintained in the computation of daily average temperature. In the continental United States, for example, there are four time zones, so that daily averages calculated from data taken from local midnight to local midnight are offset by one to three hours. Calculations of global averages can lead to difficulties involving arbitrary conventions and definitions of the "day." By international convention and agreement, meteorological observations are tied to Coordinated Universal Time[4] (UTC). The time with respect to the solar cycle, however, must be taken into consideration when computing daily average temperatures. In addition, daily average temperatures calculated from maximum/minimum thermometer observations taken in the morning will differ from those taken in the evening, and both will differ from averages computed from hourly, three-hourly, and six-hour intervals starting at local midnight (Byrd, 1985).

Daily temperature observations at many sites are limited to the maximum and minimum, and their average, the daily mean temperature. These observations are typically made for a 24-hour period, using an instrument that records the maximum and minimum since the last reading, but may be taken at varying times of day at different locations. The daily maximum and minimum temperatures, and therefore the daily mean, are influenced in subtle ways by the time of day that the observations are recorded. Schaal and Dale (1977) and Janis (2002), among others, have shown that the value of the daily mean temperature computed from observations taken at some fixed time will differ from the mean computed from observations occurring at any other fixed time during the 24-hour period defined as the day the observation was made. Maximum/minimum thermometers that are observed and then reset late in the afternoon will systematically show higher mean temperatures than those reset early in the morning (Vose et al., 2003). We show an illustrative example in Box 6.1. In the analysis of daily mean temperature, it is important to have documentation of the observing time and how it may have changed over the period of the observational record. Many of the historical records of temperature from the Cooperative Network in the United States have had to be adjusted for biases introduced by systematic changes in observation time from late afternoon or early evening to early morning (Karl et al., 1986; DeGaetano, 2000).

An analysis of the observed temperature, and other climate variables, over an area will reflect the local microclimates. The climate at a particular location is influenced by topography, proximity to large bodies of water, surface land characteristics, degree of urbanization, and other factors. Finally, analyses of climate may require spatial averages on the scale of a few tens or hundreds of kilometers in order to be useful in comparisons to numerical model analyses and forecasts. The proper formation of these spatial averages provides a whole suite of challenges to analysis that are discussed further in Chapter 11.

While the 2 m temperature has been, and continues to be, the most widely measured and used temperature, there are a host of additional temperature measurements of interest. These include: sea surface temperature (discussed in Chapter 8), "skin" or "land surface" temperature (generally measured by remote sensing from satellite or aircraft, discussed in Chapter 10) and upper air temperature (temperatures from the surface through the depth of the atmosphere, discussed in Chapter 3).

[4] Coordinated Universal Time (UTC), previously, and sometimes still, referred to as Greenwich Mean Time (GMT), or Z, is the time at (or very near) the prime meridian, which is longitude 0°. Eastern Standard Time (EST) is five hours behind UTC.

Box 6.1 Time of Day Temperature Bias

Many weather observations are taken once a day. This is true for the U.S. Cooperative Observing Network volunteers who record maximum and minimum temperature, precipitation totals, and snowfall at a fixed time every day. Many observers prefer to record these observations in the late afternoon or early evening. Others record their entries in the early morning before starting the work day. When a cooperative observer shifts their observing time from evening to morning, the result can be an artificial trend in the temperature record for the reasons illustrated in Figure 6.2 below.

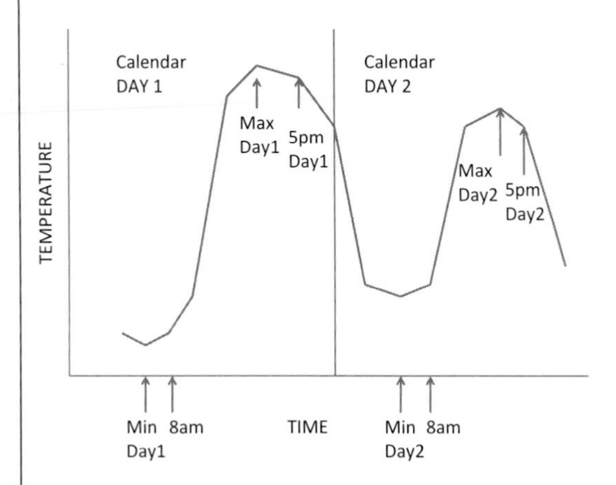

Figure 6.2 A schematic example illustrating how estimates of the daily maximum, minimum, and average temperate obtained from a maximum/minimum thermometer, reset once in a 24-hour period, depends on the Time of Day (TOD) the thermometer is reset. In this example, the temperature in Day 1 had greater amplitude with a lower minimum and higher maximum than Day 2. This might happen when a clear, sunny summer's day is followed by a cloudy day. If a maximum/minimum thermometer is reset at 8 am on both days the minimum temperature for Day 2 will be closer to the minimum value on Day 1 rather than the actual minimum for Day 2. If the maximum/minimum thermometer is reset at 5 pm on both days then the maximum on Day 2 will be closer to the maximum temperature on Day 1 than the actual maximum for that day.

While we use temperature as an example, other climate observations and datasets are influenced by the frequency and time of the observation. See, for example the discussion on snowfall in Chapter 9.

Because of its thermodynamic importance and spatial and temporal variability, the moisture component of the air is often captured in a supplemental set of temperature measurements. These include the dew-point temperature, wet-bulb temperature, virtual temperature, relative humidity, specific humidity, and others discussed in Chapter 3, none of which has traditionally been emphasized in climate studies aside from a few exceptions in specialized applications. This is changing as modern instruments and numerical model assimilated moisture data become more readily available (e.g. Mather and Voyles, 2013).

6.3 Temperature Scales

Temperatures in meteorology and climate science are measured in Celsius and recorded to the nearest 0.1°C. The Celsius scale defines 0° as the melting point of ice and 100° as the boiling point of water, both at a standard atmospheric pressure of 1013.25 hPa (14.696 pounds per square inch in English units). The internationally agreed unit of temperature, however, is the Kelvin, abbreviated as K. The size of the Kelvin, or temperature increment, was defined to agree with the Celsius degree in the range of temperatures encountered in meteorological measurements. The Kelvin thermometer is calibrated at the triple-point of water (where the three phases of water – liquid, ice and vapor – co-exist in equilibrium) at standard atmospheric pressure. The triple point of water has a value of 273.16 Kelvins. This results in a difference of roughly 0.01 degree between the Kelvin and Celsius thermometer at the melting point of ice, a value smaller than the precision of standard meteorological instruments. For historical and traditional reasons, the United States continues to use the Fahrenheit scale in practical everyday weather observations, although this scale is not used in scientific work. In the Fahrenheit scale the freezing point of water is 32°F and boiling point is 212°F, spanning 180° in contrast to the 100° span on the Celsius scale. Care is needed in conversion of temperature data from Fahrenheit to Celsius, (see Box 6.2 Units Conversion). Some historical temperature data may be recorded in units of measurement that are no longer used.

Box 6.2 Units Conversion

The Celsius scale is used in scientific meteorological studies, but many of the older temperature records were originally recorded in different scales. The United States continues to use the Fahrenheit scale for day-to-day temperature measurements. Temperatures are recorded to the nearest tenth of a degree Celsius or Fahrenheit in some observational datasets even though a degree in the Fahrenheit scale is only slightly more than half (5/9) of a degree on the Celsius scale. In most cases, meteorological temperatures are measured and recorded in degrees and tenths of a degree Celsius. In the conversion from Fahrenheit to Celsius units some choices need to have been made on how many significant digits to retain and whether the resulting values are rounded or truncated. Users of these datasets must be aware of any changes in the temperatures scales used in the original observations and the conventions applied; for instance, truncation versus rounding and the number of significant digits carried to arrive at the historical record. In the units conversion, the significant digits retained should not exceed the precision of the temperature instruments.

In the United States, routine measurements of atmospheric temperature did not become available until the mid-to-late nineteenth century. Some isolated records exist back to colonial days; notably George Washington, Thomas Jefferson, and Benjamin Franklin recorded meteorological observations. Early temperature

Box 6.2 (*cont.*)

observations were generally limited to those taken at some Army posts and by a few private citizens. It wasn't until 1890 that a regular network of observations was developed by the Weather Bureau, predecessor to the National Weather Service (NWS). The temperature records taken at NWS sites have been augmented by a network of cooperative observers (volunteers) since 1930 and extended back to the 1890s by retrospective analysis. The cooperative network is discussed in more detail in the following section.

Some regular temperature observations were taken at a few locations in Europe as early as the 1600s, shortly after the invention of the thermometer. Regular air temperature observations became common in most inhabited portions of the world by the mid-nineteenth century. Instrumental temperature records prior to the mid-1800s usually require careful analysis to relate them to the modern observations.

6.4 Observational and Instrumental Considerations

Data in the historical records that have been adjusted for changes in instrumentation and observational practice are often available from the major archive centers. For example, the Global Historical Climatological Network (GHCN) dataset is available from NOAA's National Centers for Environmental Information (NCEI). NCEI also provides the unadjusted data and metadata explaining why and how the data were adjusted. The discussion in this section outlines some of the issues addressed in these kinds of adjustments to provide an appreciation for the care that must be considered in dealing with "raw" datasets.

6.4.1 Environmental Considerations

How representative is a temperature measurement at a particular location? Temperature at a location is influenced by the local environment, with factors such as the degree of urbanization, the nature of land use, the complexity of nearby terrain, and the presence or absence of surface water all relevant. Before historical temperature observations can be meaningfully analyzed the nature and magnitude of the changes in the observing environment must be assessed. The relative importance of changes in the immediate environment depends to a significant extent on the purpose of the analysis. If, for instance, temperature data are being examined to determine the sign and magnitude of multi-year (say decadal and longer) climate variability and trends, then the constraints on absolute accuracy of the instruments and stability of the observational environment are considerable. If, on the other hand, the purpose of the analysis is to compare temperature anomalies (differences from some base point or average) of a particular month or season to previous months or seasons over the past several years, subtle changes in the observing environment may not be as critical. This is because the magnitudes of year-to-year temperature anomalies for a month or a season are generally much larger than the changes associated with a slowly changing observing environment.

However, changes in the environment cannot be completely ignored, even for analysis on the shorter term. Movement of the temperature gauge to another location can strongly influence the temperature observations, as discussed in Chapter 2 and in Box 6.3. Some typical examples of the kinds of location changes that can produce large

Box 6.3 Backyard Temperature Observations

Changes in Instrument Location

A neighbor of one of the authors (CFR) has been making daily observations of maximum and minimum temperature at their home in suburban Washington DC and recording those values for over 40 years. He asked whether there would be interest in using these records for climate research. After a brief conversation with the neighbor, it became clear that there were several reasons that, while these observations are interesting to look at and their history provides a rough estimate of climate conditions in our neighborhood, they cannot be used as a research-quality climate record for climate change studies without careful examination of the record and application of bias corrections and error estimates. In fact, these data serve to illustrate the challenges in trying to analyze climate using uncorrected data that have only minimal metadata. This box discusses some of the everyday issues that arise.

The first issue is that even though my neighbor has lived at the same street address for the entire 40 years of record, the thermometer has not been in the exact same location over the entire period. For the first dozen or so years of the record the instrument, a maximum/minimum recording thermometer, was mounted about 1.75 meters above the ground on the northwest support post of a small porch. At the end of this initial period, the observer built a much larger porch with a surrounding deck. The new porch required moving the instrument, on a date that unfortunately was not specifically recorded, to a support post, at eye level, on the newly constructed deck. This new position is about 2 m above and 8 m northeast of the original instrument location. The changes in instrument height and location on the deck are likely to have introduced a bias in the observed mean temperature that will have to be taken into account if these data are to be used in climate studies. Bias corrections to the data would entail finding the exact dates when the change in instrument location took place followed by comparisons of the temperature records at nearby observation sites (e.g. Dulles International and Reagan National Airports). Temperature comparisons between those sites and the backyard records before and after the thermometer was moved could identify relative bias changes to quantify the magnitude of the bias in the backyard temperature before and after the instrument move.

Box 6.3 *(cont.)*

The analysis would also have to take into account any systematic changes in the temperature records at the nearby airports.

Changes in the Environment

A major environmental change occurred to our neighbor's backyard observing site when a 60-year old maple tree came crashing down, just missing the thermometer, on the NW corner of the deck in a January snow and ice storm. The loss of the tree did not noticeably influence the temperature readings during the winter but, as the first year without the nearby tree progressed, the backyard morning temperatures appeared to be systematically higher than those recorded in earlier years. In addition, they seemed to be higher than at either of the two nearby airports. Prior to the tree fall, backyard morning temperature values were generally between temperatures at Dulles to the west and Reagan to the east. My neighbor and I noted that the morning sun reached into the backyard only from mid-spring through mid-autumn. Without the tree the morning sun still did not directly strike the thermometer but did shine on the southeast face of the post supporting the instrument. The observed morning temperature increases suggested that the tree, when in leaf, had shielded the southeast face of the support post resulting in lower morning temperatures than were subsequently observed. The seasonally varying temperature bias would have to be taken into account if these data were to be used in climate studies. Thus, the answer to the question of whether the neighbor's backyard temperature data may be useful depends on how much effort the analyst is willing to expend to identify artifacts in the record as well as on the nature of the climate studies.

changes in the observed temperature are changes in elevation, surrounding surface (for instance by moving the instrument platform from grass to blacktop), exposure to sunlight, changes in the observing time for daily maximum and minimum temperature (discussed in Section 6.2), and changes in instrumentation (see Box 6.4).

6.4.2 Instrumentation Changes

The instruments used to measure temperature have changed greatly over the past hundred or more years. The weather services in the United States and many other countries have moved away from recording temperature with traditional mercury-in-glass thermometers in Cotton Region shelters and now use a thermocouple aspirated with a fan to keep air moving over the thermometer. The modern devices have many advantages over the traditional measurements, including much faster response times and the ability to continuously record the temperature. These advantages, however, can introduce systematic differences from those temperature values that would have been obtained using the traditional instrumentation. Ideally, the old and new instruments would have been used in parallel through at least one annual cycle to obtain estimates of the differences (bias)

BOX 6.4 Backyard Temperature Observations

Changing the Thermometer

Here we discuss how changes in instrumentation limit the usefulness of the neighbor's suburban backyard temperature observations and the reasons why they cannot be taken at face value for many climate studies. In the 1990s work circumstances dictated that my neighbor could not make the daily weather observations during the week. He prevailed on his family to make the daily observations during his many work-related absences. This they did happily until the temperatures started to fall below freezing. Wandering out to the deck at 10 PM in the face of a stiff breeze, precipitation, and, often, freezing temperatures to manually re-set a mercury-in-glass maximum/minimum thermometer was, for them, more than they had signed up for. The family solution was to install an electronic thermometer that could be read remotely from inside the house. The electronic thermometer was installed adjacent to the old thermometer. The instrument worked well and was so convenient that even when my neighbor was able to resume the observational duties he continued to rely on the remote readings for temperature. No records were made of the maximum/minimum temperatures from the old, traditional, instrument to form a comparison to establish the bias between the instruments. There is a notation in the record of the date when the instrument change occurred, which could be useful if it were ever decided to digitize and analyze the record, but without further analysis we don't know the magnitude of the bias between observations taken before and after the instrument change. If we want to use these data to look at long-term temperature trend, we would have to examine ways to compute the instrument bias. If, on the other hand, we are only interested in the magnitudes of seasonal temperature anomalies we could simply compute the anomalies based on the means from the period that the traditional max/min thermometer was being used in computing anomalies for the first part of the record and anomalies based on the means computed from data from after the 1998 instrument change for the second part of the record. Similar issues influence climate observations at many observation sites.

between the new and old values associated with varying weather conditions. (See Box 6.4.) Parallel observations were taken in some, but not all, Weather Service locations in the United States. For those locations, the results of the comparisons were documented and archived even if the actual data that were recorded are not available.

The examples described in Boxes 6.3 and 6.4 may seem idiosyncratic and even trivial – after all, climate scientists don't actually have to rely on backyard observations by amateur enthusiasts. However, they are in fact quite illustrative of the kind of problems that do face climate scientists in trying to construct global datasets from thousands of instruments with many millions of individual observations taken all over the world, often in very poorly recorded circumstances. The amount of detailed analysis required by many scientists to create these global datasets is often underappreciated.

6.5 Datasets Based on Station Data

Climate analysis often requires data over extended regions, such as an entire country or the entire globe. Such studies need data from a large number of observing sites to support the analyses. In many countries, the national hydrometeorological service maintains a network of observing sites to support their operational requirements. Data from these weather observation and other networks form the backbone of many national climate studies. In the United States, data from existing networks, including the Cooperative Observer (COOP) Network (Williams et al., 2006) and the National Weather Service observing sites, form the basis for state and national temperature averages. For example, data from the Cooperative Observer Network is averaged into 344 Climate Divisions (Figure 6.3) in the 48 contiguous states with up to 10 Climate Divisions per state (Guttmann and Quayle, 1996).

These Climate Division data have been used in a multitude of climate studies as well as for monitoring monthly and seasonal climate in real-time (see examples on the CPC/NOAA and NCEI websites).

For studies that cross international borders, data availability can be a limiting factor. In many countries, fewer observations are freely available to users outside of the country than are found within the national data archives. This occurs because many

Figure 6.3 Map showing the 344 climate divisions in the contiguous United States. Average monthly temperature and precipitation estimates are available for each climate division starting from 1895.
Source: NOAA/NCEI.

nations view their weather observations as having national security or commercial value and limit availability to the set of locations identified in World Meteorological Organization (WMO) data exchange agreements (see the URL for WMO Resolution 60 in the reference list). The meteorological community is justly proud of its long history of collaboration and data sharing in support of weather forecasting. The amount of monthly data freely exchanged through WMO agreements, however, is less than 10% of the 12,000 or so routine observations made several times each day by national hydrometeorological services (Figure 6.4a and b).

Additional data may be available for purchase, or in some cases, for use in non-commercial (academic) research with the proviso that the data not be passed along to third parties. In some cases, more locations and variables are exchanged regionally than with the greater international community.

In addition to monthly summaries, many weather services transmit data hourly, or 3-, 6-, or 12-hourly, together with a Summary of the Day (SOD). The SOD contains the daily mean, maximum, and minimum temperatures, precipitation amount, and other variables. As mentioned above, there are roughly 10 times as many observing sites supporting real-time weather analysis as provide data summaries for climate studies. The more frequent weather data are routinely used at global weather forecasting centers to define initial conditions for numerical weather prediction. While these data have long been available to support real-time weather forecasting efforts, the systematic archiving of the hourly, 3-, and 6-hourly data for later use in climate studies has become common only in the past two or three decades. Even though much of the data is now being

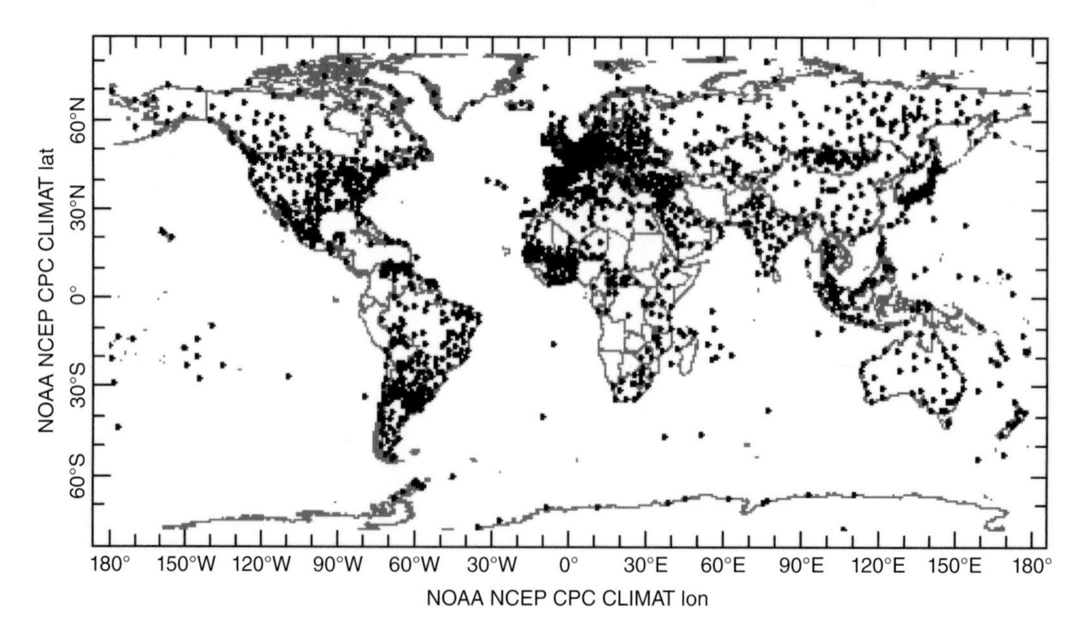

Figure 6.4a Map showing the typical number of locations that exchange monthly summaries of climate data (CLIMAT data) under international agreement.
Source: NOAA/CPC.

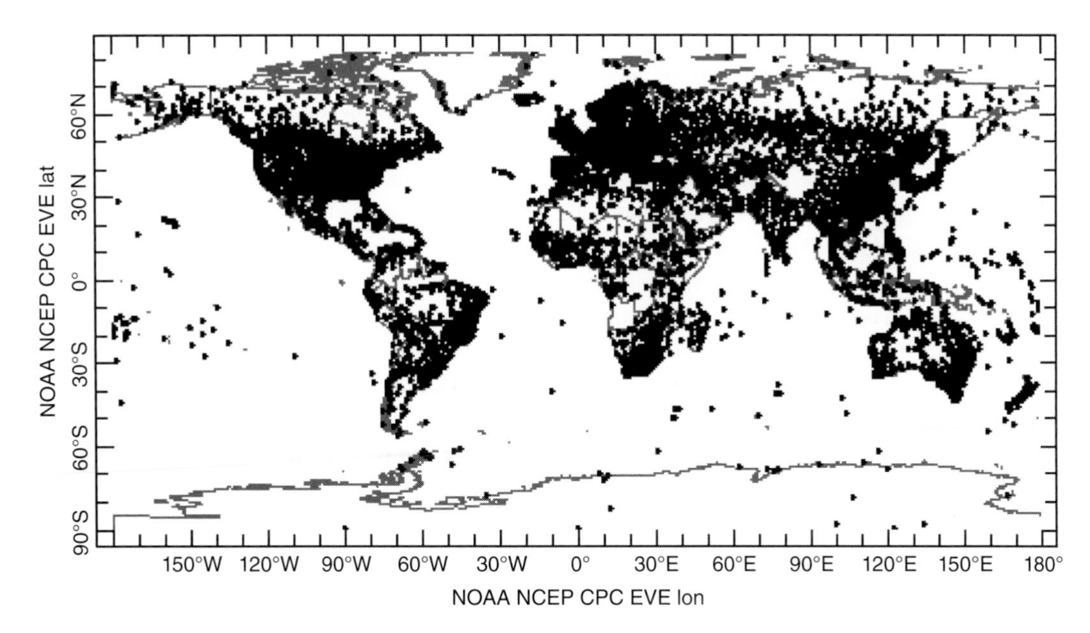

Figure 6.4b Maps showing the typical number of locations that available for daily weather analysis and forecasts over the Global Telecommunications System.
Source: NOAA/CPC.

archived and made available, for example at the NCEI, their use in climate studies still presents many challenges. Many observing sites lack comprehensive metadata documenting changes in instrumentation, location, and observing practice, and the data for any given location are often incomplete. To further complicate matters, even though 10,000 or more observations appear daily in the networks supporting weather forecasting, these data are not necessarily from the same set of observing sites at each observation time. Recently, efforts have been launched to encourage the routine archival and exchange of daily and sub-daily observations (Rennie et al., 2014; Thorne et al., 2017; WMO Resolution 60). In addition to encouraging the exchange of more real-time data, these efforts also encourage the digitization and archival of pre-1950s hard copy data.

The data exchanged via national and international networks operated by hydrometeorological services and the WMO constitute the core data for many regional, continental, and global temperature analyses. Many nations, however, maintain additional observing networks that may, or may not, appear in meteorological data archives. It is only recently, for instance, that the NCEI in the United States has been able to include access to observing networks maintained by other US government agencies such as the Departments of Agriculture, Interior, and Energy. In some cases, the data archive centers have merged observations from two or more networks into a single dataset to create a more complete product.

A number of research questions and practical applications require values of temperature at a location that doesn't have any historical temperature data. Very often the nearest available observing site is in a compromised environment, such as an airport thermometer located near an active runway. In such cases, the data may provide qualitative guidance rather than quantitative estimates for the desired location. In other situations, the observing site with the most complete data record may not be the site closest to the location of interest. The decision as to which dataset to use in a specific study must weigh the advantages of using data from one of the available sites against using data from some combination of sites.

In some cases, the temperature data that are available at a particular location may not yet have been subject to careful scrutiny, as in the observations discussed in Boxes 6.3 and 6.4. In such cases the metadata are critical. The metadata may be detailed enough to permit the necessary bias adjustments to the data, or at least to identify the limitations of the data from a particular observing site or combination of sites. Other issues like periods of missing or incomplete data are discussed in Appendix C.

Increased interest in global warming in recent decades has inspired the collection of long records of air temperature over land (and ocean, as discussed in Chapter 8) from as many sites as possible. Time series of temperature observations adjusted for changes in environment, instrumentation, and observing practice are now available from a number of data centers such as the NCEI. Datasets comprised of such observations are exceedingly useful in studies that require a history of the actual temperatures at a site. The temperatures recorded at a number of nearby sites will, however, often differ from each other because of the differences in local environment or other factors discussed in this chapter. For instance, the temperatures in the city will generally be greater than those in the nearby suburbs (a phenomenon termed the urban heat island). In examination of long-term temperature trends, we need not be concerned with biases among stations that remain constant over time. Rather, we monitor the year-to-year changes in temperature at each site with respect to its historical mean computed for the same base-period at all sites.

6.6 Gridded Temperature Datasets Based on Observations

Gridded temperature data that are complete in both space and time are very attractive since they lend themselves easily to statistical analysis of data fields, for example with Empirical Orthogonal Functions (EOF) as discussed in Appendix A, and can more easily be compared to output from numerical models and satellite observations. In addition, all of the data issues discussed earlier will presumably have been addressed in forming the gridded datasets. As with the data from individual sites, however, the documentation should be consulted to aid in understanding the limitations of these data in the context of the study.

6.6.1 General Characteristics of Gridded Temperature Datasets

Gridded temperature datasets are available from several data centers at various spatial resolutions. Early gridded temperature datasets based on station data provided monthly averaged data on relatively coarse grids measuring 5° latitude by 5° longitude (for example, Jones et al., 1982). More recently, much more ambitious daily gridded temperature datasets based on statistical interpolation techniques have been produced at resolutions of 0.5° by 0.5° and less (e.g. Fan et al., 2008). The most popular of the gridded datasets have been those that are updated in near real-time, making possible extensions of the global temperature time series and comparisons to model or satellite estimates of temperature.

In its simplest form, a gridded dataset might be calculated using the arithmetic average of the temperature observations within each grid area as the value for that area. Since observing sites within a grid have elevation and other environmental differences, and since the mix of observations within a grid will almost certainly vary over time, averaging actual temperatures to create a mean value over the grid area is challenging. Since an important use for such datasets is the analysis of anomalies, which are the differences of observed (or estimated) temperature from the mean value at each observing site, an often better option is to compute anomalies at each site and average them over the grid area to estimate gridded temperature anomalies. Some more sophisticated gridded temperature datasets estimate the temperature differences associated with elevation differences among observing sites in the same grid by incorporating estimates of the mean decrease of temperature with height into the computation of the grid area-mean temperature. For a dry atmosphere, the temperature deceases about 10°C/km or 5.5°F/1000 feet. Moist atmospheres have smaller temperature decreases with increasing height. These datasets may also use complex statistical models to adjust the elevation corrections for different weather conditions and other factors (e.g. Parameter-elevation Regressions on Independent Slopes Model – PRISM, Daly et al., 1994; and the PRISM website). The resulting gridded temperatures are estimates from a statistical model using existing observations to provide the initial conditions for the model and thus are somewhat analogous to the numerical models that use physical and dynamical laws to constrain the analysis.

Most gridded temperature datasets are spatially and temporally complete, but that completeness is likely to have been achieved through spatial, temporal, or statistical interpolation. For some grid areas, the temperature record may consist entirely of estimated values. The documentation associated with the gridded data must be consulted to determine how the datasets were constructed and how grid values were obtained when no observed data exist within the grid area. The way the estimates were obtained when no data are available, whether from data within a grid area, by interpolation, or through some other statistical process, will determine the limitations on the use of these data.

6.6.2 Relating Gridded Averages to Temperature at a Location in the Grid

Some analyses require the comparison of temperature data from a single location within a grid to the gridded value. This involves the classic problem of having to compare data at a point to data that comes from an average over an area. These two different kinds of data should be highly correlated but they will not be identical. The character of the temperature histograms for a specific location within a grid area and for the gridded values will differ from one another in systematic ways. The magnitudes of the grid area temperature extremes will generally be less than at individual stations and the point data will usually have more variance. The temperatures for a grid area may, in some cases, be significantly different from temperatures at a specific fixed location within the area. In general, this is less of a problem in the analysis of temperature anomalies, but will often be a significant issue, for instance, in the analysis of the number of days that the temperature exceeds some threshold. Presumably, grid area values will be used in an analysis when there are no site-specific data for the location of interest. The relationship between the gridded value and the temperature at a specific location may vary with time because the number of stations going into the grid average may vary considerably for data spanning several decades (Figure 6.5).

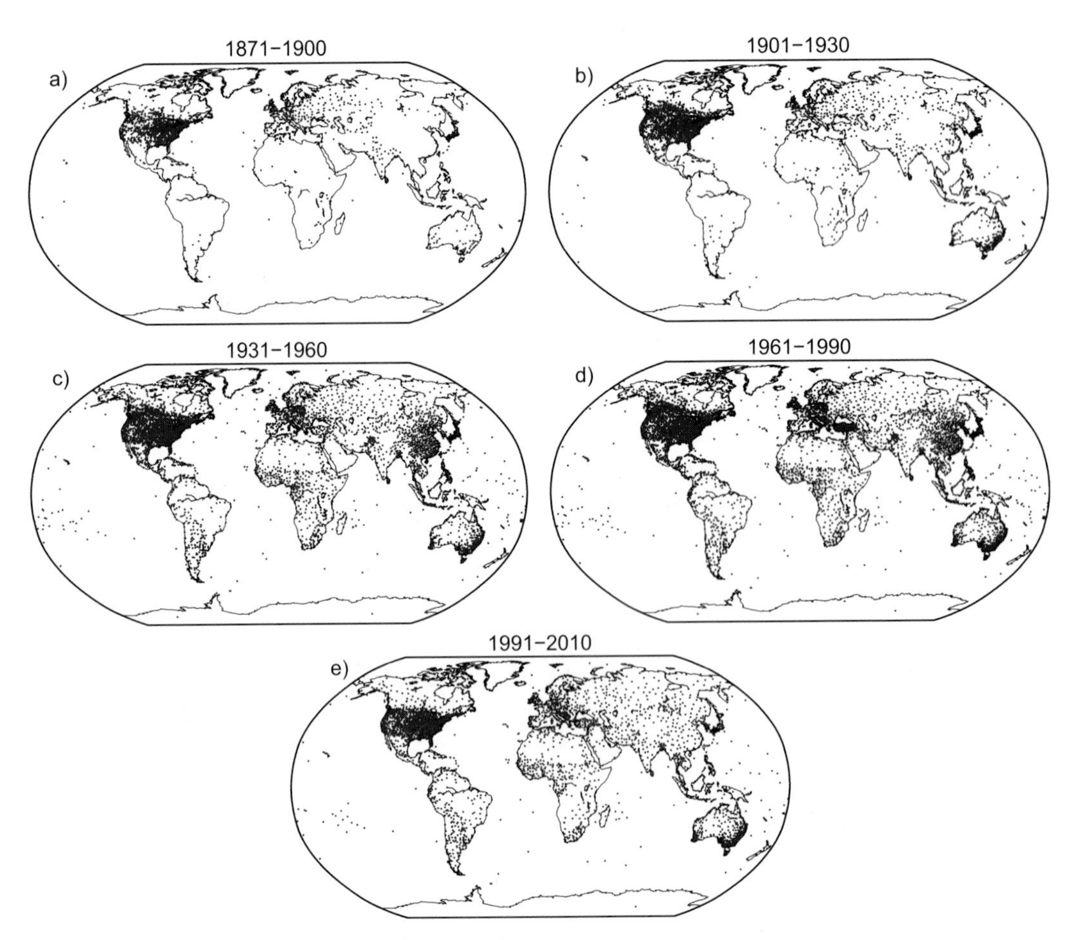

Figure 6.5 Map of locations with monthly temperature data in the Global Historical Climatology Network (GHCN-M) for various times in the past.
Figure from Lawrimore, after Lawrimore et al., 2011.

6.7 Other Estimates of Surface Air Temperature

Temperature data from global numerical models of the atmosphere are also available from several reanalysis datasets produced during the last couple of decades and discussed in detail in Chapters 3 and 11. Reanalysis refers to the analysis of historical observations using a current generation, state-of-the-art, numerical model of the atmosphere and/or ocean. The model can be thought of as a data interpolation method that is constrained by the laws of physics that govern the atmosphere or ocean.

The datasets from most reanalyses contain gridded near-surface air temperature fields at the spatial resolution of the model. Some of the current reanalysis data assimilation systems ingest surface temperature observations and others do not. In all cases the surface temperature in the reanalysis datasets refers to the model surface rather than the actual earth's surface. Nonetheless, the surface temperatures from the reanalysis are generally highly correlated with estimates based on surface temperature observations (Chelliah and Ropelewski, 2000). The surface temperatures from reanalyses can be very useful in comparisons of the statistics of observed temperatures with the statistics generated by model estimates that have not assimilated surface temperature observations. These kinds of comparisons provide some estimate of how well a model can replicate past surface temperature patterns and can provide some estimate of the uncertainties in temperature patterns from model projections into the future.

The advent of instrumented satellites has revolutionized many meteorological observations. Satellites can provide relatively high-resolution data, globally, at regularly scheduled observation times. Attempts to provide a satellite-based equivalent to the standard 2 m air temperature over land have had limited success. There are several limitations, but the main difficulty lies with the variability of surface emissivity[5] over land at very small spatial scales (Becker and Li, 1990). Currently, the most fruitful avenues of success have been through the use of numerical models in conjunction with satellite estimates (Siemann et al., 2016). The situation is not so dire over the global oceans because of the relatively smaller spatial variability in the emissivity over water (see Chapter 8).

In this chapter, we have discussed many of the challenges involved in building climate datasets from a collection of individual observations, using air temperature over land as an example. Although the details will differ, each of the issues we have discussed related to changes in the environment, instrument type, and observational practices apply to all of the observations commonly used in climate analysis. Likewise, the importance of metadata describing the time history of observations is critical to interpreting datasets for climate analysis. In the following chapter, we raise some of the same issues for the analysis of precipitation datasets.

[5] Emissivity refers to the ratio of the actual thermal radiation from the surface that is emitted toward space to the theoretical maximum radiation that would be emitted by a perfect blackbody. A perfect mirror would have an emissivity of zero, while a perfect blackbody has an emissivity of one.

Suggested Further Reading

Classic Examples of Inhomogeneities in Climate Datasets etccdi.pacificclimate.org/docs/Classic_Examples.pdf.

Automated Surface Observation System (ASOS) User's guide (1998) – PDF can be downloaded from www.nws.noaa.gov/asos/pdfs/aum-toc.pdf.

Comparisons of gridded temperature datasets (NCAR/UCAR Climate Data Guide) https://climatedataguide.ucar.edu/climate-data/global-temperature-data-sets-overview-comparison-table.

Vose and co-authors, 2011: NOAA's Merged Land-Ocean Surface Temperature Analysis, *Bulletin Of the American Meteorological Society*, **93**, 1677–1685. (A detailed description of one of the major global surface temperature datasets).

Questions for Discussion

6.1 Suppose a time series of daily mean temperatures for a 91-day season has gaps amounting to 10% of the total length of the dataset. How would you compute the differences in the seasonal mean and standard deviation if (a) the gaps are ignored in the computation, (b) the gaps are filled with the mean value computed by ignoring the gaps, (c) the gaps are filled with values selected at random from the time series?

6.2 What if the data gaps in 6.1 represented (a) 20% of the data? (b) 5% of the data?

6.3 The mean annual cycle will exert some influence on a 91-day time series of daily mean temperatures, especially during the fall and spring seasons. How would you take this into account?

6.4 Discuss methods that might be used to relate station temperature data and gridded temperature data from a numerical model.

6.5 Where on the planet would you expect the magnitude of the mean annual temperature cycle to be greater (or smaller) than the diurnal mean amplitude?

7 Precipitation

Combining *In Situ* and Remotely Sensed Observations in Constructing Climate Datasets

7.1 Introduction

Rain and snow are among the most commonly discussed elements of weather and climate. When we complain that the weather forecast was wrong, we are much more likely to be griping about a failure to accurately predict the occurrence of precipitation than an error in the temperature or amount of cloud cover. Long lasting anomalies in precipitation lead to droughts and floods, with consequent dramatic impacts on vegetation, wildlife, and society. The measurement of precipitation is as challenging as it is necessary, due to the extraordinarily high spatial and temporal variability that it can exhibit. *In situ* observations of precipitation at several locations in close proximity may disagree, adding to the difficulty in building representative precipitation datasets based on measurements at individual locations. In this chapter, we will illustrate that, in contrast with land temperature datasets, precipitation datasets used in climate analysis rely on the combination of information from many sources.

The hydrological cycle, or water cycle (Fig. 7.1), consists of the reservoirs of water in gaseous, liquid, and solid form in the Earth System and the exchanges among those reservoirs. The amount of water vapor in the atmosphere changes locally as weather systems evolve, but is relatively stable over the atmosphere as a whole. Precipitation occurs when water vapor condenses and forms raindrops or snowflakes that fall to the surface. That process would totally desiccate the atmosphere within a week or two unless the moisture was replenished by evaporation from the land and ocean. Over the course of a year and averaged over the globe, evaporation and precipitation very nearly balance one another.

Of course, this long term, large-scale balance does not reflect a well-behaved uniform distribution in time and space. Precipitation in particular is very challenging to measure and to predict because it is one of the few properties of the atmosphere that is discrete on the macroscale. Most atmospheric components and properties are quite well mixed, with values that vary smoothly over short distances. Precipitation, on the other hand, is composed of objects that are quite large compared to atoms and molecules, large enough to see and feel individually. Precipitation begins when water vapor condenses into small droplets or ice crystals, but does not actually occur until these particles exceed a certain size threshold. As a consequence, the amount of precipitation actually reaching the surface can vary from zero to large amounts over extremely short distances and time spans.

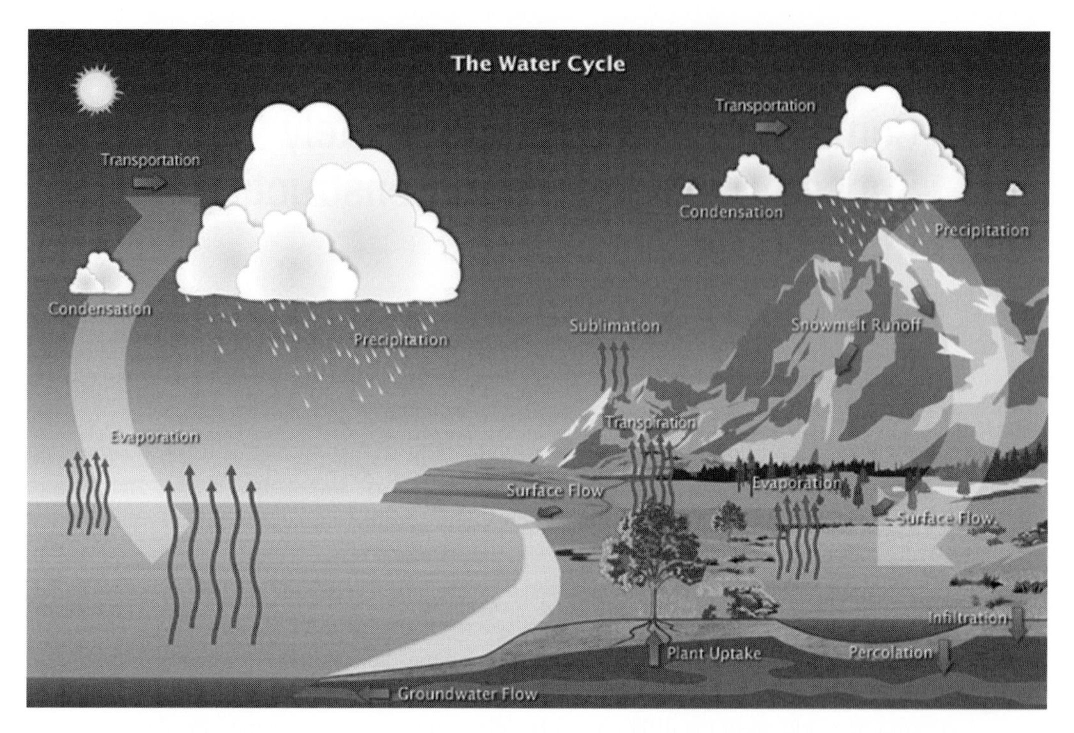

Figure 7.1 This schematic of the hydrologic (water) cycle illustrates the ways that precipitation processes interact with many components of the climate system including the atmosphere, ocean, cryosphere and land components.
Source: NASA.

Measurement of precipitation has a number of very practical applications. For example, a certain amount of water is required to ensure the growth and maturation of various crops. Precipitation observations can enable a farmer to know when enough moisture is available for planting, and whether precipitation is insufficient during the growing season and must be supplemented by irrigation. Reservoirs are often used for multiple purposes, such as providing water for human consumption and irrigation of crops while at the same time reducing the risk of river flooding. Measurements of precipitation in the catchment from which the reservoir receives water can tell the operator whether releases are required to avoid overtopping and damaging the dam, or whether the runoff into the reservoir must be retained to ensure adequate supplies in later dry periods.

Precipitation observations also have significant scientific benefits for understanding and predicting the weather and climate. Perhaps most critically, since model predictions of precipitation are based on a large number of approximations and simplifications, substantial systematic errors are common. One of the best ways to reduce and/or correct such errors is to compare them to observations for the same time/space domain. This process can enable model developers to identify errors and create improved versions that better predict or simulate precipitation, and can also facilitate the development of statistical models that improve upon the model predictions.

While the preceding chapter discussed the challenges in constructing climate datasets of temperature from many individual observations, here we face the somewhat different problem of creating datasets from a wide variety of observations that, while related to precipitation, are actually of a number of different parameters. This presents a set of complex challenges, as we shall see.

We begin this chapter by describing methods for observing or estimating precipitation at specific times and locations, including both *in-situ* and remote-sensing techniques. We also discuss other approaches that permit the inference of precipitation variability, such as atmospheric or land surface budgets, observations of land/ocean properties such as soil moisture or salinity, or the assimilation of observations into an atmospheric model. Next we discuss methods through which such observations, estimates, and/or inferences are combined into spatial analyses of precipitation covering large areas and time spans. Since many of the data used are available only since about 1980, we further discuss statistical methods that enable the analysis of earlier periods.

7.2 Measurements, Estimates, and Inferences

Precipitation occurs when water vapor condenses or freezes and the resulting liquid/solid particles fall to the surface. This statement vastly oversimplifies the process: The immediate result of condensation/freezing is the formation of clouds, and further complicated events must transpire before particles large enough to fall to the surface are formed. These atmospheric processes are well covered in a number of places (see, for example, Fleagle and Businger, 1980; Pruppacher and Klett, 2010; Wang, 2013) and will not be discussed extensively here.

For convenience, we will refer to rain and snow gage observations as measurements, values derived from satellite or radar observations as (remote sensing) estimates, and values derived by atmospheric models from observations of other atmospheric values as inferences. Precipitation can also be inferred from observations of other properties of the land or ocean, such as soil moisture or salinity.

7.2.1 Rain Gage Observations

Many people are familiar with the use of gages to measure precipitation. Here we refer to any collecting device that captures precipitation as it falls and permits it to be measured using a ruler or similar device (weighing is equivalent) as a gage. There are mentions of gages being used, generally in connection with planning for agricultural activities such as planting, from India, China, Korea, and the Middle East for millennia, but specific detailed records are sparse until the Renaissance. Records of precipitation accumulation from individual gages over extended periods began to be available in the late 1600s, and sufficient geographic coverage to permit meaningful spatial analyses over large parts of the Northern Hemisphere was achieved by the beginning of the twentieth century (Strangeways, 2011).

A wide variety of devices have been used as gages. (A full description of the Standard Rain Gage can be found in the references under websites.) Most often, falling rain or solid precipitation is captured over some period of time and measured periodically. The measurements are most often of volume, usually interpreted as height or depth of rainfall over the surface area of the gage. Sometimes the captured precipitation is weighed, which avoids variations in volume due to temperature and the small errors associated with wetting of the sides of a storage gage. Solid precipitation presents a number of difficult challenges, and a great many techniques have been developed to measure snow, especially in high latitudes. Typically, measurements of total accumulated precipitation are recorded daily at a specific time. While the time of measurement is generally consistent within countries, it often varies quite widely across national boundaries, introducing significant complications when trying to create analyses over continental or global domains.

While gages are the most direct method of measuring precipitation, they are subject to a wide variety of errors. The goal is always to obtain the amount of precipitation that would fall onto a specific small area in the absence of any disturbance of ambient air motion. Since a gage is a physical structure that modifies the wind flow in its vicinity, and since the trajectories of precipitation particles are strongly influenced by the motion of the air through which they fall, the amount of precipitation that falls into a gage is generally different than would reach the ground in the absence of the gage. Furthermore, objects such as structures or vegetation, particularly trees, that are near the gage cause distortions in airflow that change the amount of precipitation that falls into the gage. Once precipitation enters the gage, other errors can become significant. In the case of very light rain or drizzle, drops can adhere to the side of the gage (wetting), while, especially in heavy rain or in the case of large drops, some water can splash out of the gage. In gages that accumulate rainfall such that the accumulated liquid is exposed to the atmosphere over time, especially during non-precipitating periods, evaporation can reduce the apparent amounts. Since the content of most gages is determined by comparing the water level to a scale impressed or printed on the gage itself, errors can be caused by incorrectly identifying the surface level, especially for large amounts in gages with sloping sides (used to exaggerate small amounts to make them easier to read). Other gages measure the amount of rain entering the gage by allowing a small bucket to empty once filled (tipping bucket), or by recording the increase in weight of the accumulated precipitation. Such gages are more difficult to maintain, and tipping bucket gages can underestimate heavy rainfall since rain that falls while the emptying is underway is not recorded. Measurement of solid precipitation, freezing rain, or drizzle is particularly difficult, and conventional gages used for rainfall, such as the Standard Rain Gage or a tipping bucket gage, are poorly suited for such situations. Techniques and tools have been developed for measuring snow (Rasmussen et al., 2012) and are used in high-latitude nations. However, errors in accumulated precipitation are greatly increased by the occurrence of solid precipitation, and national differences in methodology can lead to systematic differences across national boundaries.

Generally speaking, the most crucial element in ensuring accurate measurement using gages is the siting of the instrument. The U.S. Cooperative Observer Program offers the following advice:

Precipitation gauge siting: *The exposure of a rain gauge is very important for obtaining accurate measurements. Gauges should not be located close to isolated obstructions such as trees and buildings, which may deflect precipitation due to erratic turbulence.* ***To avoid wind and resulting turbulence problems, do not locate gauges in wide-open spaces or on elevated sites, such as the tops of buildings.*** *The best site for a gauge is one in which it is protected in all directions, such as in an opening in a grove of trees. The height of the protection should not exceed twice its distance from the gauge. As a general rule, the windier the gauge location is, the greater the precipitation error will be.*

Since the gage itself distorts the wind flow in its vicinity and makes the amount of precipitation captured different from that which would have fallen in the absence of the gage, a great deal of effort has been invested in developing gages that minimize that error (see Nespor and Sevruk, 1999, and references therein). Furthermore, calculations of the error as a function of wind speed have been used to develop statistical corrections (Folland, 1988). Corrections for errors related to wetting, splash, and evaporation are not generally employed, since the errors are small compared to those due to wind.

Particularly when properly exposed and corrected for errors related to wind, gage observations are considered quite accurate. However, they can only represent the precipitation at their location, and so using them to estimate areal averages or as the basis for analyses (maps) of precipitation requires consideration of the spatial sampling issues involved (see Section 7.3).

7.2.2 Estimates from Remotely Sensed Data

Precipitation can be estimated from observations of visible, infrared, and microwave radiation emitted or reflected by raindrops, ice particles, or clouds associated with precipitation. Unlike gages, which provide direct measurements of the precipitation that falls at a specific location, remote sensing provides estimates of precipitation over some larger area, or more accurately, of falling precipitation over some volume of the atmosphere. Most remote estimates of precipitation are derived from surface-based radars or from observations from instruments on Earth-orbiting satellites, although some come from instruments on aircraft. In the majority of cases (referred to as *passive*), the observations are from instruments, called radiometers, that measure the radiance emanating from or reflected by a part of the Earth and atmosphere, and the principal challenge in deriving useful estimates is to infer the precipitation that resulted in the observed radiance values. This is a complicated problem, since many other properties of the land surface and atmosphere have their own effects on the observed radiance. Some instruments (radars) improve upon this through the more *active* process of emitting a microwave beam toward a potentially precipitating target and measuring the reflected return. The various remote-sensing techniques yield widely varying accuracies and coverages, and are discussed later in this section.

More specifics about applications of remote sensing to meteorological observations can be found in Kidder and von der Haar (1995).

Visible and Infrared Methods

In order for remote sensing of precipitation to be feasible, the instruments used must detect radiation at wavelengths where information can propagate far enough so that the signals from precipitation can reach the satellite or other platform. In practice, this limits opportunities to two bands of frequencies, visible/infrared and microwave, since it is only in these window[1] regions where the sensing of such distant objects can occur. Since instruments located at the surface or on aircraft have a limited field of view, essentially all such estimates are derived from instruments on satellites. In the visible region of the spectrum, instruments sense the location and, to some extent, the thickness of clouds, while in the infrared the observation is of the temperature of the top of the cloud. (While we will often refer to clouds, we actually mean collections or systems of clouds since the spatial resolution of the instruments is generally inadequate to fully resolve individual clouds). Both of these provide some limited information on the location and intensity of precipitation, and algorithms that use visible/IR data to estimate precipitation do so by identifying precipitating clouds and assigning a precipitation rate to them.

Visible and infrared (IR) instruments are carried on satellites in two kinds of orbit, distinguished by their altitude as low (LEO, for low Earth orbit) and geostationary (GEO – see Chapter 3, Figure 3.3). LEO satellites typically orbit at altitudes between 750 and 1500 km, where a complete orbit takes 1–2 hours (the duration of the orbit is largely determined by the altitude). Meteorological satellites are often placed in LEO orbits that pass over the poles, referred to as polar orbiting, and a careful choice of altitude makes it possible to ensure that each orbit passes over the equator at the same local time as all the others. Such orbits are referred to as sun synchronous. The other commonly used orbit is much higher, at about 37,000 km. At that altitude, a complete orbit takes exactly 24 hours, and when the satellite is positioned over the equator it orbits permanently above the same point on the Earth's surface at all times. These orbits are called geostationary, since they appear stationary relative to the surface. Instruments on these satellites afford the ability to observe atmospheric properties as they evolve, whereas the instruments in polar orbits view the atmosphere at the same time of day all around the world. Both of these have advantages for estimating precipitation, but neither is ideal for all purposes. Obviously, other orbits are possible, and versions that are used specifically to improve precipitation estimates will be described later.

The most obvious advantages of polar orbiting observations are greater spatial resolution, since the instruments are closer to the Earth, and better observations of high latitudes, which appear foreshortened from geostationary orbit. Polar-orbiting satellites were available about 15 years before geostationary satellites, and the first estimates, based on simple assessments of the frequency of thick or cold clouds, began to be available in the mid-1960s. Since precipitating cloud systems move and evolve

[1] A band in the electromagnetic spectrum that offers maximum transmission and minimal attenuation through a particular medium with the use of a specific sensor (Glossary of Meteorology, 2012).

relatively rapidly, and since visible and IR-based estimates are less accurate that those derived from microwave observations (next section), the limited temporal sampling available from polar orbits means that visible/IR estimates from those instruments are little used at present.

Up to the present, precipitation estimates from geostationary satellites are limited to those based on visible and IR observations. The spatial resolution of such observations ranges from 1 to 10 km, with an interval between successive images of 10–60 minutes (the most modern geostationary meteorological satellites have the capability of more frequent imagery, with intervals as short as one minute, but over limited areas). Since these instruments are always located over the equator, their field of view is limited by foreshortening as viewing angles increase away from nadir, which is the point directly below the satellite. Typically, precipitation estimates can be calculated for points that are up to 50° away from nadir.

Estimates of precipitation based on visible/IR satellite observations were first produced in the mid-1960s, soon after the observations began to be available from polar-orbiting satellites. While the methodology varied quite substantially among the various products (see, for example, Lethbridge, 1967 and Barrett, 1970, among others), the fundamental information underlying each was essentially the same: The precipitation at a point was proportional to the frequency of occurrence of clouds with cold and/or bright tops. For some applications (flash flood detection, for example), qualitative indices were adequate. For others, quantitative estimates were required and were generally obtained by regressing an index against gage observations for the same region.

Visible/IR-based precipitation estimates improved when data from operational geostationary satellites became available beginning in 1979 due to the greatly improved temporal sampling. Simple threshold algorithms were developed that related precipitation to the frequency of cold cloud (Arkin, 1979; Arkin et al., 1994), along with more complex algorithms that included information related to the temporal evolution of cloud patterns (Griffith et al., 1978) and the spatial gradients of brightness temperatures (Adler and Negri, 1988).

Visible/IR observations continue to be widely used for a great number of applications, and continuation of the current combination of low orbit and geostationary satellites is quite likely. A wide variety of visible/IR methods are still in use, and improvements continue. These methods yield precipitation estimates that are quite useful for many purposes, such as flash flood warnings. However, the reliance on the properties of the top of clouds, when the actual precipitation is always within or below the clouds, is a handicap impossible to overcome with those data alone. Appreciation of this fact has led to the development of algorithms based on observations at microwave frequencies, and to the use of methods that combine visible/IR with other information, as will be discussed later.

Passive Microwave Methods

The second main "window" through which radiation can propagate without being absorbed and reradiated by the gaseous components of the atmosphere and can thus carry information on precipitation to satellite instruments is in the microwave region for frequencies ranging from a few to a few hundred Gigahertz (GHz). In this spectral region, unlike the visible/IR domain, radiation propagating in the atmosphere is minimally affected by the small particles that make up clouds but interacts relatively strongly with larger, precipitation-sized particles. This has the important practical advantage that estimates derived from microwave observations are largely uninfluenced by the clouds that obscure the precipitation in visible/IR observations, thus removing a significant source of systematic error.

Instruments that record the microwave radiation emitted by or reflected from precipitation are referred to as passive, in contrast to those that emit a beam of radiation and record the reflection (active radar, to be discussed in the next section). Passive microwave observations are nearly unaffected by the clouds that always accompany precipitation. However, the relatively long wavelengths require much larger antennae than are needed for visible/IR measurements to obtain adequate spatial resolution. Microwave instruments on polar-orbiting satellites, given the size constraints inherent in building and operating such systems, can provide spatial resolution at the surface of 6–50 km. At the present time, passive microwave instruments are not practical on geostationary satellites.

The limitation of microwave observations to polar orbit greatly limits the temporal sampling that can be obtained. In particular, since most of the satellites that carry such instruments are sun-synchronous, most estimates derived from microwave observations from an individual instrument are for two times each day separated by 12 hours (at the equator) and thus provide poor sampling of the diurnal cycle (see Chapter 1), which can be quite significant in many areas. Two approaches can ameliorate this problem: observations from satellites in precessing orbits, where the local time at which the instrument observes changes from one orbit to the next, permitting more complete sampling of the diurnal cycle over some time period, and combining of observations from multiple satellites in polar orbit but with different equator crossing times to provide better diurnal sampling even for short time periods. Both of these approaches have been used and will be discussed further in this section.

Microwave radiation interacts in complex ways with large (precipitation-sized) water and ice particles in the atmosphere, and these interactions determine the techniques used to estimate precipitation. Microwave radiation upwelling from the Earth's surface can be scattered by the particles, or it can be absorbed. In addition, the precipitation particles themselves emit microwave radiation. When the particles scatter radiation, the satellite-based instrument will observe less radiance from an area where precipitation is occurring than from regions without precipitation, and the degree of reduction will be related to the number and size of the precipitating particles. Particles that absorb upwelling radiation will themselves emit radiation, and the amount of radiation emitted will increase as the density of precipitating particles increases. The degree to which particles scatter and emit microwave radiation changes with frequency, and differs for water droplets and ice particles.

At relatively low frequencies, below 50GHz, emission is the dominant effect, and can be used to estimate precipitation over oceans. At these frequencies, the thermal emission from raindrops in the field of view of the satellite instrument leads to an increase in radiance observed that is proportional to the number, size, and volume of raindrops within the column. This increase is detectable over oceans, where the water surface has low emissivity and thus appears cold to the satellite. Over land, where the surface emissivity is higher and more variable, the increase due to raindrops is less evident and emission algorithms are not usable. Other issues can affect the utility of such methods over oceans: Surface ice, or even cold water, can degrade estimates, and radiation from the smaller droplets comprising the clouds from which the precipitation falls contributes some enhancement for which semi-empirical corrections are required.

At higher frequencies, scattering due to large ice particles, particularly in convective clouds, becomes more important. Such scattering reduces the radiation reaching the instrument, and the reduction is proportional to the number and size of the ice particles. In convective systems, the density of such ice particles is closely related to precipitation intensity and so can be used to estimate accumulated precipitation. This approach works over land and water, and so can be used globally. However, surface ice and snow introduce large errors, and estimates for such areas must be eliminated. Precipitation resulting from non-ice processes (often referred to as warm rain) does not produce a scattering signal, and thus can lead to underestimation of total precipitation. At even higher frequencies, estimates of the intensity of falling snow are possible and are beginning to be used in precipitation products (Meng et al., 2017).

Both emission and scattering approaches have an additional error resulting from the relatively coarse spatial resolution of the instruments. Since the relationship between the emission and scattering signals and precipitation intensity is not linear, a nonhomogeneous distribution of precipitation within the field of view will result in an erroneous estimate of the mean intensity over the full pixel. This is sometimes referred to as beam filling bias, and is impossible to correct without observations or assumptions of the fine scale distribution of precipitation.

Passive microwave observations suitable for estimating precipitation have been made since 1987 when the first Special Sensor Microwave/Imager (SSM/I) instrument was flown on the Defense Meteorological Satellite Program (DMSP) F-08 satellite.

Estimates based on passive microwave observations, while widely used and clearly superior to those based on visible/IR alone for individual values, do have shortcomings. Since these instruments can only be used on low orbit satellites so far, their temporal sampling is limited, and temporal and spatial averages are subject to significant errors. At present, about 10 passive microwave instruments are available on a combination of operational and research satellites, and the resulting sampling of less than three-hour interval for 90% of the day is adequate for many applications., More frequent sampling would make the estimates more useful, but continuation of even the current level of observations is not assured. An additional challenge results from the relative lack of information over land and for snow or ice-covered surfaces. Estimates of orographic and high-latitude precipitation are particularly lacking.

Radar (Active Microwave) Methods

The fact that metallic objects often reflected radio waves was noticed early in the twentieth century. Devices using this fact to detect ships during foggy weather were created soon after, and radar (**RA**dio **D**etection **A**nd **R**anging), which was intended to both detect objects like aircraft and ships and measure the distance to them, was developed for military purposes in the decade preceding World War II. During the war, radar operators noticed that weather phenomena, including precipitation in particular, were also detectable, and following the war meteorologists began to investigate and develop practical applications. The U.S. National Weather Service began to use radars operationally to detect severe thunderstorms and tornadoes in the 1950s, and many nations around the world now operate comprehensive networks of advanced weather radars that can estimate precipitation as well as other aspects of severe weather such as winds.

Estimates of precipitation are derived from radar observations using mathematical models that infer hydrometeor density and type from the reflected energy. In order to sample a representative volume around the site, the radar beam is usually rotated and elevated in steps in association with pulses of radiation. In combination with radiative transfer models and assumptions about relevant properties of the precipitation, this permits the calculation of the amount of liquid water or ice contained in precipitating hydrometeors in a set of radially oriented volumes.

Surface-based radars, which comprise the vast majority of available instruments, are subject to a variety of errors related to the trajectory of the radar beam. Since the beam and its reflection travel in an approximately straight line, the curvature of the Earth's surface results in increasing altitude at increasing distance from the radar location for a given elevation angle. Even at the lowest practical elevation, typically 0.5°, the beam reaches altitudes well above the surface within a short distance from the radar site. This means that the radar observation is essentially never exactly the desired surface rainfall, and in addition that a horizontal (relative to the surface) map of precipitation must be approximated by taking values at different distances from different elevation angles. It is quite typical to find range-related artifacts in rainfall maps derived from individual radars. The conversion of reflected energy to surface precipitation also depends on the type of precipitation, its phase (liquid, solid, or a combination), and how much evaporates as it falls through the layer beneath the radar beam.

Radars do not have to be in a fixed location at the surface. Many research radars are portable and are transported where they are needed, either for field experiments or to obtain detailed measurements of specific storm systems. Radars have been placed on ships and aircraft as well. In general, data from these radars are less useful for climate studies due to their limited duration of sampling. Radars on low Earth-orbiting satellites have been available since the launch of the Tropical Rainfall Measuring Mission (TRMM) Precipitation Radar (TPR). The TPR was on the TRMM satellite,[2] which was launched in late 1997 and continued to collect data until April 2015. The TPR provided a unique perspective on precipitation, with limited sampling in time and space, but with the ability to detect the vertical profile of precipitation in a wide variety of

[2] The TRMM and GPM satellites and their instrument complements were and are a collaboration between the U.S. National Aeronautics and Space Administration (NASA) and Japan's Japan Aerospace Exploration Agency.

systems. A more advanced Dual-frequency Precipitation Radar (DPR) is contained on the Global Precipitation Measurement (GPM) Core Observatory, which was launched in 2014.

Most radar-derived estimates of precipitation have proven difficult to use for climate studies. The spatial coverage of such data is relatively limited even now, with regular observations available only in a few nations. Creating maps of rainfall over hours or days from individual radars is challenging, since the retrieved values have errors that can change rapidly in time and since approximating a surface value from the elevated measurements is difficult. Experience has shown that the use of surface rain gages to calibrate the radar retrievals is necessary. To the present, no climate analyses of precipitation based solely on surface radar networks, even gage-corrected, have been created. However, the 17-year long TRMM mission has permitted the creation of a number of research datasets based on the TPR that have been helpful in improving the understanding of the role of tropical precipitation in climate variability. These datasets are being extended and improved using data from the DPR.

7.2.3 Estimates from Remotely Sensed Data

The challenge of creating datasets that describe the variations of precipitation in time and space for the whole Earth is daunting. Direct gage measurements and estimates based on remote sensing individually are severely limited, but in combination, as will be discussed in the next section, can provide near-global analyses. However, there are many other observations of the atmosphere and the land and ocean surface that offer the possibility of inferring precipitation independently, and thus either augmenting and improving the analyses, or verifying their accuracy and estimating their errors. These methods can be grouped into two main categories: observations of changes in surface properties that result from precipitation, and using atmospheric observations of winds, temperatures, pressure, and moisture to estimate condensation and precipitation.

Surface Properties

Precipitation, whether falling as rain or snow, produces detectable changes on the surface onto which it falls. Some of these changes occur promptly and offer information on short time scale events, while other responses are slower and provide evidence regarding integrated precipitation over time. Precipitation amounts and variations can be inferred from observations of these changes. To provide the sampling and coverage necessary for climate analyses, satellite observations are essential; here we describe some of the observations that have been used.

For land areas, certain observations provide evidence of obvious consequences of precipitation. For example, rainfall leads to wet and muddy soil, and even to puddles on the surface. Satellite observations can identify such conditions and provide pretty clear indications of where rain has fallen, as well as some information on intensity. Visible/IR observations are relatively ineffective for this purpose, because precipitating clouds obscure the surface and make the time history unclear, and the signature of wet surfaces is not unequivocal. Observations of snow-covered surfaces are more effective since, especially in the visible spectrum during daylight hours, the signature is clear. Passive microwave observations are not impeded by clouds and so provide a better record of the surface changes, although it has proven more effective so far to use microwave information to estimate falling hydrometeors.

Estimating the total change in soil moisture over time can be useful, and when compared to estimates of evapotranspiration can be used to estimate accumulated precipitation over the period. The most difficult problem in applying such an approach is often in obtaining realistic quantitative estimates of evapotranspiration. For example, changes in vegetation are clearly related to changes in precipitation and evapotranspiration. Since chlorophyll has a distinct signature in visible and short IR wavelengths, changes in vegetative growth can be estimated using an index computed from such observations. The NDVI (Normalized Difference Vegetation Index) is one example that has been available since the 1970s. While it is difficult to quantify the specific changes in vegetation that the index represents, the enhanced growth that results following rainfall events in many regions is quite evident. It has not proven possible so far to derive useful precipitation analyses from the NDVI, but the relationship between vegetation and precipitation continues to be the subject of research. The NDVI is discussed further in Chapter 9.

Longer-term changes in precipitation can be inferred from changes in the gravity field. Changes in soil moisture and ground water lead to changes in the distribution of mass and in consequence to changes in the gravity field of the Earth. The GRACE (**GRA**vity recovery and **C**limate **E**xperiment) satellite mission provides observations of such changes on continental scales, and data from GRACE have been used to estimate seasonal and interannual changes in precipitation accumulation over regions such as the Amazon basin and Greenland.

Other observations may provide information about precipitation. For example, Aquarius is a satellite mission that provides estimates of sea surface salinity from passive microwave observations. When combined with estimates of evaporation and internal oceanic transports of fresh water, these can be used for an estimate of precipitation independent of the satellite-derived estimates discussed earlier.

While all of the approaches mentioned here offer the potential of providing some information about the temporal and spatial distribution of precipitation, none is likely to be as accurate as the gage observations or remote sensing estimates discussed earlier. Their main benefit is that they are to varying degrees independent of the more accurate methods and thus can be used to confirm results.

Atmospheric Model Forecasts

The remaining method by which precipitation can be estimated is through predictions from mathematical models of the atmosphere. Such models are commonly used to provide weather forecasts based on observations of winds, temperature, pressure, and moisture combined with the physical laws that govern atmospheric behavior. While weather predictions are not perfect, they have improved greatly over time and at very short time ranges compare very well to observations for many atmospheric variables. The physical processes governing precipitation are extremely complicated, and precipitation forecasts are not as accurate as forecasts of winds and temperatures. Nevertheless, there are situations and locations where model forecasts of precipitation may be useful in the construction of climate analyses due to the relative lack of other usable observations.

Several issues challenge the use of forecasts in precipitation analyses. Weather models are relatively poor at predicting convective precipitation, and have limited quantitative skill in the tropics and in midlatitudes in the warm season. Furthermore, the use of model forecasts in climate analyses is handicapped by the frequent changes made to models in order to correct errors. This issue is of course pervasive in climate science, and atmospheric reanalyses (see Chapter 3) have been performed to deal with this. The precipitation fields from reanalyses may eventually prove to be both usable and useful, as will be discussed later.

7.3 Analyses

Climate analyses of precipitation can be constructed from a subset of available observations, such as rain gages alone, or from combinations of different observations, estimates, and inferences. As discussed in Chapter 3, this is an analysis problem where the goal is to construct a complete field with values at all points of a regular grid. For precipitation, there are benefits to be found from a variety of approaches and here we discuss several. We will also discuss methods that can be used to extend analyses farther into the past using a combination of observations and analyses.

7.3.1 Gage-Based Analyses

Analyses based on gage observations have been available for more than a century. For most of that time, the analysis process was subjective: Precipitation totals were plotted on an appropriate map and an analyst drew isopleths[3] according to his or her best judgment. Such subjective analyses represented the most advanced state of the art through the end of the 1970s. Since usable gage observations are limited to land areas, most such analyses covered continents.

The preparation of these analyses required decisions regarding which data to use, both in terms of period of record and choice of stations, and how to deal with questionable values. These issues are common to all climate analyses, but can be particularly difficult to handle in the case of precipitation due to its large spatial and temporal variations, which can lead to very large errors in cases where precipitation events are undersampled or missed entirely. The selection of stations and time span depends upon the specific objectives of the analyst: If very high spatial resolution is required, then the maximum number of stations should be used, even if they represent subsets of the total time span. If average values for some subinterval of the total record

[3] In map analysis, isopleths are lines of equal value. After the Glossary of Meteorology (2012).

are required, then only stations with consistent observations throughout the period are used. In some cases, temporal consistency is most crucial, in which case only stations with long observational records are used. In all cases, quality control of the observations is crucial, since distinguishing extreme events from errors is challenging. This continues to be the case today, even though subjective analyses have been superseded for most climate studies by various methods of objective analysis.

Once adequate computing power began to be available during the 1960s and 1970s, methods of using mathematical algorithms to generate a set of equally spaced values from irregularly distributed observations were developed. Such analyses were much more difficult to apply to precipitation fields than to most meteorological parameters, since the spatial gradients are greatly exaggerated compared to, for example, wind or pressure fields. While these gradients are smoother in time averages, they are still significant, even in long-term climatologies. In the preparation of time series of analyses for months or seasons, spatial variability presents a significant challenge and is one of the principal reasons for utilizing other sources of information such as satellite-derived estimates.

Several algorithms for interpolating gage values were developed and applied in the 1970s and 1980s (see Xie et al., 1996 for background). In all of these, the value derived for a specific point on a regular grid is some form of weighted average of gage observations selected for their relevance. Weights are generally inversely related to distance between the station and the grid point, and various approaches are used to reduce the influence of gages that are nearly in line with others but more distant from the grid in question. Some authors have suggested that analyses based on anomalies relative to some climatological distribution are more successful. The strong influence of elevation and wind direction on precipitation (Daly et al., 1994 – see PRISM discussion) has led to many attempts to include topography as a factor in precipitation analyses.

Several precipitation analyses based on rain gage observations and covering nearly all land areas of the globe are currently available. They include those produced by the Climatic Research Unit (CRU) of the University of East Anglia, the Global Precipitation Climatology Centre (GPCC) operated by the German Weather Service (Deutscher Wetterdienst) and the Climate Prediction Center (CPC) of the National Oceanic and Atmospheric Administration (NOAA). The GPCC produces a near real-time analysis (called "first-guess"), a monitoring analysis available with a two-month lag, and a "full" analysis that includes more data and better quality control but whose availability lags by a few years. The CPC analyses are produced and published online regularly for individual months with a time lag of a few months. Both GPCC and CPC analyses are used in the creation of global integrated analyses (discussed later). The CRU analysis is based on a smaller subset of the available observations, updated less frequently but with more extensive quality control, and is intended specifically for the validation of climate model simulations and projections.

These analyses are all relatively coarse in spatial and temporal resolution, typically about 2° in latitude/longitude and monthly, and make use of fewer observations than are potentially available. All of these groups face two important challenges: how to identify incorrect data values, and what to do in regions with few or no observations. The various quality control methods for identifying and correcting errors in observations are well understood and will not be detailed here. Correction or removal is generally possible for significant errors, but subtle errors can be very difficult to identify, making the use of complementary data such as satellite-derived estimates helpful.

Handling gaps in temporal or spatial coverage is both crucial and challenging. The objective of such analyses is to calculate an estimate for the areal average of precipitation over the area surrounding each grid point. When that area contains a number of observations, an estimate of the average is relatively straightforward. However, when there are few or no values, and even more so when neighboring areas themselves are sparsely covered, estimation of values becomes much more difficult. This difficulty is exaggerated when the topography of the area is complex, since precipitation is strongly influenced by the interaction of elevated terrain and winds.

For well-observed regions such as the United States, many more rain gage observations can be obtained, making possible analyses on finer spatial and temporal scales. For such fine scales, it is both possible and necessary to take into account factors such as wind flow and elevation, and an example of this approach is PRISM (**P**arameter-elevation **R**egressions on **I**ndependent **S**lopes **M**odel). PRISM precipitation values are calculated on a grid with 4 km intervals for individual months, using available gage observations in combination with information about elevation, direction of sloping terrain and prevailing winds, distance from coasts and climatological gradients. The PRISM precipitation analyses are very highly resolved and have been extensively used, but are not available globally. The PRISM documentation cautions against using this data set for studies of decadal climate variability.

While gage-based analyses are important, they are not adequate by themselves for most climate analysis applications. In the United States, Western Europe, and some other regions, gage-based analyses such as PRISM and the CPC operational analysis are considered as accurate as any other product. However, even these have errors, generally of uncertain magnitude, in regions with sparse observations, complex topography, or both. In fact, the PRISM analysis illustrates a difficulty that arises from an approach that computes values at every point of a dense grid in the absence of actual observations – unusual events that are missed by the gages will be omitted from the analysis, but the analysis will appear to be fully detailed nonetheless.

7.3.2 Remote-Sensing Based Analyses

As discussed previously, remote-sensing estimates of precipitation are made from satellite observations and from surface radars. Analyses based on satellite-derived estimates are more often used for climate applications, since they provide broader coverage and are less dependent on local advanced technology. In essentially all cases, gage observations, or analyses derived from gage observations, are important to reduce biases over land, and methods for doing so will be discussed in the following section.

Radar-based analyses can only be produced where an adequately dense network of rain radars capable of producing quantitative precipitation estimates is available. Until recently, this was possible only in the United States, Japan, and some countries in Western Europe, but recently a number of other countries have begun to implement their own integrated networks. For the most part, radar-only analyses are not yet used for climate analysis and so we focus here on analyses based on satellite-derived precipitation estimates.

Many of the early estimates derived from satellite observations were spatially and temporally complete, and thus could be used as analyses. The earliest of these were based on the occurrence of cold or bright cloud tops, and were derived simply by accumulating or interpreting estimates onto a regular latitude-longitude grid over a time period, typically a month. The only such product still used in climate analysis, primarily in qualitative applications, is outgoing longwave radiation (OLR) estimated from LEO IR observations. Its value arises from its use in a great variety of diagnostic studies and its lengthy time series based on relatively consistent processing. It has been used to derive quantitative estimates of precipitation in diagnostic and monitoring studies and as one component of combined analyses (Xie and Arkin, 1998).

Quantitative estimates of precipitation suitable for climate analyses began to be available shortly after estimates began to be derived from GEO imagery in the 1970s. The first approach to be applied broadly, generating long time series of analyses that covered a large fraction of the globe, was the GOES[4] Precipitation Index (GPI, Arkin and Meisner, 1987). The GPI, like the approaches mentioned earlier, is based on the relative frequency of precipitating cloud but is derived from geostationary imagery that provides both good sampling of the diurnal cycle and complete coverage in the tropics and midlatitudes. Estimates were based on the statistical results of Arkin (1979) and Richards and Arkin (1981), who found that the area-averaged precipitation estimated from radar observations over the western tropical Atlantic Ocean was highly correlated with the fractional coverage of cold cloud when averaged for time periods of a few days or longer and large enough areas. The GPI gave results that, for climate scales, were comparable to more complex algorithms based on the same data and was simple enough to be applied routinely to all available imagery. Since the GPI is calculated for a regular latitude-longitude grid, no further processing is required to derive an analysis. The GPI was used in a number of studies of precipitation variations associated with the annual cycle (Arkin and Meisner, 1987), the diurnal cycle (Meisner and Arkin, 1987), and the El Niño/Southern Oscillation (ENSO; Arkin and Meisner, 1987).

A number of analyses based on individual microwave algorithms have also been produced. A comprehensive list of such single-source analyses is available in Table 3 of the International Precipitation Working Group's (IPWG) description of precipitation datasets (the IPWG is listed in the references under websites). The relatively sparse sampling of the microwave observations prevented useful analyses for periods shorter

[4] GOES stands for Geostationary Operational Environmental Satellite, the label given to the series of US geostationary meteorological satellites that first became available in the 1970s and continue to be used.

than a month and areas smaller than 2.5° latitude × 2.5° longitude. An analysis based solely on the TRMM Precipitation Radar (TPR) estimates may be the most-used of all single-source analyses at the present. The extremely sparse sampling of the TPR, due to its narrow swath and limited orbit, requires accumulation over relatively long time periods. However, TPR-based estimates are considered more accurate than any other remote-sensing based estimates for its area of coverage (35°N–S), and thus analyses derived from them are greatly valued for insight into the aggregate details of precipitation.

This plethora of single-source precipitation analyses might confuse a climate analyst. When should any one of the available products be used, and for what purpose? The answer depends on the goals of the analysis, and the constraints to which it is subject. For real-time monitoring of current events, the OLR product is the most quickly available. For research, if an integrated analysis (see later in this chapter) is not available or usable, it is wise to use several single-source products. Much useful information can be inferred from areas of agreement and disagreement among the various products. However, future climate analyses based on estimates from individual satellite sensors are likely to be used only when based, like the TPR, on instruments with unique characteristics.

IR-based analyses are subject to significant errors related to the indirect relationship between cloud top temperatures and precipitation, and the relatively sparse temporal sampling of passive microwave observations leads to substantial uncertainty in analyses derived solely from microwave-based estimates. Since the strengths of the two approaches – good sampling in geostationary IR estimates and relatively good accuracy in microwave-based estimates – complement one another, analyses based on the combination of estimates and/or information from a variety of sources have become the choice of most scientists. The Global Precipitation Climatology Project (GPCP; Arkin, 1989) was begun in 1986 with the objective of providing global precipitation analyses for research conducted within the World Climate Research Programme (WCRP). Based on several studies that intercompared estimates from a variety of data sources using different approaches, for example the Precipitation Intercomparison Project III (PIP-3; Adler et al., 2001), the GPCP concluded that analyses based on a combination of estimates were more accurate than products based on single sensors.

Modern climate analyses of precipitation can be separated into two groups: a first generation with relatively coarse spatial and temporal resolution that relied on the combination of analyses based on separate sensors, and a second generation that uses methods that combine individual estimates at much finer scales. The first generation consists primarily of the merged analysis produced by GPCP (the analysis is also referred to as GPCP; Huffman et al., 1997; Adler et al., 2003) and the **C**limate Prediction Center **M**erged **A**nalysis of **P**recipitation (CMAP; Xie and Arkin, 1996, 1997). These analyses provide time series of monthly and five-day (pentad) global analyses with spatial resolution of 2.5° latitude by 2.5° longitude beginning in January 1979. The microwave-derived inputs to both GPCP and CMAP are not available before mid-1987, and the GPI is based on limited data prior to 1986. These satellite-only versions of GPCP and CMAP are rarely used, since their values over land are not necessarily well calibrated with available gage observations.

As satellite remote sensing technology improved during the 1980s and 1990s, improved methods of estimating precipitation on finer spatial and temporal scales came into use. The launch of the TRMM satellite in late 1997, along with the availability of data from the Earth Observing System satellites in the same timeframe, began to make possible the production of precipitation analyses with much higher resolution and generally more useful characteristics. These second-generation analyses have begun to be used for a wide variety of scientific and practical applications, and as their length of record increases they will be used more and more as climate analyses. The multinational Global Precipitation Measurement Mission (GPM) relies on data from a Core Observatory satellite launched in 2014 together with information from a constellation of other satellites with appropriate instrumentation to build on the TRMM legacy with improved instruments and algorithms.

Examples of second-generation analyses included CMORPH (**C**PC **MORPH**ing Analysis; Joyce et al., 2004), TMPA (**T**RMM **M**ulti-sensor **P**recipitation **A**nalysis; Huffman et al., 2007), Precipitation Estimation from Remotely Sensed Information using Artificial Neural Networks (PERSIANN; Sorooshian et al., 2000) and Global Satellite Mapping of Precipitation (GSMaP; Tian et al., 2010). All of these analyses utilize the more accurate microwave-based estimates combined in various ways with the less accurate but more complete geostationary infrared data. Time series beginning in 1998 are available for both TMPA and CMORPH, and combinations of those products with rain gage information are being developed as supplements/replacements for CMAP and GPCP.

The Precipitation Measurements Mission has developed and implemented a new analysis called IMERG (Huffman et al., 2015) based on an algorithm that intercalibrates, merges, and interpolates all available satellite-derived microwave precipitation estimates, together with microwave-calibrated IR estimates, precipitation gage analyses, and other precipitation estimators at fine time and space scales for the TRMM and GPM eras over the entire globe. IMERG is produced multiple times for each observing time, providing both an initial analysis and successively more complete values as additional data arrive. Monthly gage data are used to reduce errors and to create analyses suitable for research (as discussed in Section 7.3.3).

7.3.3 Remote Sensing and Gage-Based Analyses

Precipitation analyses based on remote sensing estimates alone, because of their indirect physical relationship to surface precipitation and susceptibility to algorithm error and consequent artifacts, can be improved through error screening and bias removal. This is most readily accomplished through the use of gage observations, or gridded analyses derived from such observations. Most analyses of precipitation incorporate such corrections where gage observations of the quality and density required are available. The suitability of gage data for making such corrections is related to the gage density and distribution, and to the nature of the precipitation regime and the terrain. For climate analyses, where relatively coarse spatial and temporal resolution is required, it has been found that a gage density of 1–10 observations within a 2.5° × 2.5° grid area is adequate to constrain monthly analyses except in mountainous areas.

The production of quantitative precipitation estimates and analyses based on radar observations requires the removal of artifacts related to obstructions by structures and terrain, propagation anomalies related to atmospheric conditions, and a variety of other range-related errors. Calibration errors related to variations in the drop size distribution and the vertical variability in the precipitating column must also be corrected. The most appropriate practical method for deriving and applying such corrections is to use a network of high time resolution rain gage observations. Both areal mean and spatially varying biases can be estimated from the radar to gage differences, and experience has shown that the results are more useful for applications such as flood prediction and model verification. This approach is most easily applied to individual radars, and an analysis based on the combination of a number of radars and underlying gages is more difficult.

Global precipitation datasets require a combination of satellite-derived estimates for coverage and sampling with rain gage information for calibration. Both the GPCP and CMAP analyses described in the previous section have been combined with the Global Precipitation Climatology Centre (GPCC) rain gage-based analysis to yield global precipitation analyses beginning in 1979 and continuing to the present. These analyses are very widely used in climate monitoring and research to describe and investigate variability on monthly and longer time scales. They differ in some significant ways, particularly over the oceans where the truth is not well known, but are similar in many important respects.

Second generation satellite-based analyses, while covering shorter time spans than GPCP and CMAP, offer finer resolution in time and space. Combining such products with gage information over land is motivated for the same reasons as for the first-generation analyses, but new approaches were required to deal with the finer resolution.

The TMPA (**T**RMM **M**ulti-sensor **P**recipitation **A**nalysis) is the most used current near-global analysis with an extended period of record suitable for climate research and applications. TMPA is available on a 0.25° latitude-longitude grid from 60°N to 60°S at three-hourly intervals beginning in 1998. It is based on estimates using the TRMM passive microwave algorithm (Kummerow et al., 2001) from the TMI, and instruments on other satellites where available. Remaining gaps are filled using an IR-based estimate derived using a thresholding algorithm calibrated against a 30-day record of nearby microwave estimates. The TMPA is calibrated to match the GPCC monthly product, and a "real-time" version is constructed using historical calibrations. The relatively long record and consistency with both other TRMM products and GPCP make the TMPA quite useful for studying fine scale details of climate variability, but the approach used to incorporate the IR information introduces occasional spatial/temporal discontinuities. This approach also forgoes the use of information from IR data in areas where even a single microwave-based estimate is available during a three-hour period.

CMORPH was developed as a standalone satellite-based analysis. However, Xie and Xiong (2011) have shown that a combined CMORPH-gage analysis can reduce regional time-varying biases relative to gage observations. As with TMPA, CMORPH relies on combining information from passive microwave and IR on fine time and space scales. The fundamental premise behind CMORPH is that the instantaneous precipitation estimates from passive microwave observations are significantly better than any estimates that can be derived from IR observations. However, since the microwave-based estimates are relatively sparse, a complete analysis requires some method for filling gaps in time and space. While IR-based precipitation estimates are relatively inaccurate, long experience has shown that they can provide excellent estimates of cloud motion, and hence of the movement of precipitating systems. CMORPH therefore uses storm motion estimated from IR imagery to interpolate microwave-derived estimates in time and space to fill gaps. Analysis has shown that gaps up to several hours and hundreds of kilometers can be filled with estimates more accurate than any that can be derived from IR data directly, even when calibrated against microwave estimates. This methodology is also used in IMERG.

The CMORPH values are available from 60°N to 60°S at the resolution of the IR data used, which is roughly 8 km and 30-minutes at the equator. Since the microwave-derived estimates are substantially coarser in spatial resolution, the most useful CMORPH-based analysis is produced on a 0.25° latitude-longitude grid for hourly accumulations. The combination with gage values is accomplished through a two-step process. The first uses spatially coincident gage analysis values to derive a climatological correction to the CMORPH probability density function (pdf), which is followed by a statistical combination of CMORPH and the Climate Prediction Center gage-based analysis. This approach ensures that the final analysis has a realistic pdf and mean value over land areas. Xie et al. (2017) explains the most recent developments and datasets associated with CMORPH.

GSMaP (Global Satellite Mapping of Precipitation) is another second-generation analysis that was developed by the JAXA Precipitation Measuring Mission (PMM) Science Team with support from the Core Research for Evolutional Science and Technology (CREST) program of the Japan Science and Technology Agency (JST). Both real-time and reprocessed versions of GSMaP products are distributed by the Earth Observation Research Center, Japan Aerospace Exploration Agency. The GSMaP analysis approach and some initial results are described by Kubota et al. (2007), Aonashi et al. (2009), and Ushio et al. (2009).

7.3.4 Future Developments

Many of the analyses described here continue to evolve as newer satellite observations become available and as more is learned about error correction. Ongoing precipitation analyses, including GPCP, CMAP, TMPA, CMORPH, and IMERG, will benefit significantly from the data from the GPM core and constellation satellites as well as the availability of precipitation estimates derived from next generation operational meteorological satellites, such as GOES-R and JPSS (**J**oint **P**olar **S**atellite **S**ystem) from the United States and Meteosat Third Generation and the Eumetsat Polar System

Third Generation from the European Union. A promising area of development is the incorporation of estimates from the TRMM Precipitation Radar and the Dual-frequency Precipitation Radar (DPR) on the GPM Core Observatory.

Certain situations will remain problematic, however. The most obvious of these is in high latitudes, where few satellite estimates are available, and those that exist are highly uncertain. IR-based estimates are most skillful in convective regimes where precipitating systems can best be identified and discriminated from clear sky and non-precipitating systems. Such regimes are found in the tropics and subtropics, and in midlatitudes during the warm season, but rarely in high latitudes. Current research into the development of microwave-based estimates of falling snow and precipitation over snow or ice-covered surfaces is achieving a degree of success (Meng et al., 2017), and data from CloudSat have proven useful in estimating light precipitation over oceans in high latitudes (Behrangi et al., 2014).

One possible approach to improving the quality of precipitation analyses in high latitudes is the use of model predictions. As discussed in Section 7.2.3, atmospheric model predictions of precipitation based on atmospheric observations have been shown to be quite accurate in regions and regimes where large-scale circulation is the dominant controlling factor. On very short time scales, the errors in precipitation forecasts in high latitudes are small, and the use of such forecasts as one element in analyses seems feasible. For use in climate analyses, such predictions are best obtained from reanalyses (see Chapter 3) since temporal inhomogeneities related to changes in the analysis/forecast model are eliminated. One version of CMAP (called CMAP/A by Xie and Arkin, 1996, 1997) uses this approach to produce spatially complete analyses.

Short-range precipitation forecasts from atmospheric models continue to improve in accuracy, and their use in global precipitation analyses will continue to be the subject of ongoing research. Important issues must still be addressed, including the best method to remove relative biases between the satellite and model-based precipitation and to establish weights for the different inputs in different climate regimes.

7.4 Extending the Record – Reconstructions

The fact that global precipitation analyses incorporating satellite-derived estimates begin, at the earliest, in 1979 poses a severe obstacle to the understanding of the behavior and role of precipitation in the climate on decadal and longer time scales. Prior to 1979, our information on oceanic variations is extremely limited, and in fact somewhat inconsistent. Two main sources exist: interpolation of island gage observations (Taylor, 1973) and the inference of precipitation from ship observations of present weather (Tucker, 1961, 1962; Dorman and Bourke, 1979, 1981). In general, such information was not considered dense enough in time and space to permit time series of monthly, seasonal, or annual totals, and the analyses were limited to averages over multi-decadal periods. Thus, for years prior to the satellite era global

Box 7.1 Reconstruction Methodology

In general, the methodologies used are straightforward applications of those used for SST and SLP. Requirements include a time series of spatially complete analyses for some adequate time period, called the base, and some other information that is related to the desired variable, precipitation in this case, and that is available both during the base period and the target historical period. The base dataset for precipitation is provided by the GPCP or some other complete analysis, and a couple of options exist for the other information. Two approaches have been attempted, one based on the relationship of rain gage observations to empirical orthogonal functions (EOFs) of the base dataset, and the other on the spatial and temporal relationships between the base precipitation and fields of SST and SLP as determined by Canonical Correlation Analysis (CCA).

precipitation analyses were extremely limited, and studies of inter-decadal variations or long-term trends were limited to those continental regions and island with long consistent records.

Descriptive studies using GPCP and CMAP have shown that precipitation variability is characterized by substantial coherence on large space and long time scales. One result of this coherent behavior is that observations in some areas can be used to approximate large-scale patterns with substantial skill. Such approaches are referred to as reconstruction (see Box 7.1) and have been used with success in the production of sea surface temperature (SST; Smith and Reynolds, 2004a; Kaplan, 1997) and sea level pressure climate analyses (Smith and Reynolds, 2004b). Smith and colleagues (Smith et al., 2010, 2009a, b, 2008; Arkin et al., 2010) have used a similar approach to create extended time series of near-global precipitation analyses for decades before the availability of satellite-derived estimates.

Interest in global climate change in recent years has led to speculation regarding the impact of increasing concentrations of CO_2 and other radiatively active atmospheric constituents on the elements of the hydrological cycle, including precipitation. While the period covered by modern composite precipitation analyses, which begin in 1979, is too short to reveal trends in global mean precipitation, the century long global reconstructions that have been produced recently do seem to show a small but systematic increase in globally averaged precipitation (Arkin et al., 2010) that is consistent with model simulations. However, given the lack of certainty about many aspects of precipitation observations, it is clear that much remains to be done before these results can be considered confirmed.

Suggested Further Reading

Gruber A. and V. Levizzani, 2008: Assessment of Global Precipitation Products – A project of the World Climate Research Programme Global Energy and Water Cycle Experiment (GEWEX)

Radiation Panel, WCRP Report WCRP-128, WMO/TD-no. 1430. PDF available online under Precipitation Assessment Report.
UCAR/NCAR Climate Data Guide – Precipitation https://climatedataguide.ucar.edu/climate-data/precipitation-data-sets-overview-comparison-table.

Questions for Discussion

7.1 What characteristics of precipitation make remote sensing estimates critical to large-scale analyses?

7.2 Compare the advantages and disadvantages of gage-based versus satellite-derived precipitation estimates for studies of interannual climate variability.

7.3 Make the same comparison for studies of decadal climate variability and trends.

7.4 Satellite-derived precipitation estimates can be grouped into three categories: vis/IR, microwave, and radar. What are the strengths and weaknesses of each, and what are the arguments for using combinations of all three to create precipitation analyses?

8 Ocean Climate Datasets

8.1 Introduction

The global oceans cover more than two-thirds of Earth's surface, and are one of the major components of the climate system (Chapter 1). In the latter half of the twentieth century climate scientists began to understand that large-scale year-to-year variations in ocean surface temperature were often closely tied to year-to-year variations in the seasonal climate of the atmosphere over land as well as the atmosphere over the oceans. Most notably, important climate variations over large parts of the globe are associated with the El Niño/Southern Oscillation (ENSO), as discussed in Chapter 4. Monitoring the state of the ocean is essential to monitoring, understanding, and predicting climate on all time scales. Continuing research is expanding our understanding of the coupled climate system, and helping to unravel the influences of the surface and sub-surface structures of the ocean on climate and climate variability. Our discussion of climate as a coupled system will be extended in the next two chapters to include the role of the land surface and the cryosphere as essential partners with the atmosphere and ocean in the coupled climate system.

We consider ourselves climate scientists who began as meteorologists. As such, our view of the ocean is that it is a critical component of the climate system, particularly through its interactions with the atmospheric boundary layer and external radiative forcings. Although we are not oceanographers, we have had to understand and use ocean observations and datasets in many aspects of our research. Our focus for the bulk of this chapter, 8.2–8.4, is to provide students, scientists, and other interested readers who are not ocean scientists with some introductory material and guidance to the use of ocean data in climate studies. A more detailed discussion of ocean subsurface instrumentation, observations, and networks is presented in 8.5 for those interested. Additional detail on many of the topics presented in this chapter may be found in the cited references and Suggested Further Reading.

The history of environmental observations in the oceans is significantly different than that of weather observations over land. This difference stems primarily from (1) the difficulty in making the observations and (2) the context and motivation for such observations. Before the advent of satellite remote sensing, most ocean data came from observations made on board commercial or military ships. Some of these data, from commercial fishing fleets, for example, were closely held. Since the 1970s considerable efforts have been expended to retrieve older records going back into the nineteenth

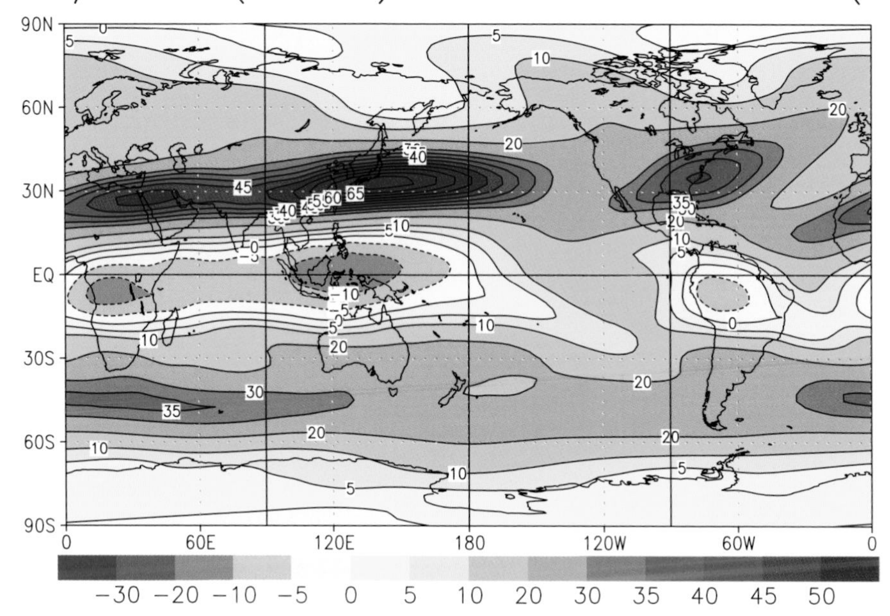

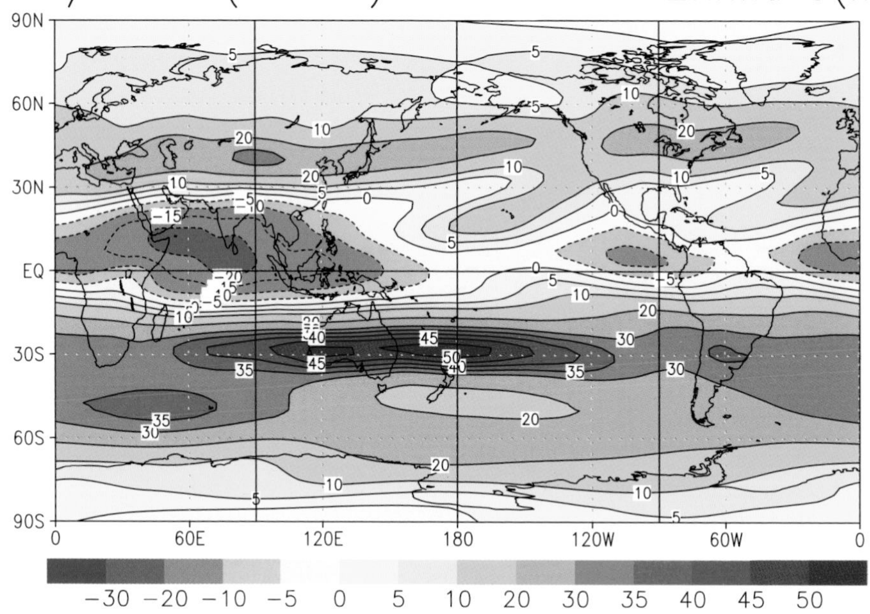

Figure 4.3 The mean zonal wind component at upper levels (200 hPa) for the 1981–2010 period based on the NCEP Reanalysis 2 (Saha et al., 2010), DJF (top) and JJA (bottom). Yellow and red shading denotes westerly winds; blue shading denotes easterlies (m/sec). The strongest mean westerlies in the winter hemisphere occur to the east of the Asian, North American, and Australian continents.

Figure courtesy of M. Chelliah, CPC.

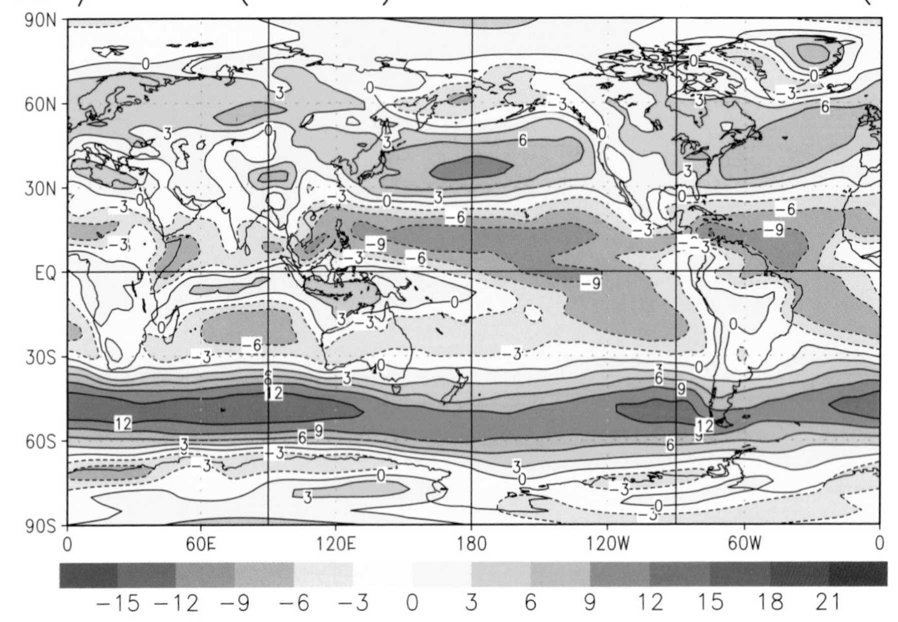

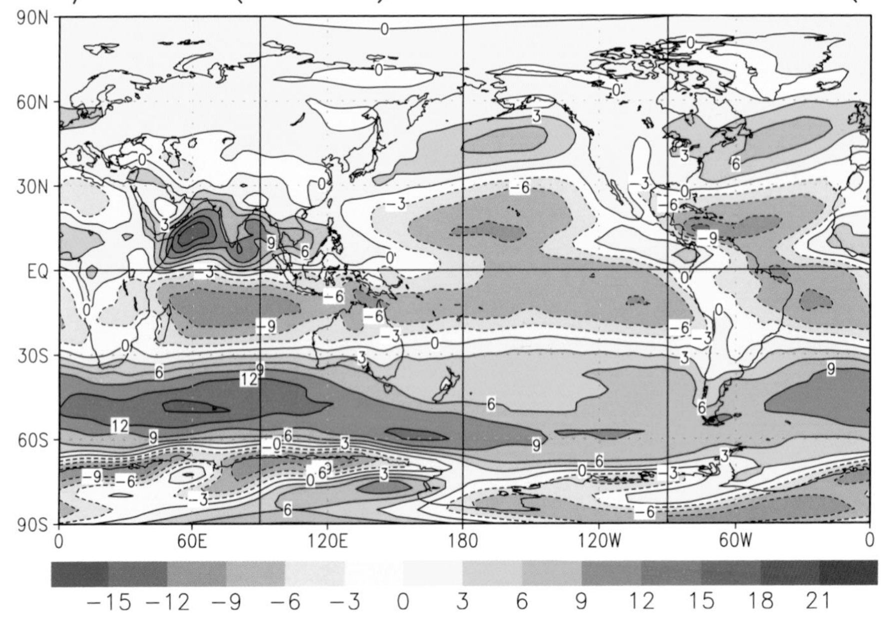

Figure 4.4 The mean zonal wind component at lower levels (925 hPa) for the 1981–2010 period based on the NCEP Reanalysis 2 (Saha et al., 2010) for DJF (top), and JJA (bottom). Yellow and red shading denotes westerly winds, blue shading denotes easterlies, (m/sec). Note the change in scale compared to the previous figure.
Figure courtesy of M. Chelliah, CPC.

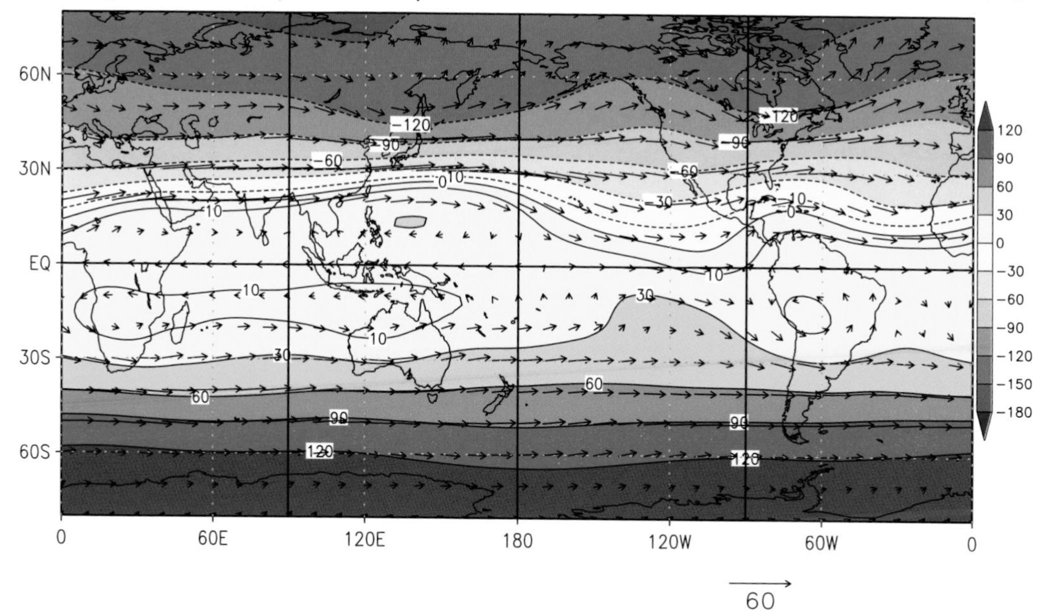

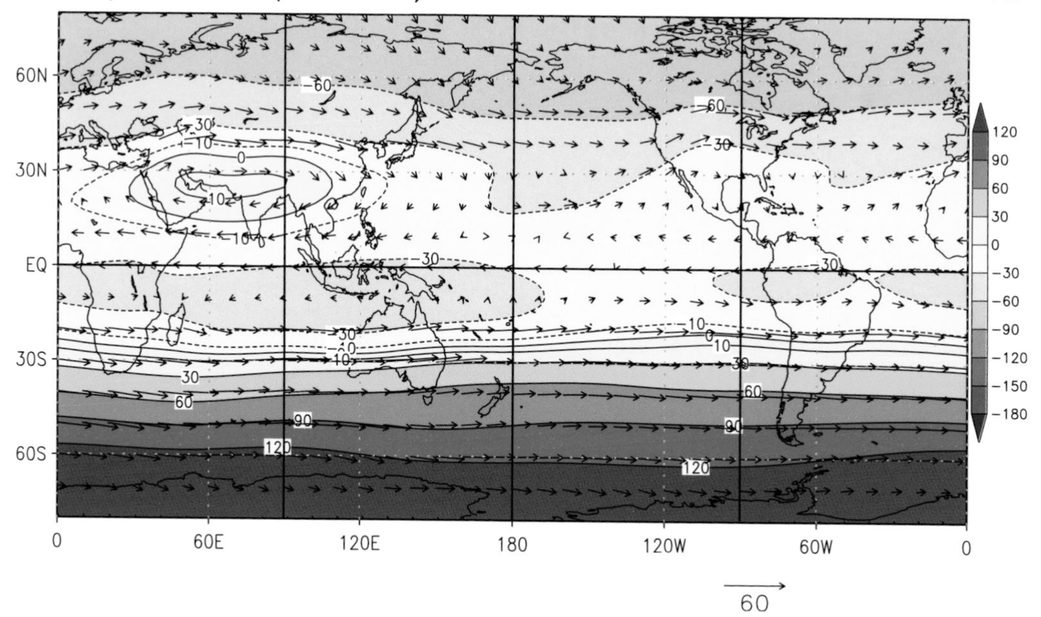

Figure 4.5 The mean 200 hPa stream function (10^6 m^2/sec), contour interval 10 units, and rotational component of the wind (m/sec) for the 1981–2010 period based on the NCEP Reanalysis 2 (Saha et al., 2010) for DJF (top), and JJA (bottom). Red shading denotes positive values, blue shading negative values.

Figure courtesy of M. Chelliah, CPC.

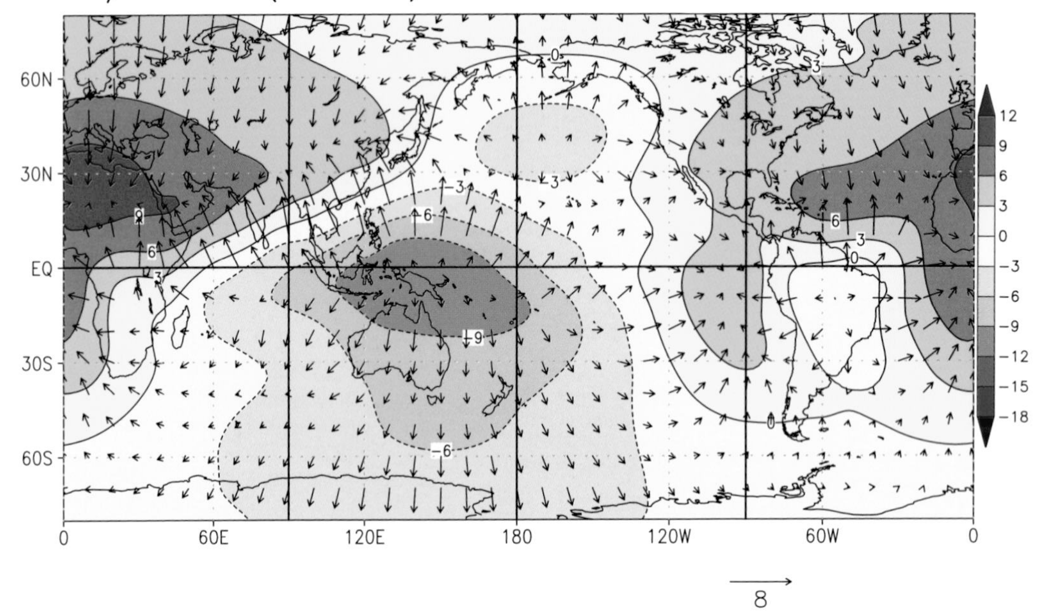

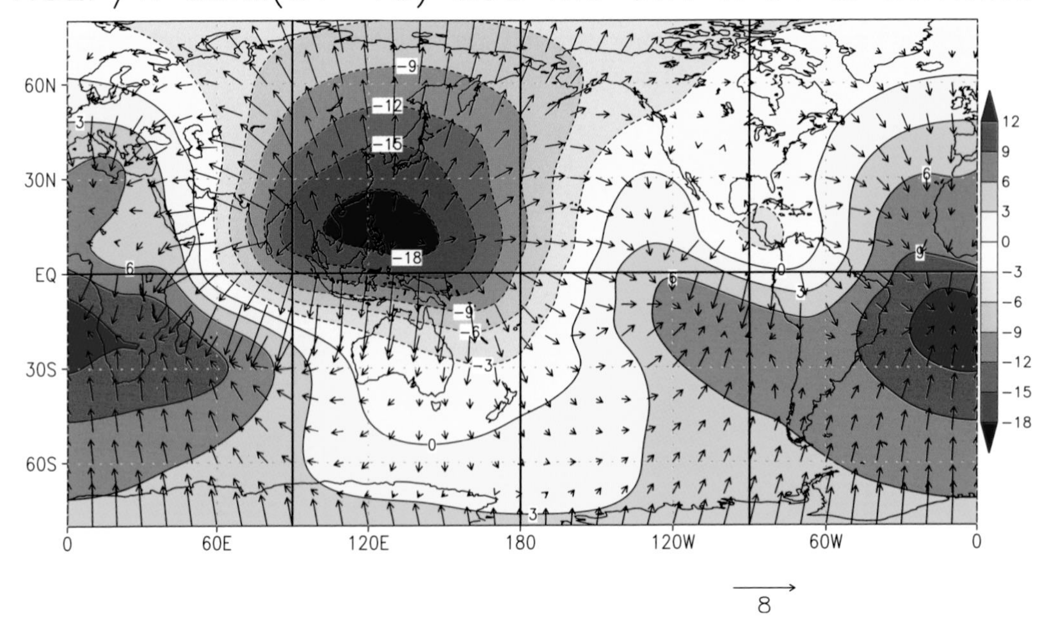

Figure 4.6 The mean 200 hPa velocity potential (10^6 m^2/sec), contour interval 1 unit, and divergent component of the wind (m/s) for the 1981 to 2010 period based on the NCEP Reanalysis 2 (Saha et al., 2010) for DJF (top), and JJA (bottom). Red shading denotes positive values, blue shading negative values.

Figure courtesy of M. Chelliah, CPC.

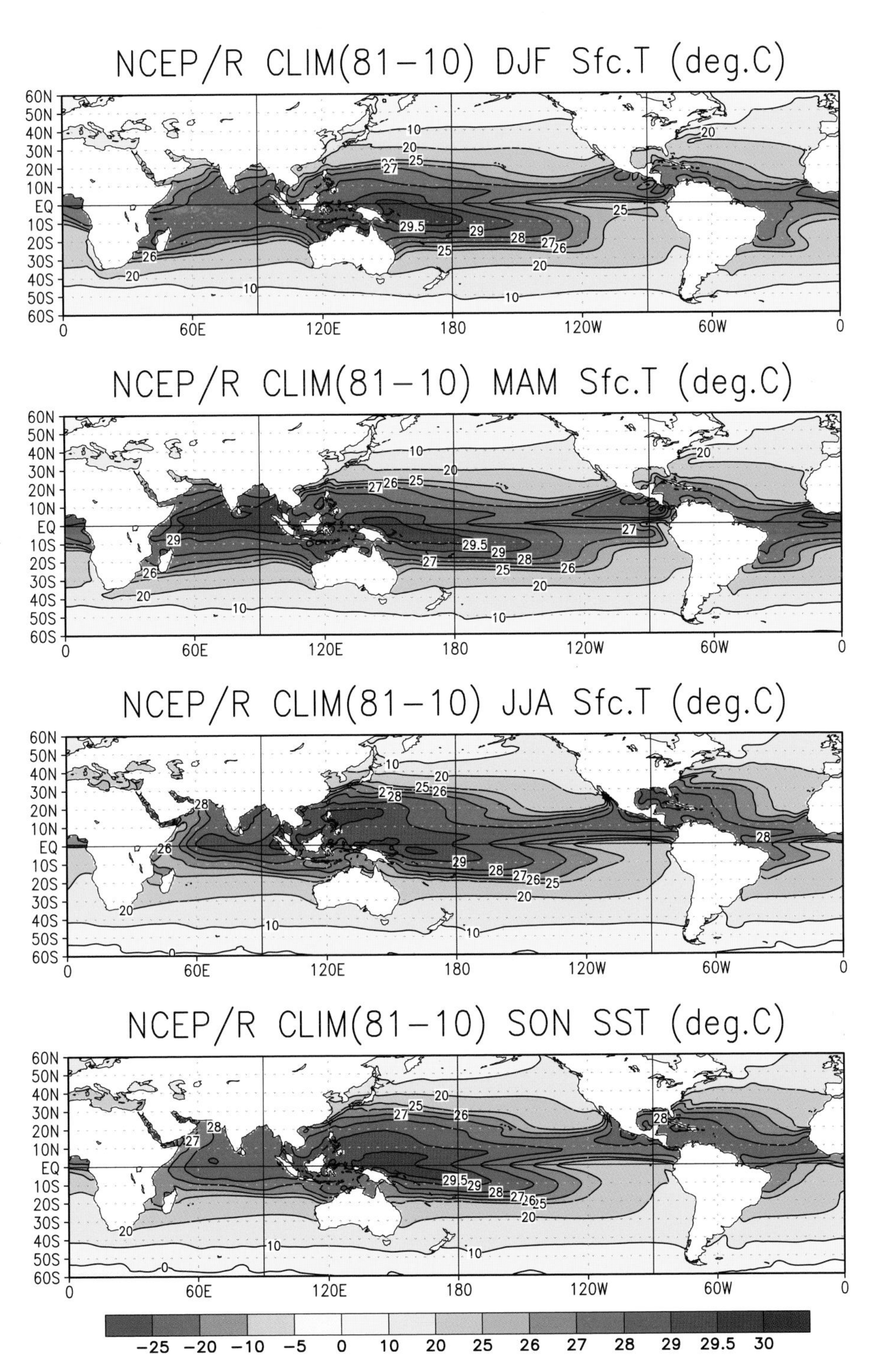

Figure 4.8 The seasonal mean (1981–2010) sea surface temperature (SST) based on satellite and *in situ* data for (a) DJF, (b) MAM, (c) JJA and (d) SON (°C). Temperatures greater than 0°C are shown in warm shades, temperatures greater than 28°C are shown in orange and red. Figure courtesy of M. Chelliah, CPC.

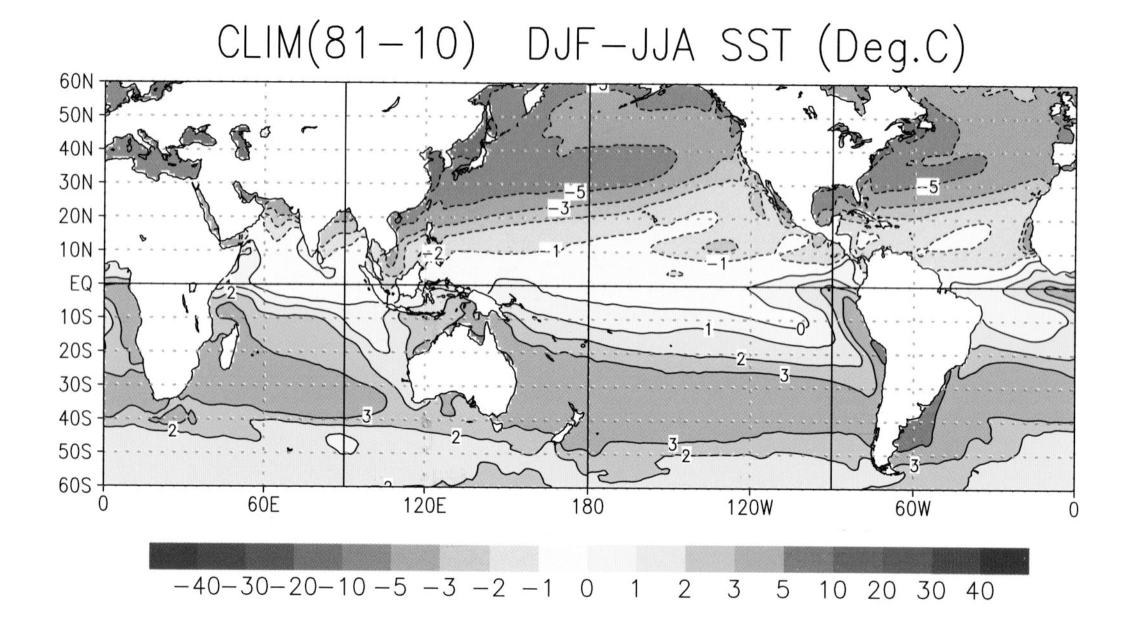

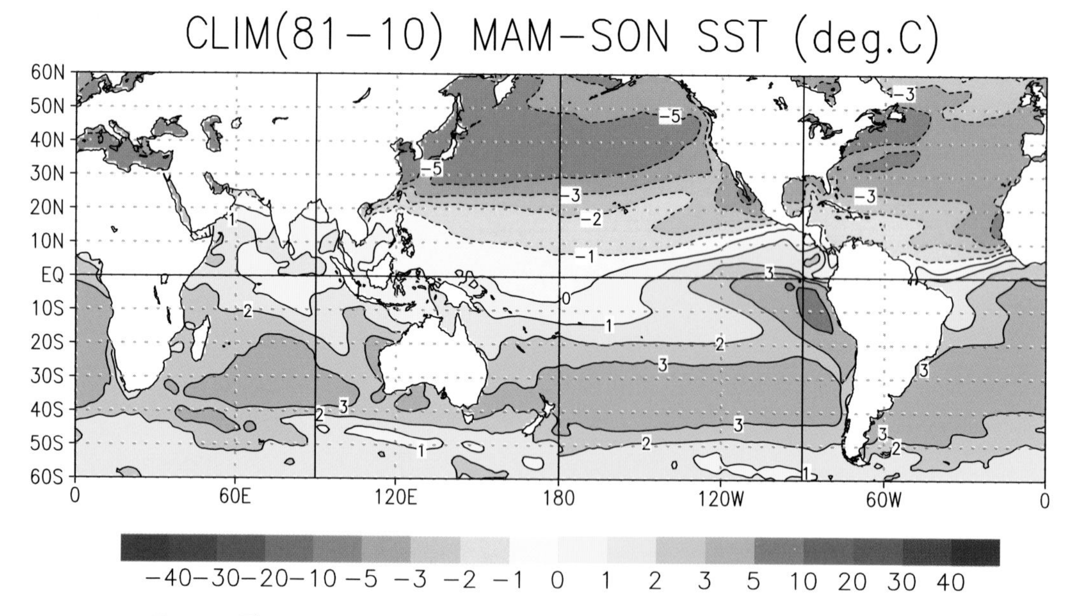

Figure 4.9 The seasonal mean sea surface temperature (SST), differences between winter and summer (DJF minus JJA, top) and between spring and autumn (MAM minus SON, bottom) in °C based on satellite and ship data for the period 1981–2010.
Figure courtesy of M. Chelliah, CPC.

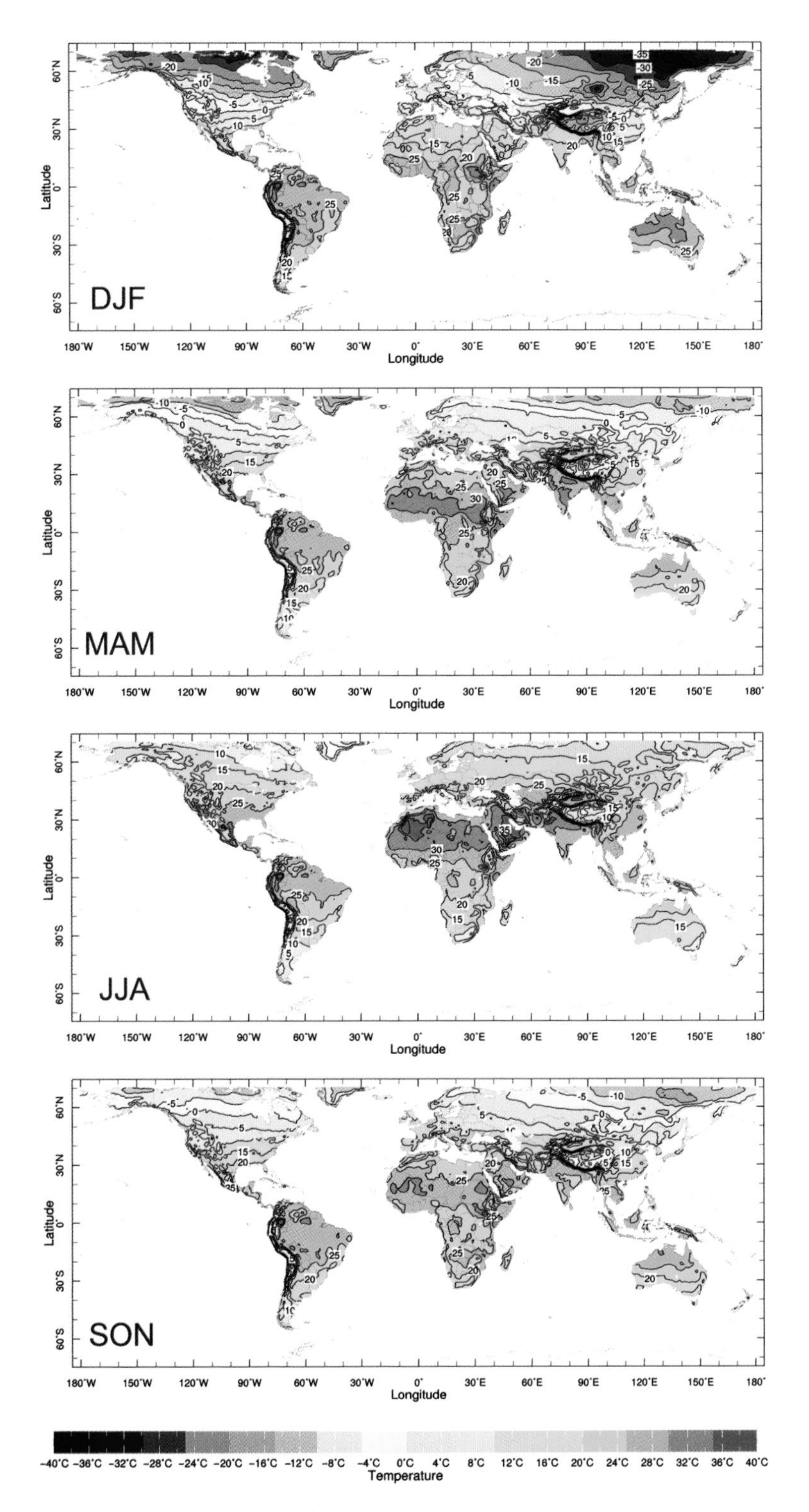

Figure 4.10a The seasonal mean surface air temperatures over land based on the GHCN-CAMS gridded monthly temperature database (Fan and van den Dool, 2008) for the 1981–2010 base period. Figure provided by M. Bell, IRI/LDEO Data Library.

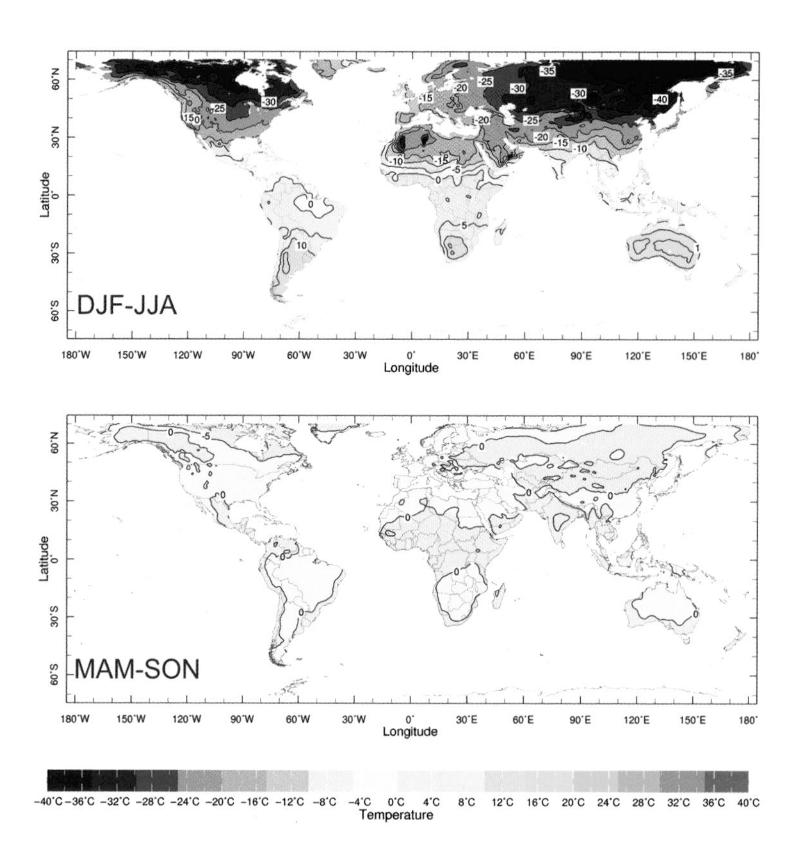

Figure 4.10b The seasonal mean surface air temperature differences over land based on the GHCN-CAMS gridded monthly temperature database (Fan and van den Dool, 2008). Northern Hemisphere winter minus summer (DJF-JJA), top, and spring and autumn (MAM-SON), bottom. Figure provided by M. Bell, IRI/LDEO Data Library.

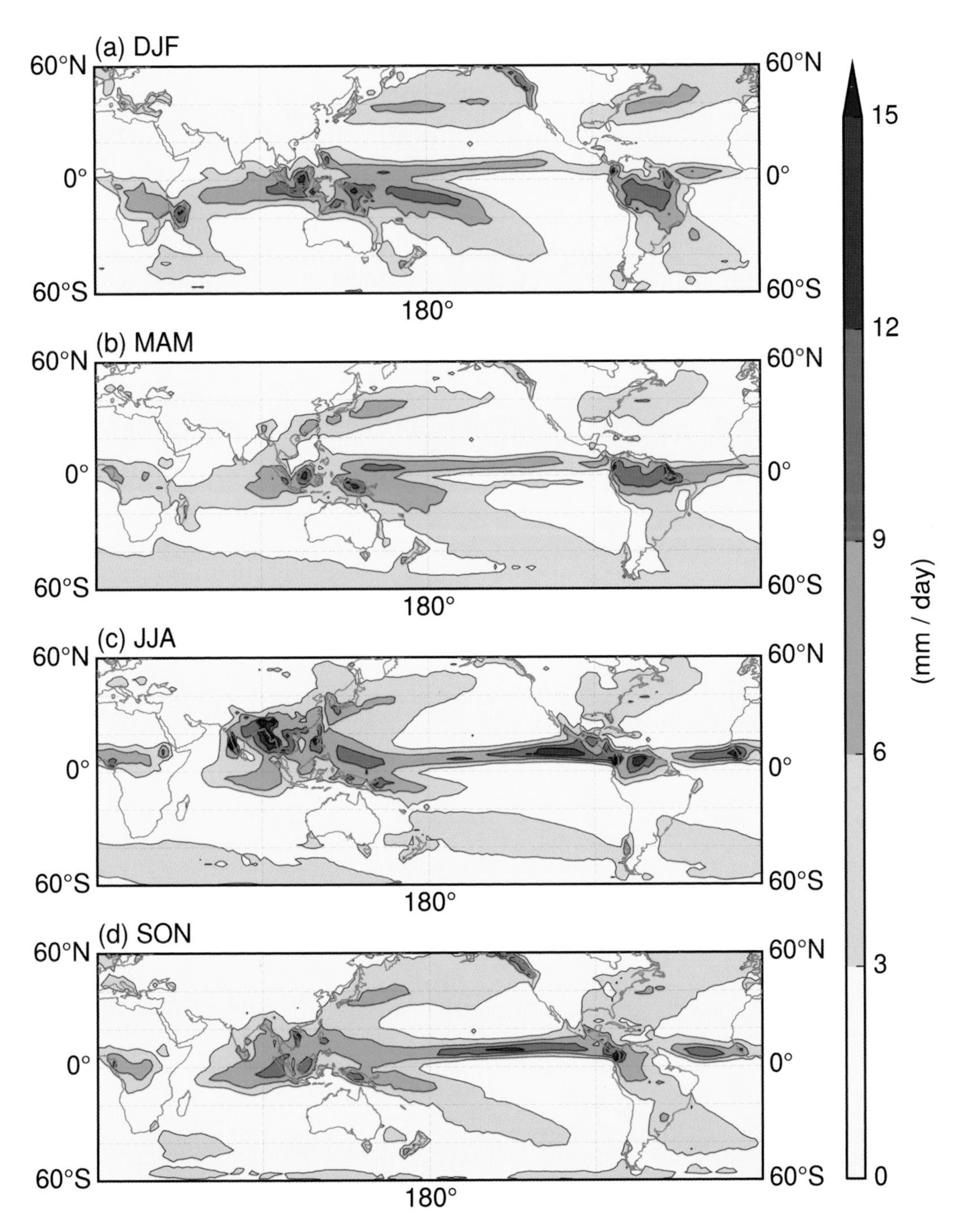

Figure 4.11 Mean precipitation in mm/day for (a) DJF, (b) MAM, (c) JJA and (d) SON for the period 1981–2010 based on satellite and surface station (GPCP) data.
Figure courtesy of G. Huffman and J. Tan, NASA.

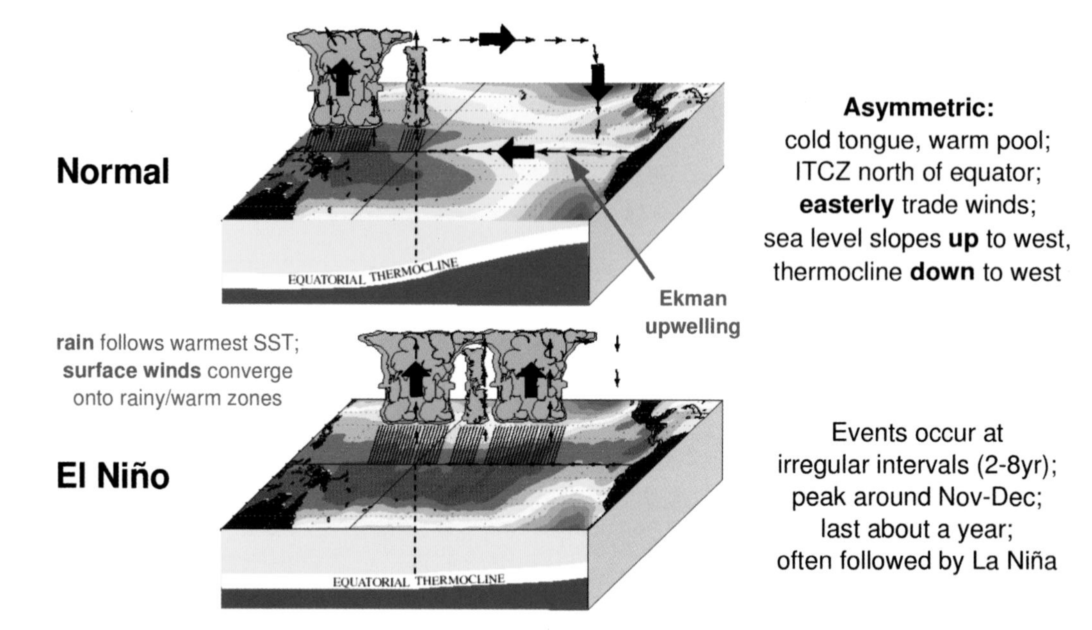

Normal
EQUATORIAL THERMOCLINE
Asymmetric:
cold tongue, warm pool;
ITCZ north of equator;
easterly trade winds;
sea level slopes up to west,
thermocline down to west
Ekman
upwelling
rain follows warmest SST;
surface winds converge
onto rainy/warm zones
El Niño
EQUATORIAL THERMOCLINE
Events occur at
irregular intervals (2-8yr);
peak around Nov-Dec;
last about a year;
often followed by La Niña

Figure 4.14 Schematic of normal versus ENSO conditions in the equatorial Pacific during northern hemisphere winter.
Source: NOAA/CPC.

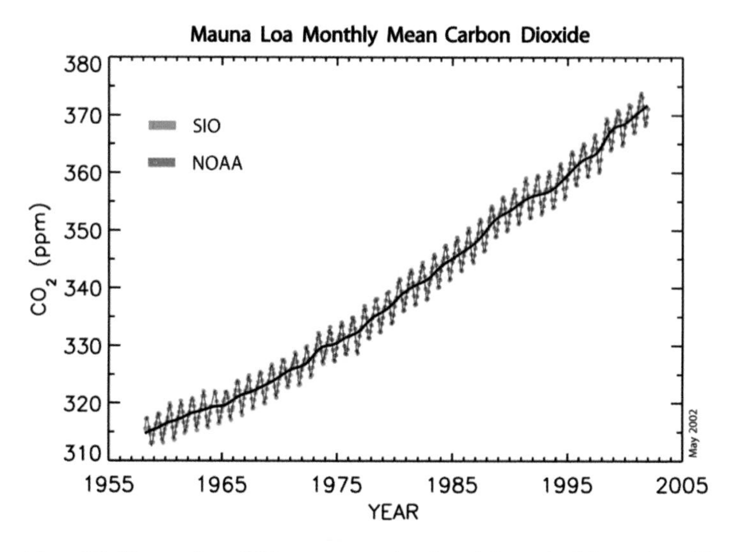

Mauna Loa Monthly Mean Carbon Dioxide
SIO
NOAA
CO2 (ppm)
380
370
360
350
340
330
320
310
1955
1965
1975
1985
1995
2005
YEAR
May 2002

Figure 5.1 Time series of CO_2 concentration (ppm) from the Mauna Loa Observatory, 1957–2005. Early data (blue) were taken by the Scripps Institute of Oceanography. The CO_2 concentration had risen to over 400 ppm in 2015 and continues to rise.
Source: NOAA Earth System Research Laboratory, Global Monitoring Division.

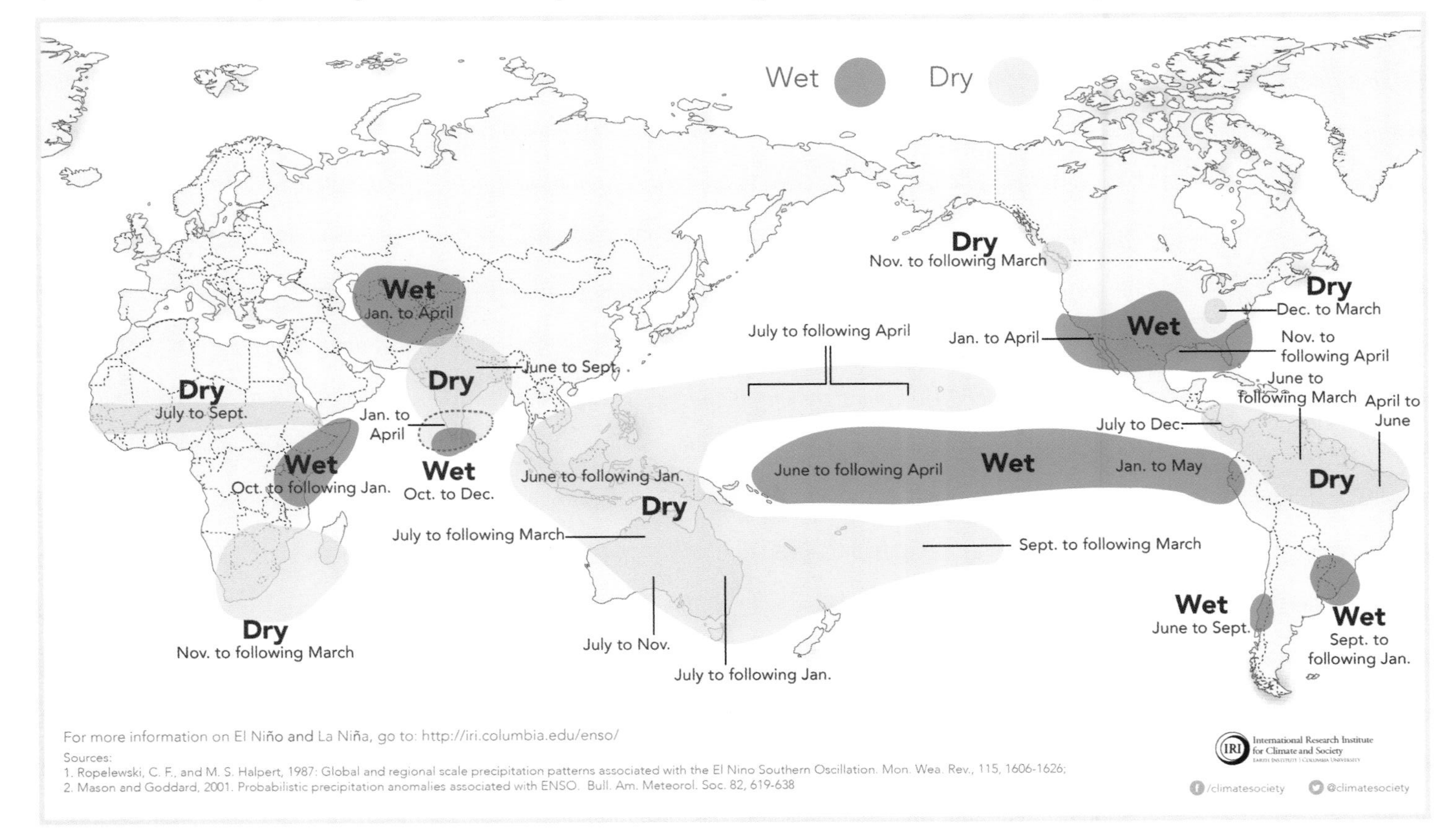

Figure 4.15 El Niño and Rainfall Schematic.
After Ropelewski and Halpert, 1987 and Mason and Goddard, 2001. Figure from the IRI Data Library, used with permission.

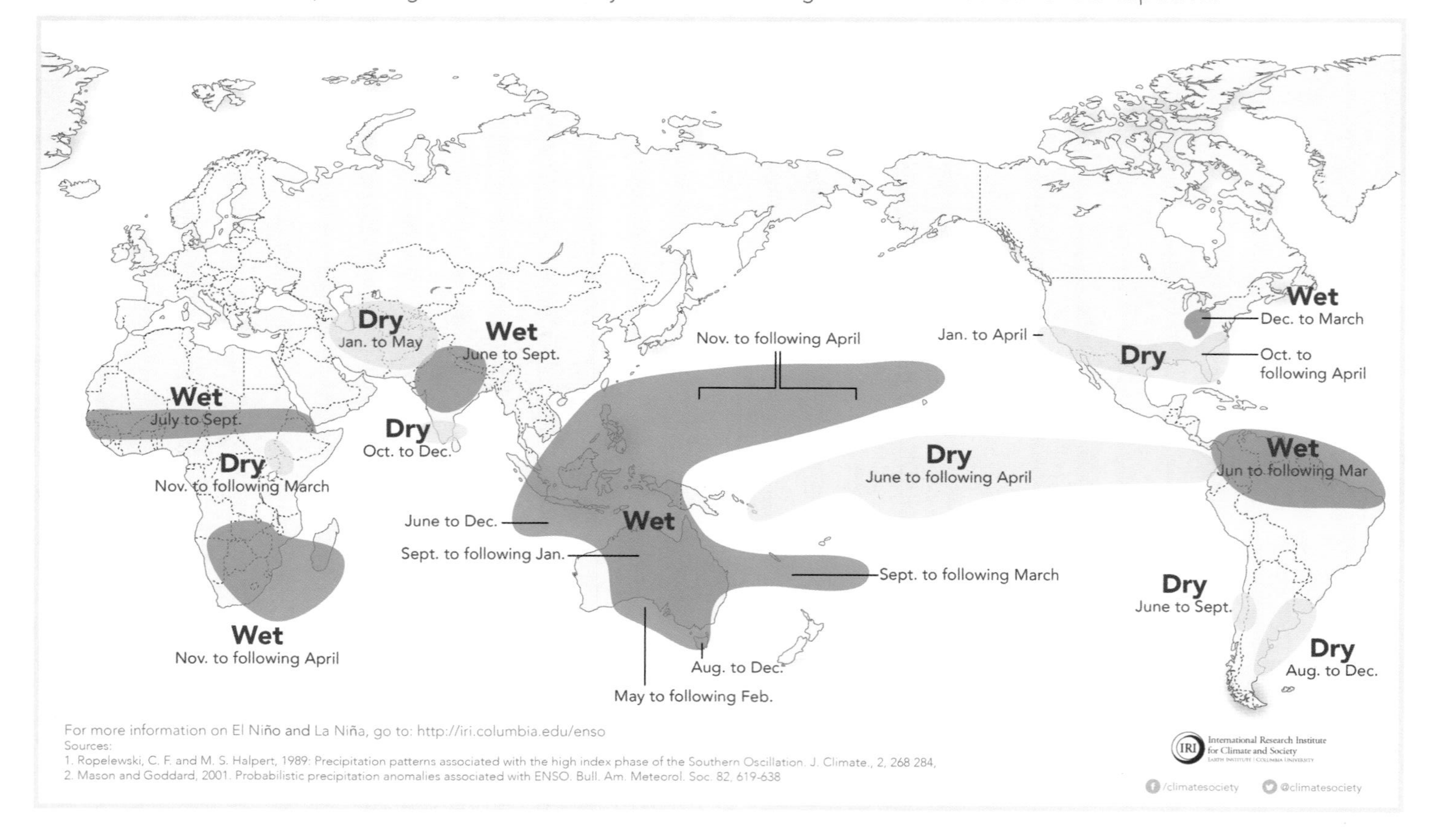

Figure 4.16 La Niña and Rainfall Schematic.
After Ropelewski and Halpert, 1989 and Mason and Goddard, 2001. Figure from the IRI Data Library, used with permission.

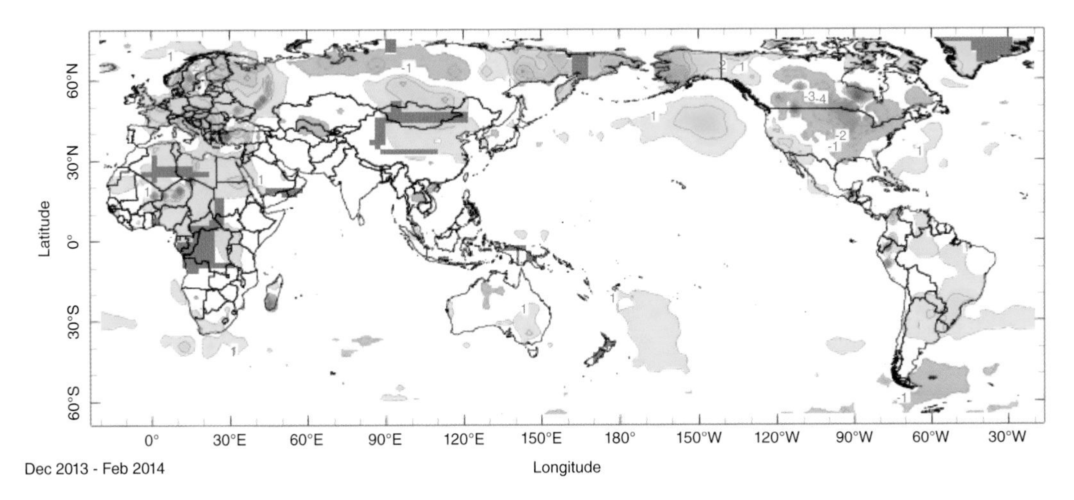

Figure 5.3a Map showing the pattern of surface temperature anomalies for a boreal winter (December 2013–February 2014).
Source: Climate Prediction Center NOAA, figure plotted using the IRI Data Library.

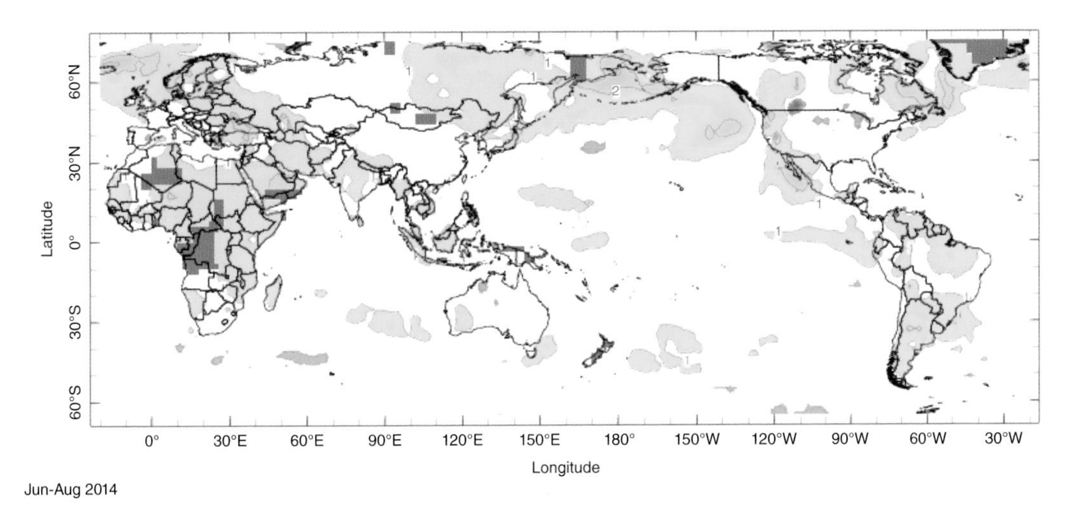

Figure 5.3b Map showing the pattern of surface temperature anomalies for a boreal summer (June–August 2014).
Source: Climate Prediction Center NOAA, figure plotted using the IRI Data Library.

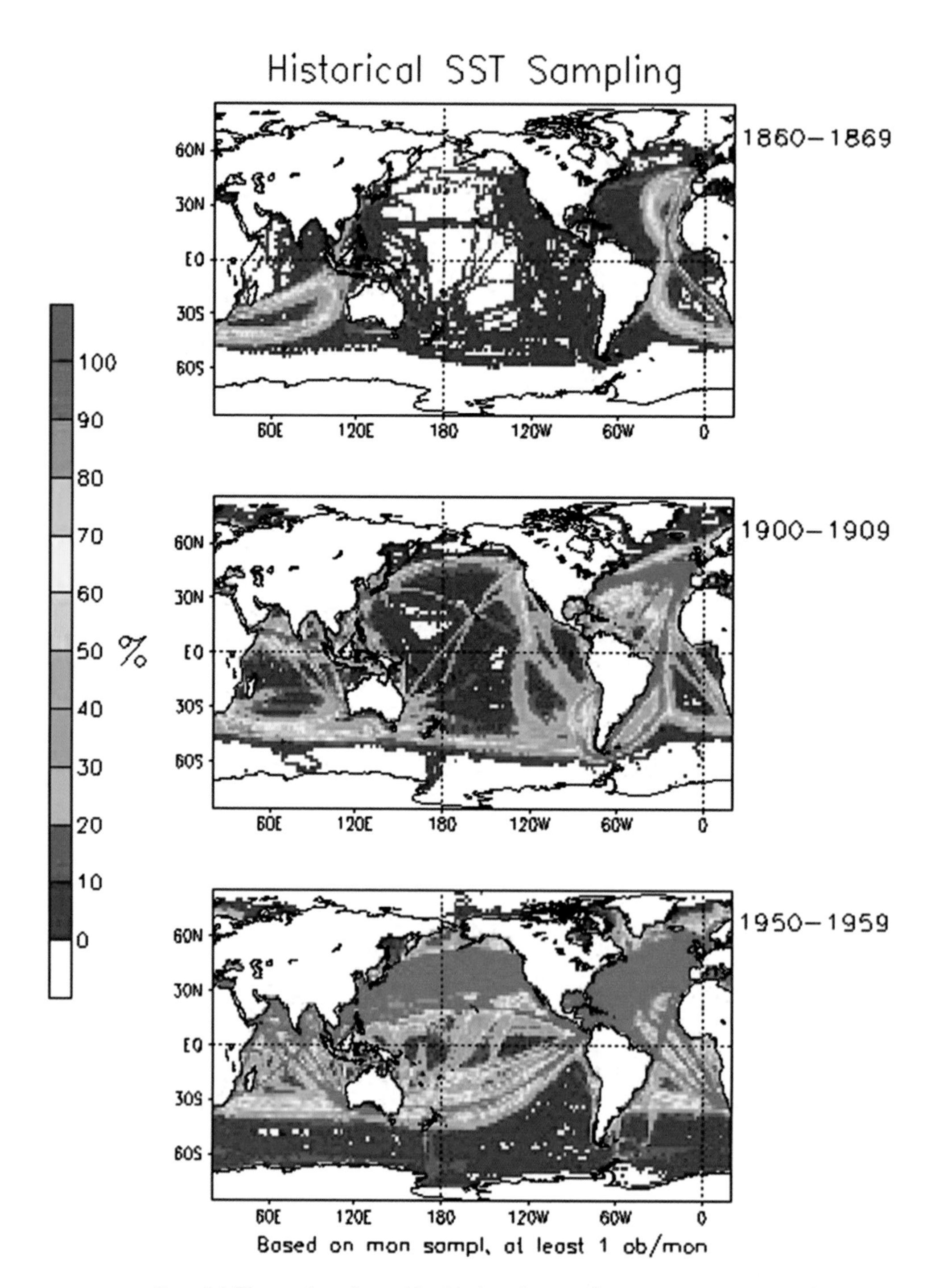

Figure 8.1 The number of monthly ship-based sea surface temperature expressed as a percent out of a total of 120 months for 3 ten-year periods. Top, 1860–1869, Middle 1900–1906, Bottom 1950–1959.

Source: R. W. Reynolds, NOAA; based on ICOADS 2 deg monthly data.

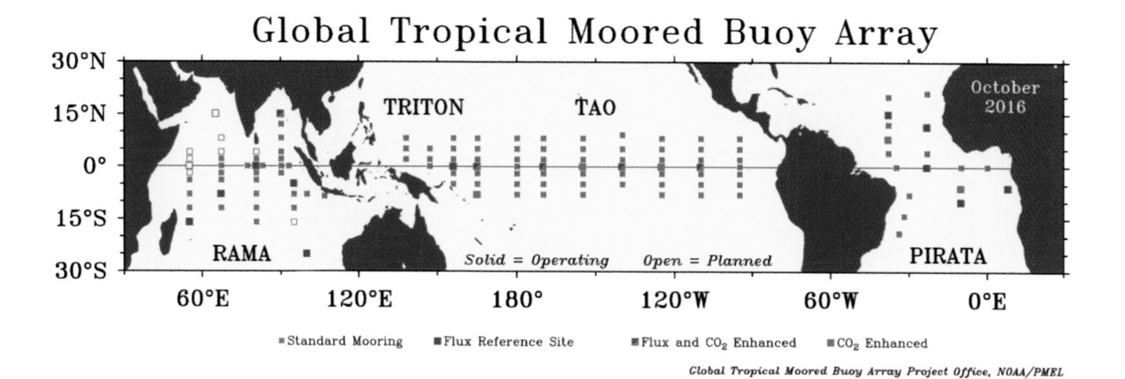

Figure 8.2 Locations of moored buoys in TAO/TRITON, PIRATA, and RAMA arrays to provide data in equatorial regions globally.
Source: M. McPhaden, PMEL/NOAA.

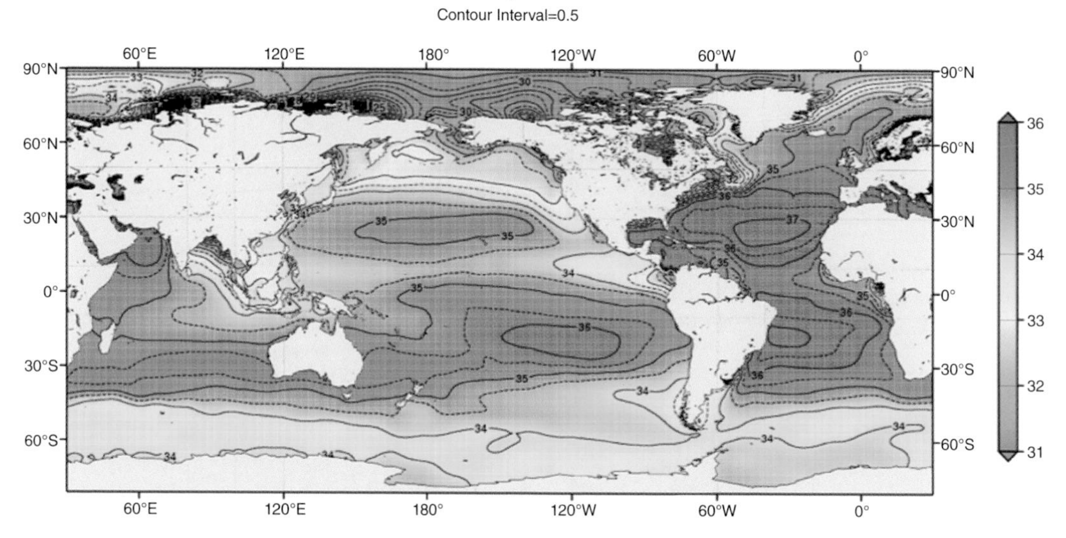

Figure 8.4 Mean annual salinity in the mixed layer of the world's oceans.
Source: NODC/NOAA, Zweng et al., 2013.

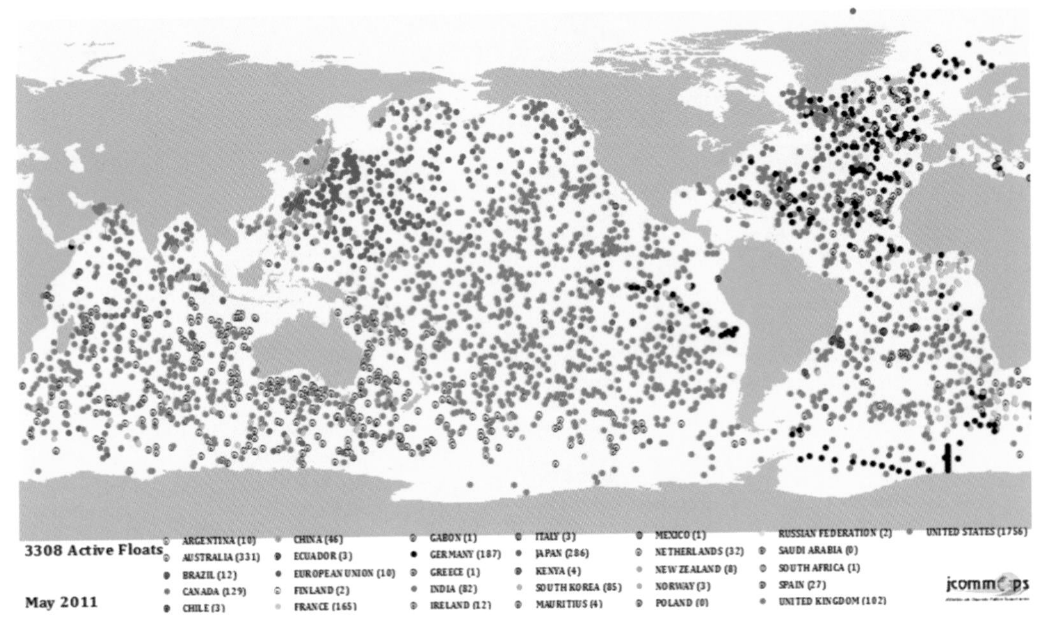

Figure 8.5 An example of the spatial coverage of Argo floats for May 2011(NOAA).

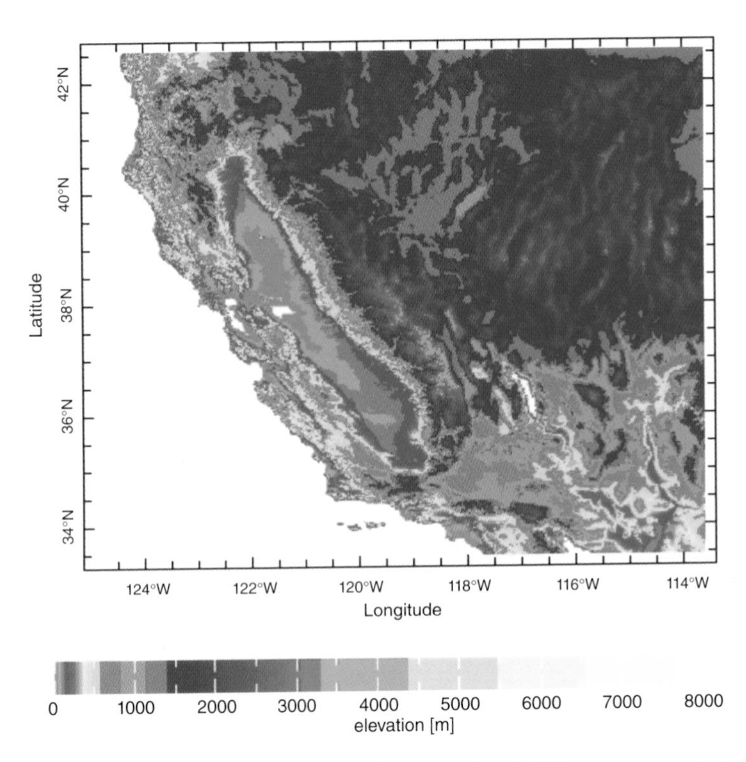

Figure 10.1 An example of a Digital Elevation Map (DEM) at 1 km resolution. This example illustrates the changes in topography from coastal California into the Central Valley and the Sierra Nevada Mountains.

Source: Hastings et al., 1999. Data downloaded from and plotted with the IRI/LDEO Data Library by author, CFR.

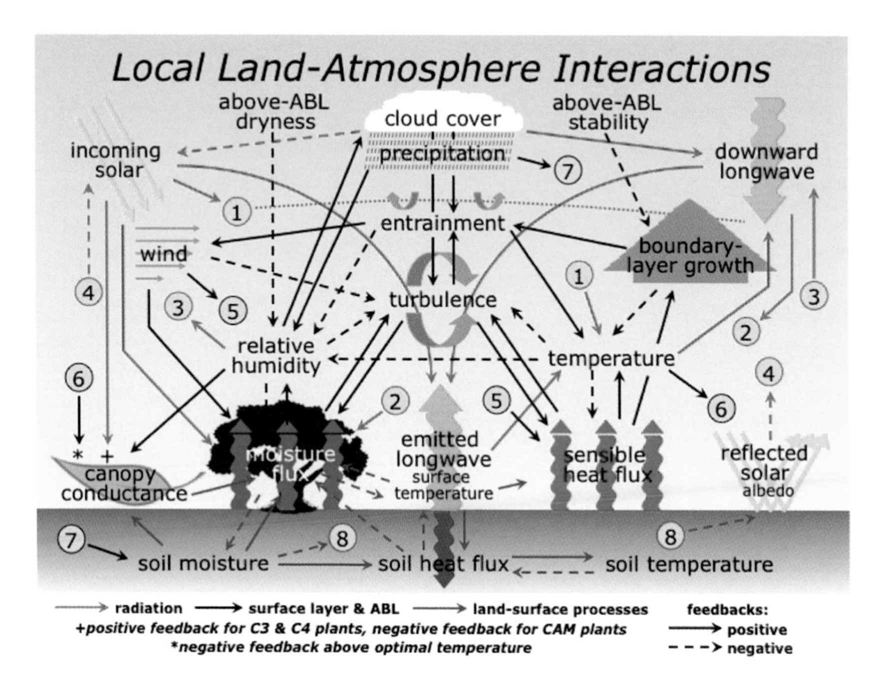

Figure 10.3 This schematic shows the components of the water and radiation budgets typically parameterized or estimated in Land Surface Models (LSMs).) The figure indicates local land-atmosphere interactions for a relatively undisturbed atmosphere, including the soil moisture and precipitation feedbacks. The solid arrows indicate a positive feedback, and large dashed arrows represent a negative feedback, while red arrows indicate radiative processes, black arrows indicate processes near the surface and lower atmosphere (termed the planetary boundary layer – PBL), and brown arrows indicate land surface processes. Thin red and grey dashed lines with arrows represent also represent positive feedbacks. The single horizontal gray-dotted line, with no arrows, indicates the top of the PBL, and the seven small vertical dashed lines, with no arrows, represent precipitation.
Source: Mike Ek, adapted from Ek and Holtslag, 2004, *J. Hydrometeorol.*, **5**, 86–99.

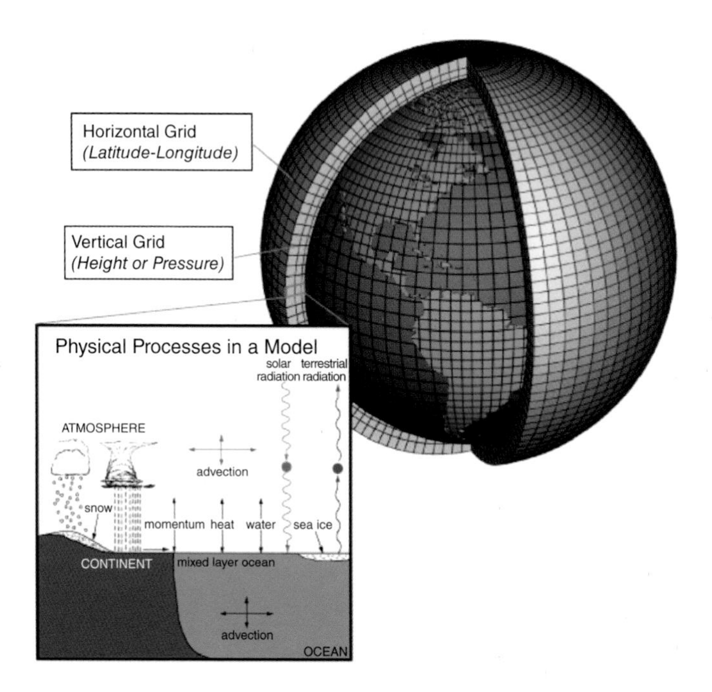

Figure 11.1 Schematic showing typical grids in the atmospheric component of climate models. Early reanalyses had horizontal grids with dimensions of a few latitude/longitude degrees and 20 or so levels. Several contemporary reanalyses have dimensions less than one latitude/longitude degree at 60 or more vertical levels.
Source: NOAA Climate.gov, courtesy Frank Niepold and Rebecca Lindsey.

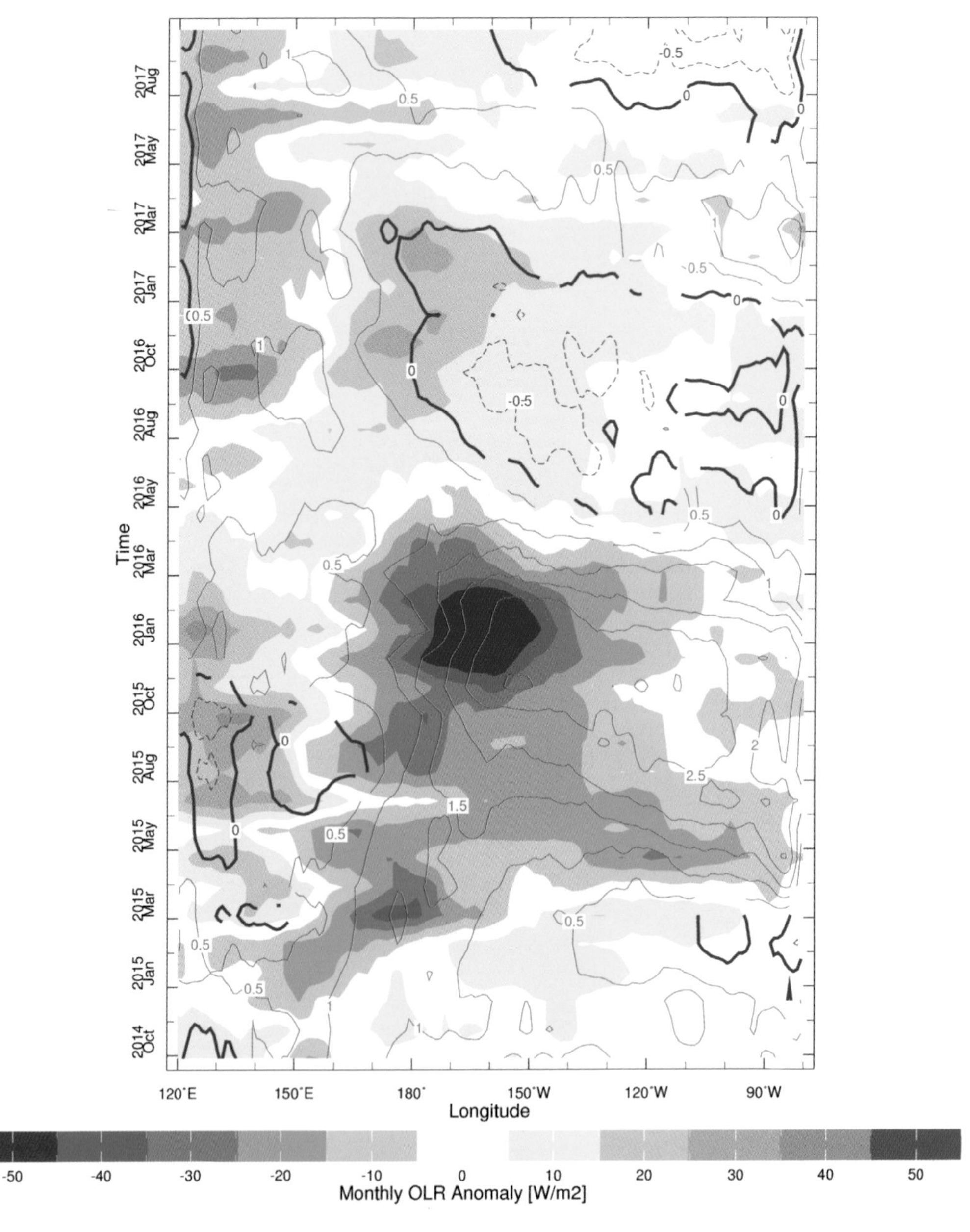

Figure 12.3 Time-longitude (Hovmöller) diagram of sea surface temperature (SST) anomalies (positive anomalies in red contours, negative anomalies in blue contours – °C) and outgoing longwave radiation (OLR) anomalies (negative anomalies indicating enhanced convection in green, positive anomalies indicating suppressed convection in brown). The data are averaged between 5°N and 5°S, from 120°E to 80°W. The earliest data are at the bottom of the diagram. Contour interval is 0.5°C. The OLR color scale is given below the diagram. Anomalies are departures from the 1981–2010 base period monthly means.

Source: data are from CPC/NOAA and figure plotted by M. Bell using the LDEO/IRI Data Library.

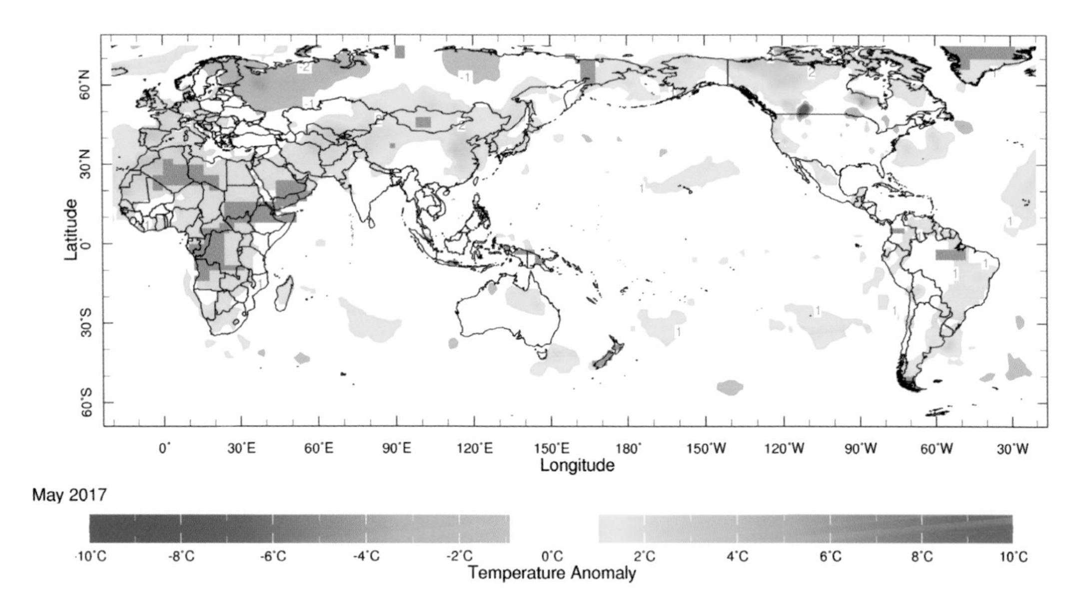

Figure 12.4a An example of surface temperature anomalies (°C). The analysis is based on station data over land and on SST data over the oceans. Anomalies for station data are departures from the 1981–2010 base period means, while SST anomalies are departures from the 1981–2010 adjusted OI climatology. (Smith and Reynolds 1998, *J. Climate*, **11**, 3320–3323). Regions with insufficient data for analysis are indicated by gray shading.

Source: data are from the CPC/NOAA; figure courtesy of M. Bell, LDEO/IRI Data Library.

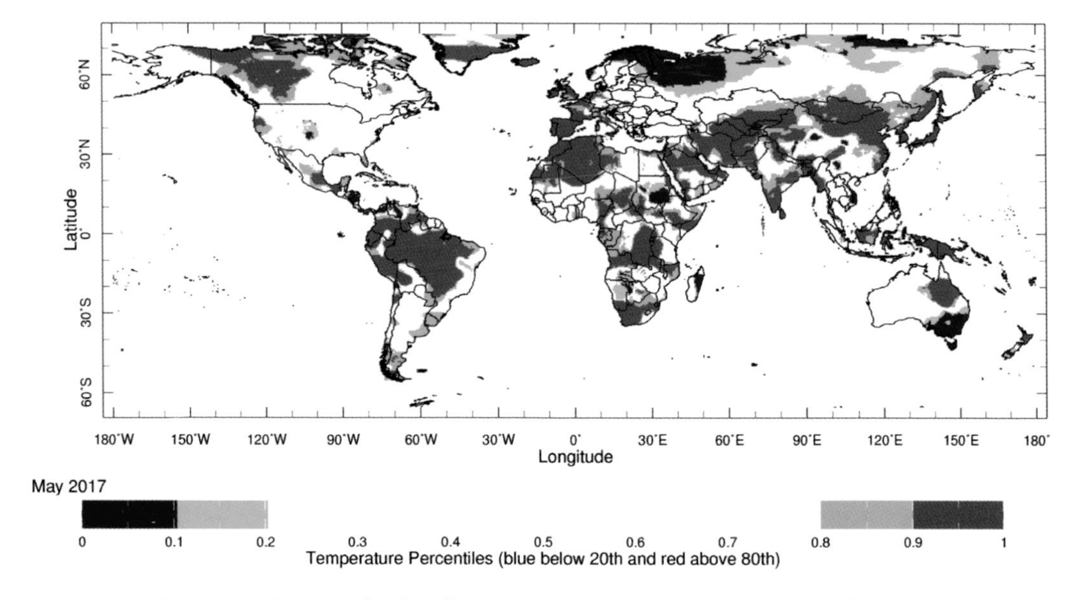

Figure 12.4b An example of surface temperature expressed as percentiles of the normal (Gaussian) distribution fit to the 1981–2010 base period data.

Source: data are from the CPC/NOAA; figure courtesy of M. Bell, LDEO/IRI Data Library.

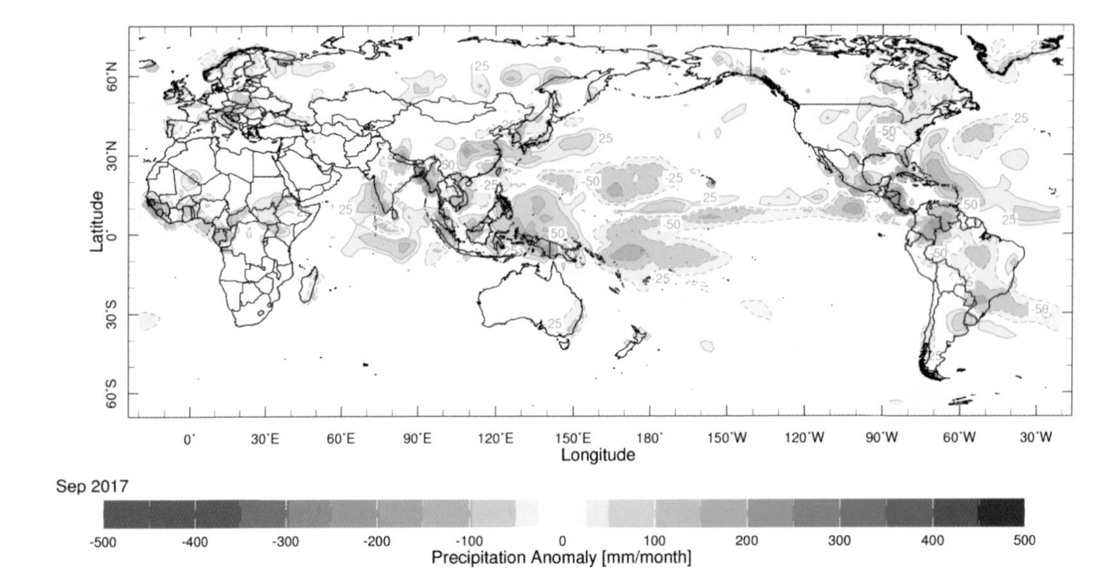

Figure 12.5a An example of anomalous precipitation (mm). Precipitation data are obtained by merging rain gage observations and satellite-derived precipitation estimates (Janowiak and Xie 1999, *J. Climate*, **12**, 3335–3342). Contours are drawn at 200, 100, 50, 25, −25, −50, −100, and −200 m.

Source: data are from the CPC/NOAA; figure courtesy of M. Bell, LDEO/IRI Data Library.

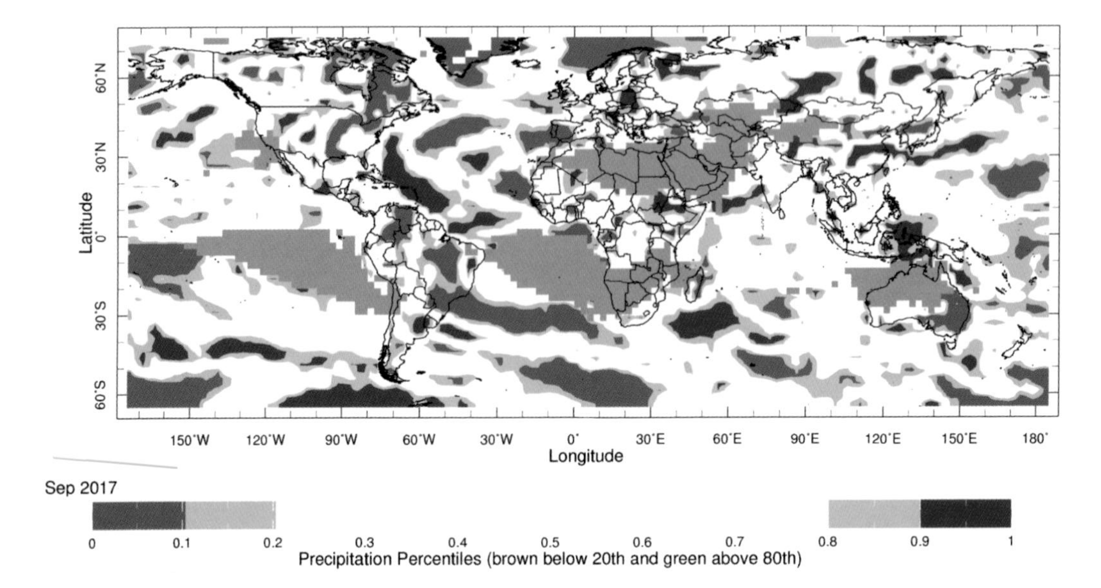

Figure 12.5b An example of anomalous precipitation expressed as precipitation percentiles based on a Gamma distribution fit to the 1981–2010 base period data. Gray shades indicate regions where mean monthly precipitation is <5 mm/month.

Source: data are from the CPC/NOAA; figure courtesy of M. Bell, LDEO/IRI Data Library.

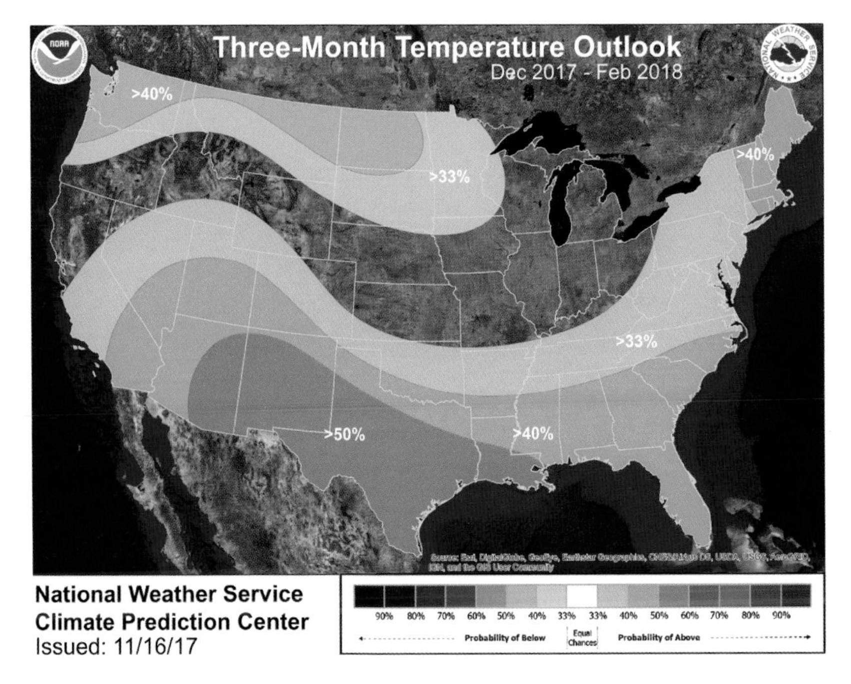

Figure 12.6 An example of a seasonal forecast from the CPC/NOAA using a tercile, or three-class (above, normal, below) system. The class limits are determined from the data in the 1981–2010 period.
Figure courtesy of the CPC/NOAA.

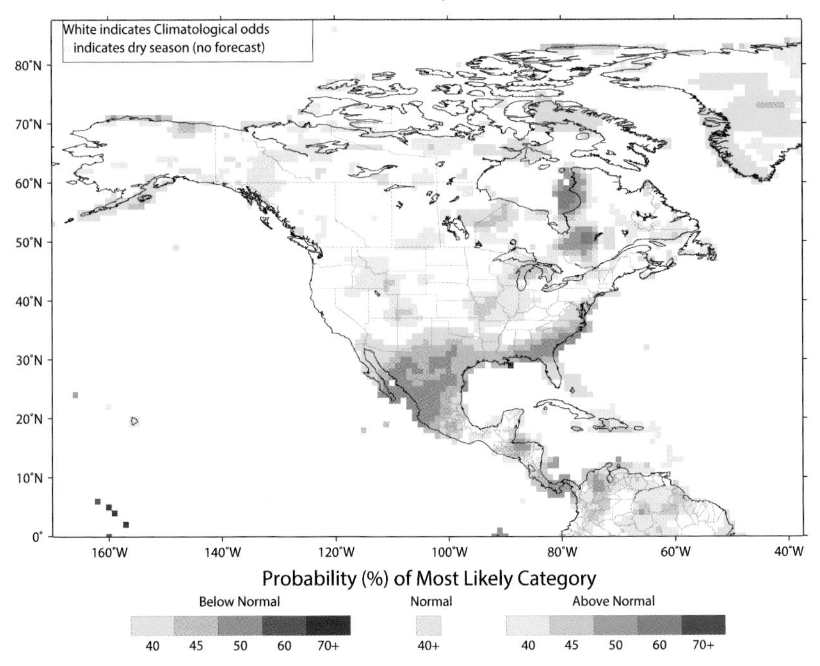

Figure 12.7 An example of a seasonal forecast from the IRI in a tercile, or three-class (above, normal, below), system. The class limits are determined from the data in the 1981 to 2010 period.
Figure courtesy of the IRI.

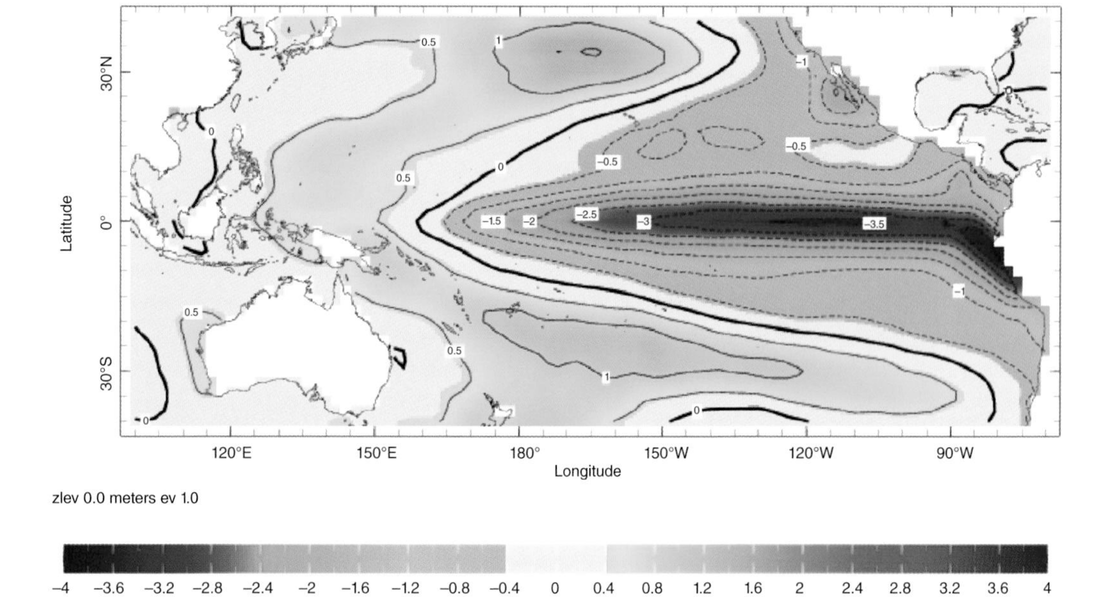

Figure A.1 Leading EOF pattern for the Pacific Ocean SST using ERSSTv3b monthly SST anomalies for 1981 to 2010 (Smith et al., 2008). This pattern accounts for 35% of the variance remaining after the annual cycle was removed. (Courtesy of M. Bell, IRI Data Library).

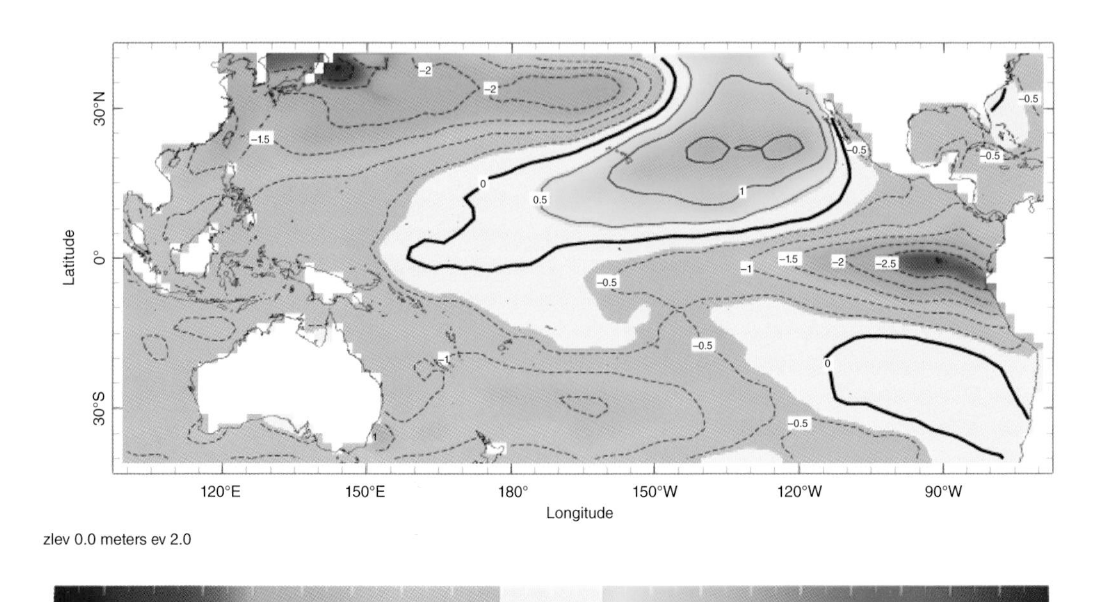

Figure A.3 The second EOF pattern for the Pacific SST accounting for 12% of the variance. (Courtesy of M. Bell, IRI Data Library).

century with the International Comprehensive Ocean Atmosphere Data Set (ICOADS) (Woodruff et al., 1987, 2011). Even to this day, however, data from individual observing experiments over the sea may not be available to the wider scientific community until some significant period of time has passed. On the other hand, due to the great interest in climate in recent decades, a series of internationally sponsored observational systems, both *in situ* from buoys and remotely sensed from satellites, as well as international data exchange agreements, have resulted in the routine availability of oceanic data from most parts of the world oceans (Reynolds et al., 2007).

It is important to note that the open ocean (beyond the continental shelf) is typically 4000 m in depth. Accurate ocean observations began to be available in the middle of the nineteenth century and were primarily surface observations. As we will discuss, the number of surface and subsurface observations increased with time. However, the subsurface observations were frequently restricted to the upper ocean and coverage decreased with increasing depth. Although the number of satellite observations has dramatically increased in recent years, these observations were restricted to the surface and could not give information on the changes in ocean parameters such as temperature with depth.

This chapter concentrates on descriptions of ocean datasets archived and accessible through established data centers. The discussions may also be a useful guide to the use of proprietary data that may be obtained through special arrangements but not available through the archive centers.

8.2 Sea Surface Temperature (SST)

8.2.1 How Is SST Defined?

To first order, the interactions of the ocean and the atmosphere are felt through the influence of ocean and atmospheric temperatures at and near the sea surface. The ocean surface temperature, usually called sea surface temperature or SST, provides the temperature boundary condition for interactions with the atmosphere over the global oceans. The SST and SST anomalies are sometimes used as proxies for surface air temperature and anomalies over the oceans. SST is not identical to the temperature of the atmosphere at 2 m height, as discussed in Chapter 6, but does provide a good representation of the atmospheric temperature patterns and their variability over the oceans.

Despite its name and common usage, *in situ* SST is traditionally based on water temperature measured at depths ranging from less than one meter to several meters. The top several meters of the ocean are generally well stirred by waves and winds, resulting in fairly uniform temperature. The mixed layer is defined by oceanographers as the region of the upper ocean where temperatures are uniform to within roughly 0.5°C. The depth of this layer from the surface is typically on the order of 100 m. The mixed layer depth varies seasonally, interannually and with geographical location. Remotely sensed SST, as measured by satellites and discussed in more detail later in the chapter, is fundamentally different from the *in situ* measurements. The satellite sensors provide estimates of the temperature at the very top few millimeters of the ocean surface, referred to as the ocean "skin temperature" (Wick et al., 2002). The SST in the topmost millimeters of the ocean, as estimated by the satellite sensors, is different from the SST measured by *in situ* measurements resulting in bias between them.

8.2.2 SST Observations and Instruments

The earliest SST observations were made in the days of sailing ships. As with all climate and weather observations, instrumentation and observing practice has evolved considerably with time, particularly from the mid-twentieth century onward. These changes introduce systematic biases in the historical record of SST that need to be taken into account when using these data.

SST from Ship Observations

The earliest archived SST measurements from ships date back to the late seventeenth century. By the mid-nineteenth century the number of observations had become sufficient to allow the estimation of near global SST patterns (Woodruff et al., 1987). Until relatively recently, most SST data were from ship-based observations. Traditionally, a wooden, canvas, or rubber bucket was lowered over the side of the ship; the bucket was then raised to the deck and the temperature of the water in the bucket measured with a thermometer. As sailing ships gave way to engine-powered vessels in the twentieth century, it became more convenient and more common to base SST on the temperature of seawater that had been piped into the engine room to cool the power plant.

Both methods for measuring SST have errors and biases that must be dealt with to construct a consistent climate record. For example, the water in a bucket may have come from different depths below the surface, depending on the size and type of ship. Different observed temperatures may result depending on how well the near-surface water is mixed. Early SST observations were taken using uninsolated buckets without strict guidance and agreement on how promptly the temperature of the water in the bucket should be measured. Thus, the water temperature in the bucket might have cooled or warmed in different ways after it was hoisted to the deck. For example, daytime bucket temperatures in the tropics may have sat in full sunlight on a hot deck for an unspecified time before the measurements were recorded. Conversely, winds may have cooled the water in a bucket, especially on overcast days or at night (Folland and Parker, 1995).

Historically, SST data have been recorded and centrally archived at many national data centers. Beginning in the 1970s, extensive efforts were initiated to develop corrections for errors and biases in the SST record. The historical data based on observations made during different eras and with different instrumentation and observing practices were made more consistent with one another through application of bias corrections. The magnitude of some of the bias corrections can be quite large, for example the United Kingdom Meteorological office (UKMO) notes a 0.5°C bias between observations taken in December 1941 and observations taken from 1942 onward. This bias has been ascribed to changes in the dominant source of SST data, from Japan early in the record to the United States in the later period (Reynolds and Smith, 1994). Digital archives of SST dating back to the 1870s are available, some of which include earlier records with a much smaller number of observations. The uncertainties associated with estimating the bias corrections contribute to the uncertainties in the historical record (Smith and Reynolds, 2005; Woodruff et al., 2011). Bias corrections for engine intake SST are not generally made and some researchers recommend treating bucket and engine intake SST separately (Mathews and Mathews, 2012). Modern SST observations by ships are taken four times daily.

Several groups of researchers began to investigate ways of creating complete historical maps of global SST based on bias-corrected data. Ship observations are not uniformly distributed across the oceans since ships carefully optimize their routes to minimize both travel time and risk. Thus, the ship temperature observations fall principally along discrete routes called ship tracks. In addition, ship tracks have changed with time due to major changes in shipping routes, notably in response to the opening of the Panama and Suez canals, Figure 8.1.

Some ship observations have not yet become fully available to the data archives because, as mentioned earlier, they are viewed as proprietary. Some geographic areas are especially poorly sampled; for example, ship observations poleward of 60° are generally sparse in both hemispheres but especially in the Southern Hemisphere. Even though the ship track data leave much of the world's oceans poorly sampled, they have been invaluable sources of data for climate studies, for example in describing the evolution of ENSO (Rasmusson and Carpenter, 1982).

SST from Fixed and Drifting Buoys

While ship data can be used for many climate analyses, others required spatially complete global SST analyses that were not possible until additional sources of observations of ocean temperatures became available. Starting in the late 1970s ship observations of SST have been supplemented significantly by observations made from drifting buoys. The Global Drifter Program (GDP) is an international effort that strives to maintain near global coverage of SST measurements with about 1,250 drifters spaced roughly 500 km apart (Legler et al., 2015). The drifters measure SST six times daily at between 0.2 m and 0.3 m. Some of the drifting buoys also measure salinity and atmospheric surface pressure.

Moored buoys, which are anchored to the ocean floor, were instituted to observe the equatorial ocean substructure and as additional sources of SST observations starting in the late 1980s. Three networks of near equatorial moored buoy arrays, the TAO/TRITON, spanning the Pacific, the PIRATA in the Atlantic, and the RAMA in the Indian Ocean (Figure 8.2), are now in place (see Section 8.5 for more detail).

These buoys observe water temperature at a typical depth between 0.5 m and 1 m and provide SST estimates that are independent of the ship observations and with greater frequency, precision, and accuracy. The moored buoys generally provide six-hour mean SST values, a frequency sufficient to define the diurnal cycle of SST. Data from moored buoys are routinely transmitted to data centers via satellite uplinks. Some buoys provide

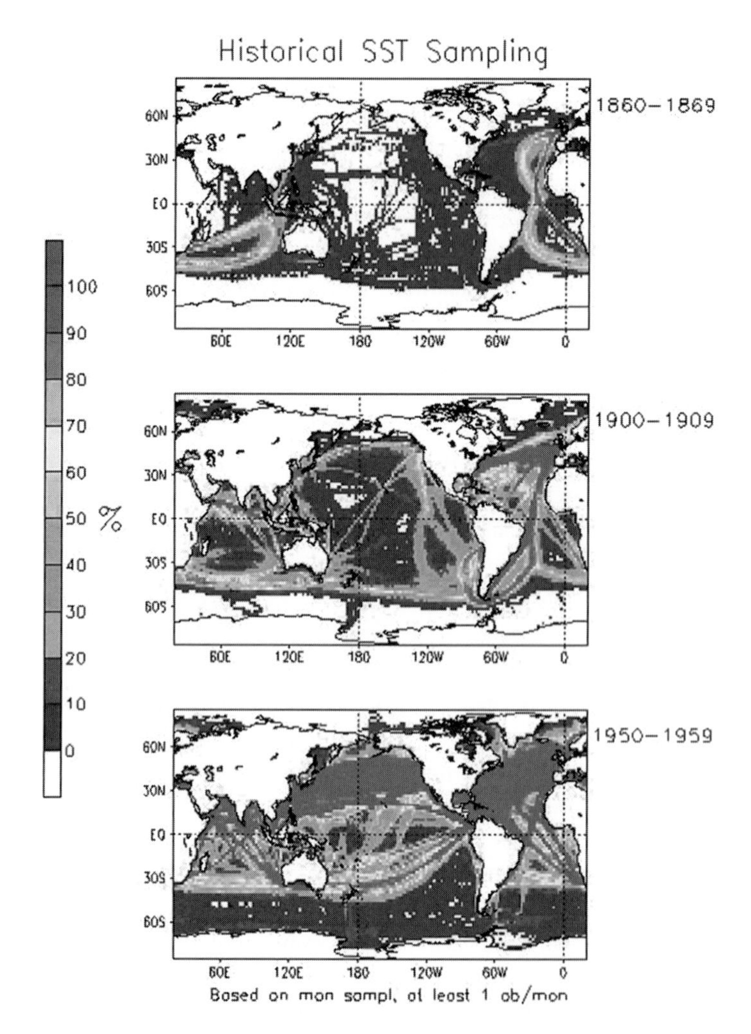

Figure 8.1 The number of monthly ship-based sea surface temperature expressed as a percent out of a total of 120 months for 3 ten-year periods. Top, 1860–1869, Middle 1900–1906, Bottom 1950–1959. A black-and-white version of this figure appears in some formats. For the color version, please refer to the plate section.
Source: R. W. Reynolds, NOAA; based on ICOADS 2 deg monthly data.

hourly SST observations but the record of these higher temporal resolution observations is not as complete. The moored buoys provide an independent benchmark dataset for comparisons with ship, drifting buoy, and satellite SST data.

SST from Satellites

Polar orbiting satellites provide estimates of SST over almost the entire ocean surface using infrared instruments (Emery et al., 2001). The cloud-free atmosphere is nearly transparent at infrared wavelengths near 11 micrometers, so that the temperature of the ocean surface can be estimated quite well by instruments sensitive at these wavelengths. The estimated ocean temperature is representative of the average SST over an area. NOAA operational meteorological satellites have made such observations since

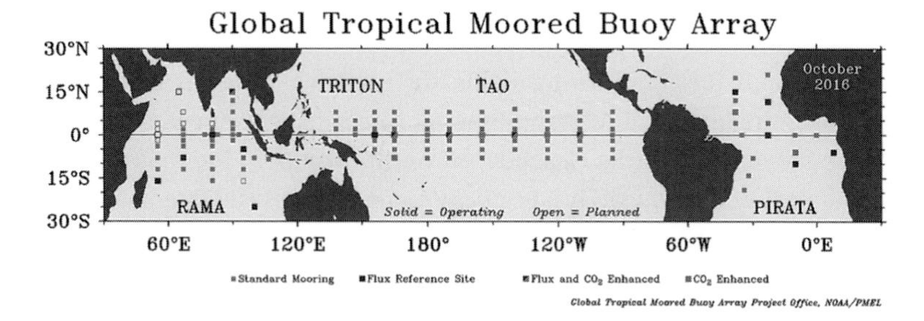

Figure 8.2 Locations of moored buoys in TAO/TRITON, PIRATA, and RAMA arrays to provide data in equatorial regions globally. A black-and-white version of this figure appears in some formats. For the color version, please refer to the plate section.
Source: M. McPhaden, PMEL/NOAA.

1979. Atmospheric water vapor influences the infrared SST estimates, but a correction for water vapor can be made by using observations at different wavelengths. The SST estimates are also influenced, more rarely but to a greater extent, by atmospheric aerosols, particularly in the stratosphere. Volcanic eruptions such as El Chichon in Mexico in 1982 and Pinatubo in the Philippines in 1992 caused significant difficulties that continue to present challenges in the historical record. The infrared satellite instruments sense the temperature of the surface as seen by the satellite. In the presence of clouds, the satellites see the cloud tops, and thus, the cloud-top temperatures rather than the SST, limiting the infrared satellite SST estimates to cloud-free areas.

In the first decade of the twenty-first century, SST estimates have become available from satellite passive microwave instruments whose observations are not strongly affected by clouds, but which cannot measure SST where there are precipitating clouds. The spatial resolution and coverage of the microwave instruments has historically been 25–50 km, considerably less than the 1 km or less spatial resolution in the infrared observations. The errors and biases in these two sets of remotely sensed sea surface temperature measurements are largely independent, thus making them useful complementary data sources. The relatively coarse spatial resolution of the microwave SST estimates, however, reduces their ability to provide SST estimates near coastlines and ice boundaries. SST analyses based solely on satellite observations are biased relative to the *in situ* SST. Identifying and removing the bias in the satellite SST estimates has been one of the major challenges in creating SST datasets for climate studies as microwave and infrared instruments have changed over time.

8.3 Creating Global SST Fields

No one set of observations can provide the unbiased estimates of SST needed for the monitoring and study of climate over the entire globe. Each of the observing systems discussed in Section 8.2, however, can contribute to production of global SST analyses that include all ocean basins. In this section we discuss how each of the observing systems contributes to providing global estimates of SST.

8.3.1 SST Fields from *In Situ* Data

The number of observations available for historical analysis along ship tracks has varied over the historical record. Shipping routes have changed as canals opened and as ships moved from strict dependence on the winds to internal power. Ship tracks also changed due to social and political circumstances; most notably, the number of ship track observations available in the historical record decreases substantially in association with each of the World Wars in the twentieth century. The number of observations along ship tracks can vary significantly over time, and indeed, some ship tracks may cease to exist. This is particularly true for the older tracks around southern Africa and South America that changed significantly with the opening of the Suez and Panama Canals. Conflicts and tensions in the Middle East and elsewhere have also altered the number of observations along some ship tracks. The changes in the number of observations along the ship tracks influence the uncertainty estimates in the SST data.

The SST values in the longer historical records have systematically shifted as observations from relatively small wooden sailing ships diminished, and eventually gave way to observations from large commercial ships. Changes in the kinds of ships taking SST observations also dictated changes in observing practices, such as insulated versus uninsulated buckets, or bucket temperatures taken on ship decks versus engine intake temperatures. Engine intake temperature measurements have evolved in time as well; for example, the engine intake temperature sensor depth generally increased as ships became larger. Because of these and other changes, SST data centers may provide analyses of the data that have time-varying bias adjustments applied to the raw SST observations (see the NCEI Website). Documentation including the magnitudes of the adjustments and an explanation of how the adjustments were made are included in the dataset descriptions. Significant efforts to increase the number and quality of ship track SST data have been made through international cooperation in ICOADS, and its predecessor COADS (Woodruff 1987, 2011). The ICOADS project seeks to capture, quality control, archive, and disseminate surface observations taken by commercial, government, and research ships. These observations include SST as well as air temperature, humidity, cloud cover, and winds at many locations.

The analysis of SST from ships is augmented by comparisons of these data with observations from moored and drifting buoys discussed above. The buoy data can also provide independent comparisons with SST estimates based solely on ship observations. Even with the addition of buoy data, the ship-only SST estimates will leave significant portions of the ocean under-sampled.

8.3.2 SST Fields from Blended Satellite and *In Situ* Data

Satellite SST observations, as discussed earlier, have the advantage of global coverage using the same instruments and provide estimates of SST patterns on large spatial scales. Satellite temperature observations are, as mentioned earlier, biased with respect to *in situ* observations. Conversely, *in situ* observations do not have the near-global coverage available from satellites. Considerable efforts have been expended to blend, or merge, both sources of SST data to take advantage of the strengths of both.

The *in situ* observations are well suited to deriving large-scale (typically 500 km or greater) bias corrections for the satellite-based SST values in areas that include observations from both. Bias corrections to the satellite SST estimates are determined from data at locations that contain enough *in situ* observations for inter-comparisons. Blended analyses based on satellite observations, to provide global coverage, in conjunction with *in situ* observations, to adjust for biases in the satellite SST estimates, were developed in the early 1980s; see, for example, Chapter 4, Figure 4.8 (Reynolds, 1988). Blended monthly *in situ* and satellite SST data at a 2° spatial resolution were introduced in the mid-1980s. Spurred by improvements in the analysis techniques and observations as well as the demands from the data assimilation systems for operational weather and climate prediction models, the blended analyses have evolved to higher and higher spatial and temporal resolutions. For example, Reynolds and Marsico (1993) and Reynolds and Smith (1994) provide daily SST estimates at 1° spatial resolution. Further improvements (Reynolds et al., 2007) have resulted in daily SST analyses on a 0.25° resolution. The higher spatial resolution comes with significant penalties in terms of the uncertainties in the estimate. The family of blended SST analyses are archived at NCEI as OISST.

Blended SST analyses using combinations of *in situ* and satellite observations have been available for over 30 years and have spatial and temporal resolutions useful in a number of studies, particularly relating to sub-monthly and sub-seasonal climate variability. These datasets have also been used to provide estimates of trends over the recent three decades. The trends in SST from these analyses are sensitive to the bias adjustments that have been applied to the satellite data. Nonetheless, global trends in SST from the blended analyses tend to have the same general character (warming) as the analyses based solely on the *in situ* observations.

Because of the relatively small number of observations in parts of the oceans, particularly at high latitudes, uncertainties there are greater in both *in situ* and satellite datasets than uncertainties in the tropics and midlatitudes. The interpretation of high latitude satellite observations is further complicated by uncertainties in the location of the sea-ice boundaries and by the interpolated estimates between the SST in open water and SST at the sea-ice boundary. The sea-ice boundary is generally defined by the percent of the area covered by sea ice; a sea-ice concentration of 50% is often used. The water temperature at the sea-ice boundary is typically set at −1.8°C (28.8°F), the nominal freezing point of sea ice. The blended satellite and *in situ* data are available from several sources including NCEI and ESRL, both listed in the References.

The global blended analyses are suitable for case studies of multi-year and decadal variability during the satellite era, and have been used for atmospheric model data assimilation, particularly in production of various reanalyses (Chapters 3 and 11) and in Atmospheric Model Inter-comparison Project (AMIP) studies.[1] Differences among the

[1] AMIP (see www-pcmdi.llnl.gov/projects/amip/ for more information) was a model intercomparison protocol developed about 1990 to facilitate comparison and evaluation of the background climates of various

blended SST analyses used in AMIP studies have been found to result in significant differences in the atmospheric circulations produced by the models (Hurrell and Trenberth, 1999), particularly in ocean regions that show the largest differences in the blended analyses. Most atmospheric reanalyses (see Chapter 3) have agreed to use the UKMO global SST analysis to minimize the differences in model-based climate analyses resulting from the use of different SST data. Whether this, or any other SST analysis, is the optimum SST analysis for use with all reanalyses has not been addressed. Nonetheless, the sensitivity of atmospheric reanalyses to relatively small differences in SST analyses points both to the importance of SST in climate studies and to the range of uncertainties in these reanalyses.

Studies of multi-decadal climate variability require time series longer than these blended analyses can provide. For these studies the analysis is required to focus on the areas of the ocean that have long periods of ship track data or use one of the reconstructed global datasets discussed in the next section.

8.3.3 Reconstructed SST Analyses

The clear need for global SST analyses and time series longer than can be provided in the era of satellite observation inspired several scientists (Reynolds and Smith, 1994; Kaplan et al., 1997, 2000) to develop and apply methods of extending global SST analyses further into the past. Their methods used pre-satellite era ship track data together with information about significant climate patterns present in the satellite era to produce reconstructions (as discussed for precipitation in Chapter 7) of historical SST. While they differ in their details, all of these methods of reconstructing a longer SST record rely on Empirical Orthogonal Function (EOF), or similar analyses (Appendix A; von Storch and Zwiers, 2002). Some reconstruction methods use EOF analysis in combination with optimal averaging or optimal interpolation techniques (Smith et al., 1996) to define the major large-scale patterns of climate variability based on the analyses of global fields available in the satellite era. The patterns revealed from these almost spatially complete satellite observations are used in various ways to project these characteristic patterns onto the earlier *in situ* ship track data and, thus, obtain more complete estimates of the occurrence of these patterns in the longer historical record. These methods have been used to construct spatially complete monthly and seasonal SST fields at spatial resolutions of several degrees of latitude well back into the nineteenth century.

The reconstructed global SST fields have been used to study the behavior of interannual climate variability reaching back more than 100 years. While the reconstructions provide an interesting resource for further analysis, they have inherent limitations. Among these limitations is that the patterns of climate variability used in the reconstructions are limited to those observed during the 30 years or so of satellite data. Projecting this variability back to earlier periods relies on the implicit assumption that

atmospheric models used for both weather prediction and climate research. AMIP is now an element of the Coupled Model Intercomparison Project (CMIP).

the SST from the satellite era captures all the important modes of (seasonal to interannual) climate variability in the longer climate record. Furthermore, the reconstructions assume that the spatial patterns of climate variability in the oceans have not changed over the longer time periods.

Another class of limitations arises in studies that utilize reconstructed SST to investigate decadal and longer variations in global SST. In these studies, the ability of the reconstructions to identify decadal or multi-decadal variability is limited by the extent to which the satellite era SST data capture climate patterns based on say, three or fewer, realizations of decadal or multi-decadal variability that have occurred during the satellite era. The reconstructions have likely not identified all the relatively small amplitude, low frequency patterns of SST variability that have occurred during the satellite era and certainly cannot provide information on patterns of variability that did not occur during the satellite era. While there is broad consensus that an upward trend in SST is captured by the reconstructed datasets, it is highly unlikely that they are able to represent regional details of that trend accurately in areas poorly sampled by ship observations. Reconstructed monthly SST datasets, dating back to 1854, at 2° latitude/longitude resolution, and identified as ERSST, are available from NCEI and ESRL and other data centers listed in the References.

8.4 Other Ocean Surface Properties

Near surface currents are significant features in the oceans (Figure 8.3) that are responsible for heat, salinity, and other exchanges between low and high latitudes in the ocean. The mean circulation of the atmosphere has a strong influence on these ocean currents.

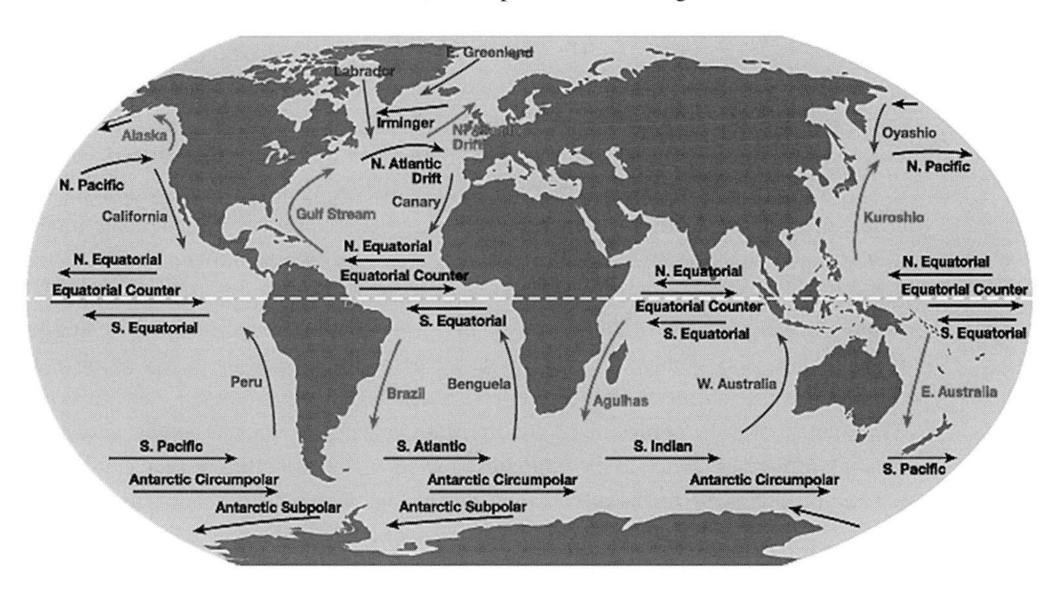

Figure 8.3 Schematic of major ocean currents.
Source: National Earth Sciences Teachers Association (NESTA), Windows to the Universe.

The equatorial easterly winds, those blowing from east to west at low latitudes of both hemispheres, drive ocean currents toward the west, called westerly currents in oceanography. In the Northern Hemisphere when the low latitude westerly ocean currents encounter the continental landmasses, North America in the Atlantic Ocean, Asia in the Pacific Ocean, they turn northward to form the Gulf Stream and Kuroshio, respectively. At mid-to-high latitudes the prevailing westerly winds help to drive ocean currents from west to east until the Pacific current encounters the west coast of North America and the Gulf Stream encounters the European continent. These currents turn to the south forming southerly boundary currents. As a net result, clockwise rotating gyres exist in the Northern Hemisphere ocean basins. Similar processes occur in the Southern Hemisphere ocean basins, including the Indian Ocean, but the gyres rotate in a counter clockwise sense under the influence of the Coriolis effect. A west-to-east, wind-driven current around Antarctica is the only continuous current not completely constrained by a continental barrier. All of these wind-driven currents exist in the ocean's mixed layers (Prager and Earle, 2000).

While the SST is generally the dominant influence of the ocean on the atmosphere in the context of climate variability and change, observations of other ocean surface properties are valuable as well. For the most part, these observations rely on estimates from satellite data, combined when feasible with very sparse *in situ* observations. Dedicated satellite missions that observe sea surface topography and height, rainfall over the oceans, and sea surface salinity, discussed later in this chapter, have already begun to produce data that is useful for enhancing climate datasets. Ocean surface topography, the variations in the height of the ocean surface with respect to the solid earth, has been estimated by a series of satellites (Topex-Posiden, Jason), called altimeters, since 1992. These satellites used nadir-pointing radars, at 5.5 GHz and 13.65 GHz, to estimate the distance from the surface of the ocean to the satellite. Microwave radiometers, at 18 GHz, 21 GHz, and 27 GHz aboard the same satellites are used to correct for atmospheric moisture and provide an overall accuracy for the height of the ocean surface of between 3 cm and 4 cm. The variations in sea surface topography arise from the tides as well the gradients in ocean temperature and salinity associated with ocean currents. In the open ocean, height gradients are used to estimate ocean currents. These observations have been used to support real-time climate monitoring efforts, notably ENSO. They are also used, in conjunction with tide gauge observations, to estimate long-term sea level changes, as discussed in Chapter 5. Other satellites, including TRMM, GPM, Aquarius, and CloudSat, provide estimates of rainfall, sea surface salinity, and clouds that are helpful in estimating the fluxes of solar radiation into the oceans as well as the fresh water flux.

The potential role of satellites continues to expand to include other variables, including properties related to the biological and chemical properties of seawater. Many of these observations are from research satellites designed to demonstrate proof of concept. Newer generations of satellites and sensors are providing an expanded suite of ocean products relevant to climate studies, particularly in the context of earth-system modeling. This context expands the concept of an earth system to include not only coupling, or interactions, among the physical components of the climate system but also the biological and chemical couplings that are part of the earth system. These include observations of ocean color (related to chlorophyll) and chemical composition as well as ocean acidification estimates. These observations are evolving as part of continuing research efforts. They are likely to form an important component of earth system analysis as more years of data become available. They will also be invaluable in guiding the development and validation of earth system models as these models include more sophisticated chemical and biological processes. Much of the data from these instruments may be obtained through the NASA Earth Observing System Data and Information System (EOSDIS) listed in the references.

The satellite instruments providing estimates of oceanic chemical composition and biological processes mentioned earlier do not directly measure these quantities. The estimates of the physical quantities, for example chlorophyll, are derived through sophisticated analysis of data from one or more sensors aboard a single satellite or in combinations with analysis of data from different sensors on other satellites. The highly derived nature of these estimates suggests that their use for climate analysis needs to be guided through close collaboration with experts in the processing and interpretation of these data.

8.5 The Ocean's Subsurface

Oceanographers have long known that, in addition to the direct influence of sunlight and winds near the surface, the SST reflects physical and dynamic processes taking place beneath the ocean surface. These subsurface processes are shaped by the geography of the ocean basins and by the fluid dynamics of the ocean. The density of the atmosphere, a collection of compressible gases, changes rapidly with height. Changes in the density of ocean water, virtually an incompressible fluid, are much subtler and depend chiefly on the temperature and salinity of the water.

8.5.1 Subsurface Ocean Structure

Three main layers characterize the ocean: the surface mixed layer, the thermocline,[2] and the deep ocean. The mixed layer is much more interactive with the atmosphere on short to medium time scales, and plays a significant role in many weather events as well as seasonal-to-interannual climate phenomena like ENSO. The thermocline is characterized by a decrease of temperature with depth, and extends from the bottom of the mixed layer down to about 1000m. In the deep ocean, temperature decreases and salinity increases much more slowly with depth, leading to small horizontal and vertical density gradients that determine the direction and strength of deep ocean currents.

[2] Strictly speaking, this segment of the vertical structure of the ocean is called the pycnocline, the region where the density increases with depth. This is because temperature decreases in the thermocline and salinity increases, in the halocline, both contributing to density increases. Climate scientists tend to concentrate on the thermocline.

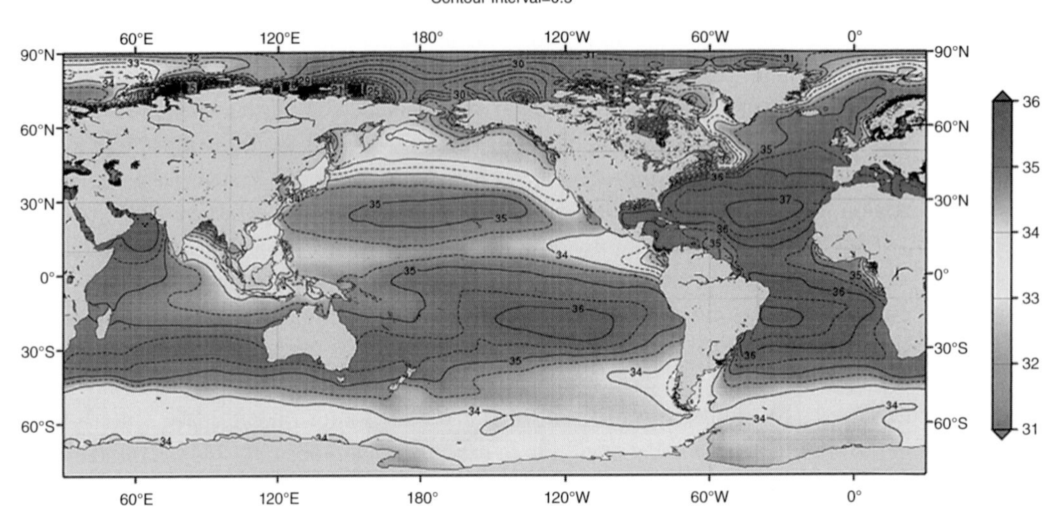

Figure 8.4 Mean annual salinity in the mixed layer of the world's oceans. A black-and-white version of this figure appears in some formats. For the color version, please refer to the plate section. Source: NODC/NOAA, Zweng et al., 2013.

The mixing in the mixed layer is driven by the winds and characterized by very small, if not always negligible, vertical gradients in temperature and salinity. The depth of the mixed layer is important for many reasons, but its most crucial role in climate variability is to increase or decrease the volume of heat available for interaction with the overlying atmosphere.

Below the mixed layer and the thermocline, the density of ocean water varies in many subtle ways. In the deep ocean salinity and temperature determine density and density gradients that, in turn, result in slowly moving deep ocean currents. The density gradients are influenced by many factors on the ocean-basin scale, but the general shape and pattern of the density gradients are determined by the geography of the ocean basin. Melting ice at high latitudes in both hemispheres provides a near-surface source of relatively fresh and cold, thus relatively dense, water that tends to descend. Water from the Mediterranean Sea has relatively higher salinity compared to the Atlantic due to high evaporation over the Mediterranean (Figure 8.4).

Thus, water from the Mediterranean Sea that enters the Atlantic Ocean at the Straits of Gibraltar is more saline, and tends to sink, leading to a deep exit flow of more salty water into the Atlantic. Such processes strongly influence the density structures in the oceans. The overall density structure of the ocean is further complicated by the existence of the western (Gulf Stream, Agulhas, Kuroshio) and eastern (Benguela, California, Humboldt) boundary currents near the edges of the continents. For instance, the Gulf Stream carries saline, but relatively warm, water to high latitudes where it cools and sinks. As discussed in the previous section, these boundary currents are a consequence of the physical boundary imposed by the continents and the Coriolis effect on a rotating earth. Other factors that alter the density of ocean water are the inflow of fresh water from the major river systems and rainfall over the oceans.

With the exception of the fixed continental boundaries all of the processes that influence the temperature and salinity vary on multiple time and space scales. Fresh water flux associated with rainfall over oceans may vary within minutes in association with the convective time scale, on seasonal and longer scales associated with ENSO, and on multi-year time scales with the waxing and waning of multi-year dry and wet regimes. In addition, seasonal and multi-year droughts and wet spells over land alter the amount of fresh water inflow from rivers. The ability to obtain a complete description of the density structure through the deep ocean has progressed slowly in the absence of adequate observing systems to measure temperature and salinity throughout the ocean volume. The need for better information below the surface mixed layer has started to be met through the implementation of the observing systems discussed in the next section.

8.5.2 Subsurface Observations and Instruments

Studies of the role of the global ocean in climate have, until the past few decades, concentrated on surface observations, particularly SST. A number of practical applications, such as sound propagation for military sonar and commercial fishing, have motivated observations through the mixed layer, including the depth of the thermocline. Until recently, however, routine measurements of deep ocean temperature, salinity, and currents were rarer, difficult to sustain, and without practical climate applications. The relatively short and spotty records have limited the use of these data principally to the production of long-term averages, or climatologies, describing the mean density and current structure at depth in the ocean basins. The historical data have been exploited by a handful of dedicated research scientists to produce descriptions of the mean ocean conditions at depth. The first estimates of density (temperature and salinity) variations from the mean at depth (Levitus and Boyer, 1994; Levitus, Burgett, and Boyer, 1994; Conkright et al., 2002) were made late in the twentieth century, but the data needed to provide definitive descriptions of the temporal evolution of subsurface ocean temperature, salinity, and currents are only now becoming available (Durack and Wijffels, 2010; Boyer et al., 2014). The motivation for such improved observations are the indications from numerical modeling and theoretical considerations that changes in the deep ocean may have significant influences on the atmosphere on the multi-year, decadal, and longer time scales. Another driver for improved subsurface ocean observations is the continuing evolution of the data assimilation systems used in coupled climate models. The use of subsurface ocean observations in coupled models aimed at forecasts of multi-year and decadal variability and long-term trends is increasing demands for observations at greater depths and with more complete geographical coverage. The instruments widely used in studies of the ocean subsurface are briefly described in this section. We do not, however, include discussions of instruments designed by individual research institutions in support of specific research programs.

Expendable Bathythermographs

Instruments called expendable bathythermographs, or XBTs, began providing experimental subsurface temperature measurements after WWII but not in sufficient numbers to enable estimates of subsurface ocean temperature structure over much of the globe until the mid-1960s. By that time observational estimates of subsurface temperatures in the first 700 m became available for roughly 20% of the ocean. The XBT sampling of temperature to this depth had risen to roughly 50% of ocean volume in the first decade of the twenty-first century. The XBT instruments can provide temperature data extending to 1500 m or more but the number of observations decreases with depth. As with all measurements, the XBT instruments have evolved with time, requiring bias adjustments in the construction of long time series.

The XBTs are not re-usable and sink to the ocean floor, where they remain, after the temperature sounding is complete. Temperature data from an electronic sensor on the XBT are sent back to the ship though wires attached to the sounder. The XBTs are deployed, primarily along commercial shipping routes, by volunteer observers. The XBT, deployed over the side while the ship is underway, observes ocean temperature as it descends. The XBT temperature versus depth profiles are constructed by assuming a standard descent rate for the instrument. Historically, XBT data have been recorded on strip charts and more recently as digital data. In either case, the data are not available until they have been received by a data center and processed. A smaller number of XBT observations are transmitted to data centers in real time.

As mentioned earlier, since there is no depth sensor on the XBT, the depth associated with any given temperature is estimated from the time elapsed from the deployment of the XBT and an assumed descent rate. Comparisons of the vertical temperature profiles taken by XBTs with observations taken from moored buoys in the late 1990s and into the first decade of this century have revealed errors in the assumed descent rates for XBTs that have resulted in a positive (warm) bias in these data (Wijffels et al., 2008). The discovery of these descent-rate biases has resulted in increased uncertainty in the amplitude of subsurface decadal variability as determined from analyses of the XBT data but estimates of total warming in the top 700 m of the global oceans remain relatively robust.

A volunteer program for making regular XBT casts by ships of opportunity is managed by NOAA's Atlantic Oceanographic and Meteorological Laboratory (AOML). Under this program, 50 or more volunteer ships make roughly 25,000 XBT casts annually. Data are available through the NOAA National Ocean Data Center (NODC). While the XBTs have provided invaluable data on the vertical temperature structure of the world's oceans along commercial shipping routes, they are much scarcer in other parts of the oceans and do not provide estimates of salinity in the oceans.

Argo Profiling Floats

In situ observations of the ocean subsurface were significantly increased in the late 1990s through addition of Argo profiling floats. The Argo floats take measurements of temperature and salinity as a function of depth and have added considerably to the

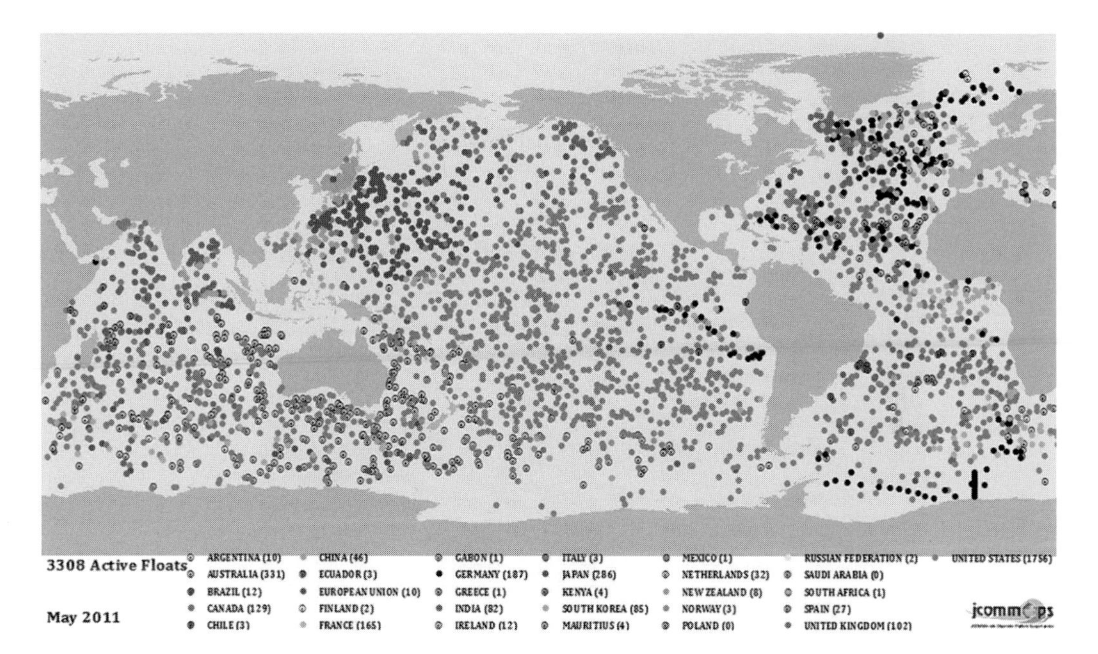

Figure 8.5 An example of the spatial coverage of Argo floats for May 2011(NOAA). A black-and-white version of this figure appears in some formats. For the color version, please refer to the plate section.

available set of ocean observations (Schmid et al., 2007; Legler et al., 2015). At any given time, over 3,000 Argo floats are deployed globally (Figure 8.5).

The confluence of technological advances in instrumentation, computer microprocessors, and satellite communications in the last decades of the twentieth century led to design, testing, and implementation of this new generation of instruments. They have an unprecedented ability to routinely sample the temperature and salinity structure in the global oceans. The international Argo program utilizes free floating, or autonomous, devices. After being launched at the surface, the Argo floats descend to a depth of 1,000 m. They then drift with the sub-surface ocean currents for 9 days. The system is designed so that at the end of the 9-day period, the floats descend to a depth of 2,000 m before beginning their roughly 10-hour ascent to the surface. Temperature and salinity measurements are made during the float's ascent. Upon reaching the surface the floats transmit the data and their location to satellites (Figure 8.6). In practice, roughly two-thirds of the Argo floats provide data to below 1,500 m and slightly under half provide data to near 2,000 m.

The Argo instruments do not measure salinity directly. The electrical conductivity of water varies with the concentration of salt and other chemicals dissolved in the water. The Argo salinity instrument measures the electrical conductivity at various depths during ascent to the surface. The conductivity observations are then converted into equivalent salinity estimates. Salinity values are available to operational weather forecast centers within 24 h of the time the float reaches the surface and communicates its observations by satellite. Early experience with the Argo salinity estimates has uncovered a tendency for the electric conductivity instruments to lose calibration with time, resulting in biased salinity values.

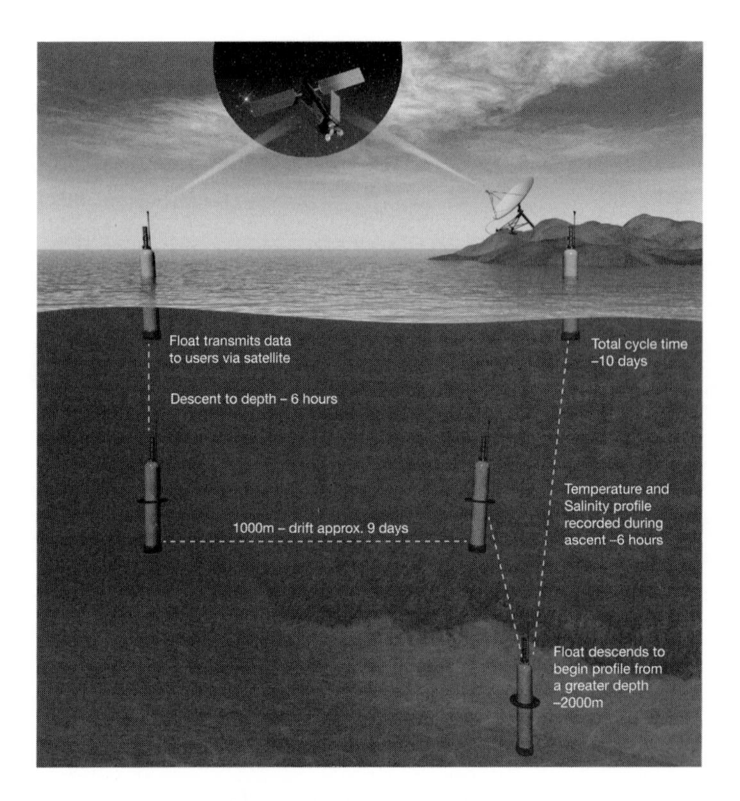

Figure 8.6 Illustration of the operation of Argo float observations.
Source: Schmid et al., 2007 / NOAA and the Scripps Institution of Oceanography UCSD, Argo Program Office.

Salinity estimates are reviewed and quality controlled at Argo international data archive centers including GODAE in Monterey CA and NOAA's AOML in Miami. Corrected salinity estimates are made available to the wider community after the data centers have processed the data and bias adjustments to the raw measurements have been made. The data centers also provide information on how the raw estimates were adjusted. Argo data may be obtained through the Global Argo Data Repository (GADR) at the National Ocean Data Center (NODC) on the NCEI website listed in the references.

The Argo observing system was designed to have at least one Argo float deployed in every 3° by 3° latitude-longitude area of the global oceans. A total of 3,000 floats are required to maintain this density of observations. The floats are designed to operate for 4 years so that approximately 750 floats need to be replaced annually to maintain the array. The full 3,000-float array was achieved in late 2007. While the period of observation is far too short to support studies of long-term climate variability based solely on these instruments, these data can provide estimates of the mean oceanic temperature and salinity structure in detail not previously available. These few years of data can also start to provide some estimates of the patterns of temporal and spatial variability, serve to check and refine theoretical understanding of the ocean structure, and serve as input data for ocean models.

The Argo floats can, in principle, provide data about 10-day mean currents in the first 1,000–2,000 m of the oceans, since the change in location of a float can be ascertained by satellite. Near-surface ocean currents can also be derived from an analysis of ocean height, either derived from models using the Argo temperature and salinity data or from use of the Argo data in conjunction with satellite estimates of ocean height structure by the Topex-Poseidon satellite and its successors.

The Argo data are being used by the data assimilation systems of some coupled ocean-atmosphere climate models. The ocean salinity and temperature profile data play a role similar to that provided by the radiosonde data in the atmosphere as discussed in Chapter 3. In addition to providing real-time data for assimilation in operational coupled climate models, the Argo data promises, in the future, to provide better descriptions and understanding of the mean structure and slow variations in the deep ocean.

Moored Buoys

The recent two to three decades have seen a dramatic increase in the number of buoys that are anchored, or moored, to the ocean floor and primarily deployed in the tropical oceans (Figure 8.2). The array of moored buoys centered on the equator in the Pacific Ocean known as the Tropical Atmosphere Ocean (TAO) array was pioneered by Stan Hayes and colleagues at NOAA's Pacific Marine Environmental Laboratory (PMEL; Hayes, 1991). The TAO buoys were designed to provide improved ocean and atmosphere observations in support of ENSO monitoring and prediction. They have been providing near-equatorial real-time observations of the subsurface ocean and surface meteorology since 1995. The array consists of a set of buoys that are moored within 6° of latitude of the equator in the Pacific Ocean. In addition to providing measurements of SST using a sensor mounted at a depth of 1m, the buoys measure ocean temperature at ten different depths. The sampling strategy is designed to monitor the ocean's mixed layer and to define the depth of the thermocline. The TAO buoys make four or five temperature observations in the top 100 m. The remaining five or six temperature and pressure measurements are spaced at wider intervals to a nominal depth of 500 m.

The TAO array buoys also monitor the lower atmosphere, providing six-hourly averaged near-surface (3.8 m) winds, air temperature, and humidity. Hourly atmospheric boundary layer observations are made at some TAO locations to provide more detailed estimates of the diurnal cycle. Precipitation estimates are available from some buoys.

The TAO array has become a standard source of research data in the equatorial Pacific (McPhaden et al., 1998). These data are also used extensively for ENSO monitoring on the Climate Prediction Center (CPC) and PMEL websites, and as input to model analyses and seasonal predictions. The TAO array formally became the TAO/TRITON array in the year 2000 with the introduction of the Triangle Trans-Ocean Buoy Network (TRITON) buoys in the western part of the array by the Japan Agency for Marine-Earth Science and Technology (JAMSTEC). Extensive side-by-side comparisons of observations from the traditional ATLAS buoy used by TAO and the TRITON buoys were performed in 1999, prior to the operational deployment of the TRITON buoys into the array. Observations from both kinds of buoys are included in the TAO/TRITON Archive. The TAO/TRITON Array is often referred to informally as the TAO array but the observations are being made by different buoys and instruments in different parts of the array leading to a potential source of systematic bias. This potential bias could present challenges in the interpretation of the historical record at locations where the ATLAS buoys were replaced by TRITON buoys. Buoy changes are documented and maintained with the data archive.

Building on the success of the TAO array, similar arrays have been established and operate in the tropical Atlantic and Indian Oceans. These arrays have been primarily designed to provide information on phenomena associated with climate variability in their basins in addition to any links with ENSO. The PIRATA array, now called the Prediction and Research Moored Array in the Tropical Atlantic (Figure 8.2), was designed to monitor climate variability in the Atlantic Ocean (Bourles et al., 2008). Of primary interest are the ocean and atmosphere boundary layers including the heat content in the upper ocean and the mechanisms responsible for variations in SST. Variations in the tropical Atlantic include changes in the north-south gradient of temperature, often referred to as the Atlantic dipole, as well as other patterns of climate variability. The PIRATA array is a joint activity among Brazil, France, and the United States. The buoys in this array are essentially the same as those deployed in the TAO array. PIRATA, initiated in 1997, covers a broader latitudinal extent (20°S to 20°N) than the TAO array and includes meteorological observations from islands, including Peter and Paul Rocks, Fernando de Noronha in the western part of the array, and Sao Tome in the east.

Data from the PIRATA array are being routinely assimilated into coupled ocean-atmosphere models used for weather prediction and for analysis and prediction of seasonal climate. Data are available from centers in Brazil and the United States, such as the NOAA PMEL Buoy Data Delivery website. The data records in the eastern part of the array are less complete than elsewhere in the array due to persistent problems with vandalism.

In the Indian Ocean, the Research Moored Array for Africa – Asian – Australian Monsoon Analysis and Prediction (RAMA) was initiated in 2004, and is the youngest of the equatorial moored-buoy networks., The 26th RAMA buoy was installed during 2016, bringing the array to just under 80% of its target size. Once RAMA is complete the equatorial ocean and atmospheric structure will be routinely monitored using an *in situ* moored buoy data network in each of the ocean basins. The RAMA array is an international collaborative effort with contributions by nations surrounding the Indian Ocean Basin including Australia, India, and Indonesia as well as from China, Japan, and the United States. RAMA provides observations of the atmosphere and ocean boundary layers north of 30°S. RAMA is part of larger efforts to obtain better understanding of Indian Ocean Basin climate variability from intraseasonal through inter-decadal time scales. Data from RAMA are expected to lead to better understanding of the atmosphere-ocean interactions including those associated with intraseasonal variability, such as the Madden-Julian Oscillation (MJO), as well as monsoon variability, ENSO, the Indian Ocean (east-west SST) Dipole, and with time, decadal variability.

The backbone of RAMA is an array of ATLAS and TRITON buoys. As with the buoys deployed in the other ocean basins, RAMA provides observations of ocean temperature supplemented by estimates of salinity as well as observations of atmospheric temperature, humidity, winds, precipitation, and downward shortwave solar radiation. A subset of four RAMA buoys is equipped with acoustical sounders to provide estimates of ocean currents on the equator starting from depths of 400 m upward to about 40 m below the surface. Nearer to the surface, acoustic current measurement observations are masked by noise from surface waves and other sources. Most RAMA data are planned to be available shortly after observation time using the same or similar technology as was developed for the TAO/TRITON and PIRATA arrays. The 400 m to 40 m current data are stored locally on a submerged buoy platform and will be available on an annual basis with the timing dependent on when the buoys are serviced.

8.6 Summary

Early climate studies often restricted their attention to examination of statistics of the atmosphere. The role of the SST in climate and the interaction of the ocean with the atmosphere on interannual time scale begin to emerge in the latter half of the twentieth century (Bjerknes, 1966, 1969). This was followed by an emerging realization that processes in the ocean subsurface are also likely to influence climate not only on interannual time scale but also on decadal time scales and climate change (Lorenz, 1970). Current generation climate analyses and climate models strive to realistically include the role of ocean as a part of a larger climate system.

Recent decades have witnessed a dramatic increase in the amount of ocean data available for climate analysis. The advent of routine satellite estimates of SST and other ocean variables, as well as the introduction of an increased number of *in situ* moored and drifting instruments, has greatly increased our ability to monitor much of the world's oceans. These data have already proved extremely useful for monitoring and predicting climate variations from the sub-seasonal through interannual time scales. As the length of record increases, these observations, particularly those in the deep ocean, promise to support greater understanding of decadal and longer time scale climate variability.

Much of the data described in this chapter is available from the data centers and websites referenced throughout. For example, climate studies requiring SST data may have the option of obtaining daily data with high spatial resolution of 1° or less (OISST) but for the relatively short satellite record from NCEI or monthly SST data at 2° latitude by 2° longitude resolution for a century or longer (ERSST) from NCEI, or (HadISST) from the Hadley Center, both listed in the references. Likewise, subsurface and other data can be accessed from the sources cited in the previous sections.

While great progress has been made, challenges remain. Many of these challenges occur at the interface between the land and the ocean, the coastal zone, and the boundary between sea ice and ocean. Climate at the interface between the land and the sea and between sea ice and open ocean is difficult to characterize with data from current observing systems. Observing systems for the atmosphere, land, and cryosphere often differ in spatial and temporal characteristics reflecting the differences in scales associated with the physical processes that influence climate in the land surface and cryosphere. In the following two chapters we discuss observing systems and data that characterize climate in the land surface and the cryosphere.

Suggested Further Reading

Gill, A. E., 1982: *Atmosphere-Ocean Dynamics*, Academic Press, International Geophysics Series, **30**, 661 pp. (A classic text giving mathematical treatment of the dynamics of the atmosphere and ocean aimed at graduate level and upper level undergraduates.)

Legler, D., H. J. Freeland, R. Lumpkin, G. Ball, M. J. McPhaden, S. North, R. Cowley, G. Goni, U. Send and M. Merrifield, 2015: The Current Status of the Real-Time In Situ Global Ocean Observing System for Operational Oceanography. *Journal of Operational Oceanography*, **8** (S2), 189–200. This paper can also be viewed at www.tandfonline.com/doi/full/10.1080/1755876X.2015.1049883

Levitus, S, J. Antonov, Z.-X. Zhou, H. Dooley, K. Selemenov, and V. Tereshchenkov, 1995: Decadal-Scale Variability of the North Atlantic Ocean. *Natural Climate Variability on Decade-to-Century Time Scales*, D. G. Martinson, K. Bryan, M. Ghil, M. M. Hall, T. R. Karl, E. S. Sarachik, S. Sarooshian, and L. D. Talley, Eds., National Academy Press, 318–324. (Includes further descriptions of ocean observations and datasets.)

Reynolds, R.W., 1988: A Real-Time Global Sea Surface Temperature Analysis, *Journal of Climate*, **1**, 75–87. (A seminal paper illustrating the merging of ship and satellite data to produce a global sea surface temperature analysis). (Available on the American Meteorological Society website).

Questions for Discussion

8.1 What datasets would be necessary for examination of the influence of the global oceans on variations of seasonal climate in the southeastern United States? What analysis tools might be employed?

8.2 What datasets would be necessary for examination of influence of the Gulf of Mexico on variations of climate in Florida? What analysis tools might be employed?

8.3 What datasets would be necessary for examination of the influence of Galveston Bay on variations of climate in Houston?

8.4 How do global SST analyses from the early twentieth century differ from those of the late twentieth century? In what ways will these differences impact studies of twentieth-century climate variability on interannual scales? On decadal scales?

8.5 Studies of the ocean's impact on climate variability and change tend to emphasize changes in SST and sea level. Does this mean that other oceanic surface parameters, such as currents and waves, have no role in the climate? Discuss your answer. What about subsurface properties of the ocean?

9 Cryosphere

The cryosphere is the part of the earth's climate system where persistent low temperatures, snow, and ice are dominant. The Glossary of Meteorology (American Meteorological Society, 2012) defines the cryosphere as "That portion of the earth where natural materials (water, soil, etc.) occur in frozen form." The Glossary also notes that the term is "generally limited to polar latitudes and higher elevations." In this chapter, however, we include seasonal snow, sea ice, and permafrost wherever they occur. Sea ice and snow cover are its spatially largest components but the cryosphere also includes glaciers, frozen lakes and ponds, and permafrost. Sea ice in the global oceans and snow cover over land areas are major interactive players in climate variability on all time scales. The state of glaciers and permafrost along with the freeze and thaw dates of lakes, ponds, and rivers are indices of climate variability on multi-year time scales (Figure 9.1). The study of the cryosphere includes multiple disciplines and branches of technical expertise, including geophysics, oceanography, meteorology, climatology, and, in recent decades, remote sensing.

In this chapter, we will first discuss the observations and instruments that contribute to the routine monitoring of climate variability in the cryosphere. These include *in situ* snow observations regularly collected at weather stations or other networks, and remotely sensed estimates of snow and sea ice made by satellites. We then turn to a discussion of ice caps and ice sheets and their role in spawning glaciers and icebergs. We conclude the observations section of the chapter with discussions of frozen water in lakes, ponds, rivers, and frozen ground, or permafrost. In the remainder of this chapter we discuss methods of observation of and climate variability in each of these components of the cryosphere.

Historically, detailed studies of the cryosphere concentrated on some element of the frozen system at one or more specific locations. Snow observations, however, have been included as part of the suite of weather observations from the late nineteenth century. In recent decades, satellite observations have provided the necessary spatial coverage to allow examination of cryospheric variability on global or near-global scales. Nonetheless, merging the satellite-based, more spatially complete data with earlier observations remains an analysis challenge.

The components of the cryosphere are important contributors to climate variability on all time scales. Variations in the extent of snow cover and sea ice alter the albedo, and thus the radiation balance, over large areas of the globe. These variations and variations in other components of the cryosphere may also influence short-term climate variability

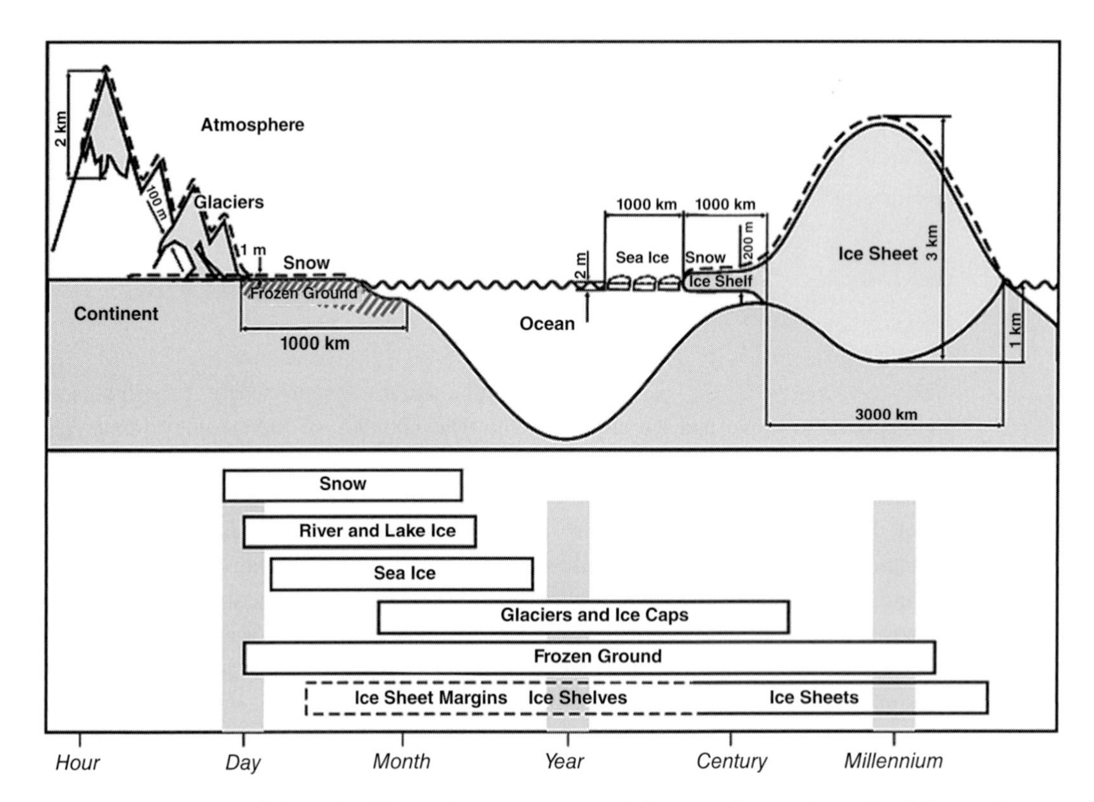

Figure 9.1 Schematic of the various components of the cryosphere and associated time-scales. Source: IPCC, Solomon et al. (2007).

by altering local wind patterns, heat budgets, and water budgets over land as well over the oceans. Local climate may also be influenced by whether precipitation falls as rain or as snow.

9.1 Snow Observations and Datasets

Until the advent of routine satellite observations, historical snowfall and snow cover data are available only from station observations over land. Before the late nineteenth century there are scant routine records of snow, with most records coming from diaries, logbooks, and accounts of extreme low temperatures and snow. Many of the early written descriptions in logs and diaries were made during the so-called Little Ice Age that has been variously described as starting anywhere from the twelfth to fifteenth century and ending in the 1800s. Comprehensive paleoclimate studies for that period have led some researchers to conclude that the "Little Ice Age" was, most likely, a regional phenomenon, centered in northern Europe (Mann et al., 2002). Early handwritten records may provide the dates that snowfall occurred, comments on the duration of snow cover and, in some cases, estimates of accumulation or snow-depth but there were no widely practiced standards for reporting snow.

More consistent observations of snow begin with the establishment of national meteorological services in the nineteenth century. Guidelines for measuring and recording snow characteristics emerged in the early twentieth century, but there remains, to this day, no consistently employed methodology to observe the physical properties of snow at all meteorological observing sites. Representative, accurate, and consistent observations of snow are difficult to achieve despite considerable efforts to design and implement instrumentation to do so (Rasmussen et al., 2012). Observations include the dates the snow falls, snowfall amount, snow depth, liquid water equivalent, and snow cover extent. Each of these is discussed in this chapter. Since the introduction of the Automated Surface Observing System (ASOS) in the 1990s, the U.S. National Weather Service (NWS) has ceased to take observations of snowfall and snow depth at most airport locations. Observations continue at some airport locations in collaboration with the Federal Aviation Administration. Snow observations also continue to be made by volunteers as part of the NWS U.S. Cooperative Network. NWS observational guidelines for airport locations and for the Cooperative Network can be found at the website listed under Snow Measurement Guidelines (2013) listed in the References. Globally, automated observing systems have also been implemented by the national weather services of many nations. The meta-data associated with historical records should be examined in the analysis of *in situ* snow observations to determine if, and when, changes in observations may have taken place.

9.1.1 Snowfall

Daily snowfall, or how much snow has accumulated during a 24-hour period, has traditionally been estimated with a measuring stick on a flat surface. Snowfall amounts are usually recorded to the nearest tenth of an inch in the United States. Where and how often that measurement is made may vary considerably by observation site. Current practice recommends the use of a snowboard, defined as a (usually plywood) board with dimensions of roughly 30–60 cm (1 foot by 2 feet), and painted white. Ideally the board is placed in an open area away from buildings, shrubs, trees, or other obstacles that may alter wind patterns or shelter the snowboard. In the United States, observational guidelines suggest that the snowfall measurements be made at least daily, but no more frequently than every six hours. The snowboards are to be cleared after each measurement and, if more than one daily observation is made, values are summed to arrive at a daily snowfall amount. The U. S. Cooperative Network observing stations generally make one snowfall observation per day. It is well known that snow tends to settle after it falls, so that the total snowfall amounts obtained by summing more frequent daily observations will be larger than the amount measured once daily. Thus, there are likely to be differences among observing sites in a region, depending on the frequency of observations. Some sites have adopted the practice of taking hourly snowfall measurements, especially during major storms, in an attempt to capture "boasting rights" for the most snowfall. While such practices are discouraged, these observing inconsistencies may remain in the historic datasets.

9.1.2 Snow Depth

Snow depth, or amount of snow on the ground, is also a standard observation. Snow depth is an integrated measurement, and includes the depth of snow from previous snowfalls. Snow depth is recorded to the nearest inch in the United States. The depth of the snow on the ground may be a more interesting quantity, from a climate standpoint, than snowfall and may be integrated with satellite estimates of snow cover, discussed later. Those who live in areas with significant seasonal snowfall are aware that fields of snow tend to decay, or melt, unevenly. As time progresses, parts of a snowfield may start to show bare ground while other nearby locations may maintain significant snow amounts. At some locations snow depth observations are taken at several different spots and averaged. If more than 50% of the ground is bare the snow depth is recorded as T, for Trace. The very meaning of average snow depth over a region, however, often becomes difficult to define and analysis of time series of snow depth must take into consideration that there may be spatially unresolved local variations.

9.1.3 Basin Snowpack

The accumulated snow depth, or snowpack, within catchment basins of river systems in mountainous regions is an important quantity used for estimates of the spring runoff that feeds rivers and streams. In the United States, the mountain snowpack is measured at several snow course locations maintained by the Department of Agriculture's (USDA) Natural Resources Conservation Service (NRCS). Manual measurements of snow depth and liquid water equivalent are taken monthly along 300 m (approximately1000 feet) "courses," as near to the first of the month as possible in the winter and spring. The first snow courses were established in 1906. In addition to the snow course data from 1,185 locations, the USDA maintains an automated network of 858 snow-telemetry (SNOTEL) observing sites in the mountainous western US. The SNOTEL sites automatically record and transmit maximum and minimum temperature, as well as estimates of precipitation, liquid water equivalent (discussed further in this chapter), and snow depth data. The length of record for both snow course and SNOTEL data varies with location. Some augmented sites also provide soil moisture data. SNOTEL estimates the liquid water equivalent of snow by measuring the pressure due to the weight of the snow on a snow "pillow," essentially a bladder filled with antifreeze. Snow depth is estimated using a sonic sensor (Schaefer and Paetzold, 2000). Climate analyses based on snow course and SNOTEL data must take into consideration the differences in instrumentation and observing practices between the two.

9.1.4 Snow Water Equivalent

As discussed in the previous section, snow depth, whether during or immediately following a storm, or during the melt season, is not a well-defined and consistent quantity. The liquid water equivalent of snow, which is the amount of water that is obtained by melting the snow without loss due to evaporation, is called snow water equivalent (SWE) and is a well-defined snow-related quantity that can be used, for example, in analyses of the water cycle. The SWE is especially important in regions where the spring snowmelt provides the water to recharge rivers, streams, lakes, and ground water. Large snowfalls may occur in very dry conditions but represent less liquid water than less deep but wetter snowfalls. Measurements of liquid water equivalent can be challenging. The standard US 8–12 inch (20–25 cm) rain gauge is typically fitted with a funnel for measuring rainfall. At observation sites where snowfall is common the funnel is removed to collect snow at the bottom of the cylindrical gauge. The collected snowfall is melted to arrive at the liquid water equivalent for the snow. If snow is not collected in the gauge, a sample may be gathered from a snowboard or other nearby location. Automated observing sites like ASOS are often equipped with heated tipping buckets to measure precipitation amounts, including the water equivalent of snowfall. Very often these measurements have been found to underestimate the liquid water associated with a snowfall by 40–50% compared to standard gauges. In the absence of measurement, the ratio of snow depth to liquid water equivalent is often assumed to be 10, that is, 10 cm of snow is assigned a liquid water equivalent of 1 cm. This assumption is generally understood as an educated guess, because the actual liquid water content depends on the moisture content of the snow. Active research on the use of satellite-based passive microwave estimates of the snow liquid water equivalent continues.

9.1.5 Snow Cover

Snow cover refers to the spatial extent of snow over a region. Before the advent of routine satellite observations, regional snow cover was estimated through interpolation (and extrapolation) of snowfall and snow depth observations recorded at observing sites and, in some cases, supplemented by anecdotal information. Estimates of snow cover extent were particularly uncertain in sparsely populated mountainous areas. Visible satellite imagery provided the basis for regular estimates of Northern Hemisphere snow cover starting in November of 1966, becoming one of the earliest satellite-based climate datasets. These early satellite snow cover estimates required subjective analysis of visual satellite images, as well as comparisons to weather maps, to differentiate between cloud cover and snow on the ground. The analysis provided estimates of snow cover extent on a weekly basis for a calendar week ending on Sunday evening. Depending on the number of days with cloud cover, the analysis could reflect snow cover on any day of the seven-day period. The snow cover was digitized onto a rectangular 89x89 grid that was used as an overlay on a polar-stereographic map of the Northern Hemisphere (Dewey and Heim, 1980). The size of the grid areas varied with latitude and was roughly190 km on edge in midlatitudes. A grid area was considered snow covered if 50% or more contained snow. Estimates of albedo in four classes were also made until it was found that variations in the albedo estimates due to differing viewing angles, from satellite image to satellite image, introduced too much uncertainty. Monthly snow cover analyses depicted the number of weeks in a month that the grid areas were snow covered. Satellite instrumentation and computer power have evolved significantly over

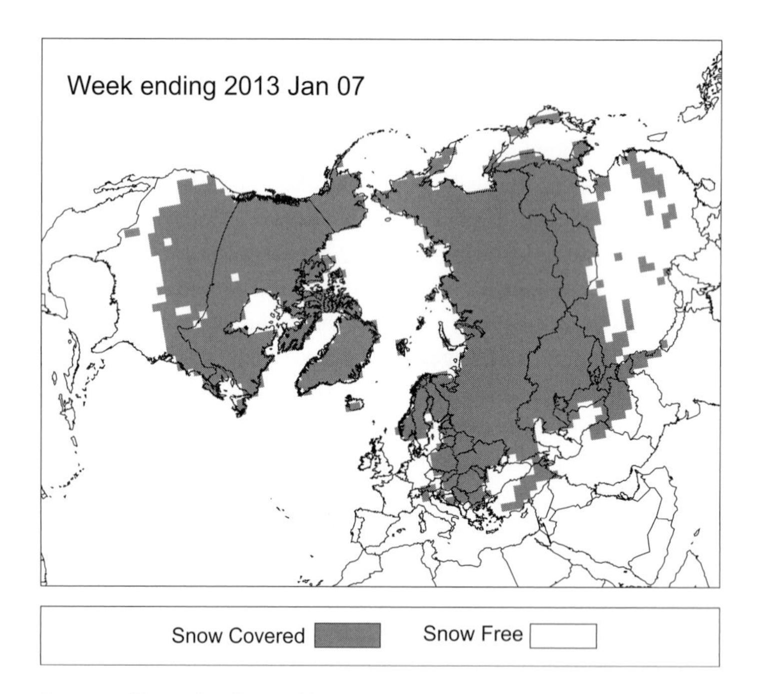

Figure 9.2 Example of a weekly snow cover map on an 89 by 89 grid for the first week of 2013. Source: Rutgers Global Snow Laboratory.

the course of the past several decades, but weekly data on the original grid (Figure 9.2) continues to be produced (Robinson, 1993) and archived on the Rutgers Global Snow Laboratory web site as well as at the National Snow and Ice Data Center (NSIDC).

While these snow cover data were updated weekly, the calendar dates of the weekly analyses varied from year to year. The weekly analysis continued until May of 1999, when daily analysis on a 24-km grid was implemented to provide initial surface boundary conditions for numerical weather forecast models. In 2004, the spatial resolution was increased to 4 km. These high-resolution daily snow cover data are archived at NSIDC. The evolution of the snow cover analyses provides an illustration of the compromises that must be made in the utilization of operational weather data for climate analysis. The longest record of data, nominally over 50 years, has a relatively low spatial resolution while the higher resolution data span only one or two decades.

9.2 Sea Ice Data and Observations

The formation, melting, and dynamics of sea ice are complex, involving ocean currents and salinity, atmospheric winds, sea and air temperatures, as well the surface radiation balance and sub-surface heat balance. One result of this complexity is that, in addition to extent, sea ice varies in morphology, thickness, sea ice concentration, and age, since ice that forms and melts within a single season (called seasonal ice) has significantly different physical characteristics from multi-year ice that lasts through one or more warm seasons. Furthermore, sea ice can consist of a homogeneous flat sheet, fragmented shards that thaw and refreeze at odd angles, and a great many other configurations (Bushugyev, 2012; NSIDC web site). Sea ice may include ice that abuts the coast, called fast ice, as well as ice in the open water that is not attached to land. Study of the nature of sea ice fields, including the complex interactions among the heat, energy, and radiation budgets, is an active area of research, and considerable efforts have been made to parameterize this complexity in climate models. In general, however, the variations in the physical quantities that relate to the characteristics of sea ice, on climate time scales, have not been well observed on regional and larger spatial scales. In the next sections, we discuss observations of the large-scale characteristics of sea ice that are typically emphasized in climate studies. These are sea ice extent, thickness, age, and concentration.

9.2.1 Sea Ice Extent

The earliest sea ice extent data are from visual observations taken from ships, on islands, and by coastal stations. As with snow cover, estimates of sea ice extent obtained from these sources are not spatially complete. Often, the location of the ice edge was estimated from observations made from some distance away. Early observations of sea ice taken by whaling ships, primarily in polar regions, and by commercial vessels globally, were documented in ships' logs and sailors' dairies. While these observations provide information about large-scale fluctuations in sea ice dating back to the eighteenth century, they are spatially and temporally limited in coverage, and rather qualitative in nature (see Box 9.1 – Historical Sea Ice Observations).

Box 9.1 Historical Sea Ice Observations

Before satellites made global sea ice datasets possible, most sea ice observations were made by ships, generally fishing and whaling vessels as well as a few exploratory voyages. These observations were only capable of recording the edge of the sea ice field. Most pre-satellite era observations implicitly assumed that the sea ice fields were continuous poleward from the observation location. Early aircraft observations noted that this assumption was not always correct and sea ice could have several types of discontinuities, including open water between landmasses and the ice field, as well as leads and polynyas, which are surrounded by ice. Thus, the ship-based observations tend to overestimate the total extent of the sea ice. Spurred by the need for sea ice data for retrospective simulations by climate models, Walsh and Chapman constructed a historical time series of monthly sea ice coverage data for the Arctic for the period 1870–2009 (Chapman et al., 2013). Their 1870–1952 sea ice estimates are based on "Climatology with increasing amounts of observed data throughout the period," while the 1953–1971 estimates were derived from a variety of sources, and those from 1972 onward are based on satellite data.

The need for consistent global sea surface temperature (SST) datasets to support operational weather prediction and climate studies (see Chapter 8) led to a need for improved monitoring of sea ice extent. Since seawater is saline, the melting or freezing temperature of sea ice is less than 0°C. It is generally taken to be around −1.8°C in analyses of SST (as discussed in Chapter 8). The need for regular estimates of sea ice extent to support global SST analyses, coupled with the requirements for climate models to include sea ice to support prediction, also spurred improvements in sea ice analysis from the 1990s onward.

As with snow cover, truly global sea ice extent data were not available until the beginning of the satellite era. In the early years of satellite observations, snow cover and sea ice extent estimates were performed independently, in many cases using the same visual images as the basis for the analysis, but adopting different conventions in resolution, frequency, and data storage formats, if indeed the data were saved after their initial operational uses. It was not until the weather modeling community required these data in common formats that some consistency was achieved between the snow cover and sea ice analyses. As mentioned previously, sea ice fields are often not continuous, ranging from completely ice covered, through mostly ice with a few open water leads, to a mix of open water and ice. An estimate of the percentage of an area covered by sea ice is referred to as the sea ice concentration. A concentration of 50% is often taken as the boundary of sea ice extent in climate models, but much lower concentrations are used for advisories to ships (discussed in a later section).

Sea ice extent has been monitored by a variety of passive microwave sensors aboard polar orbiting satellites since 1972 (Ropelewski, 1983; Pope et al., 2017), supplanting the subjective analysis of sea ice from visual satellite imagery initiated in November 1966. Passive microwave sensors are sensitive to microwave emissions from the Earth's surface. Since sea ice is a stronger emitter in the microwave band of the electromagnetic spectrum than seawater, passive microwave sensors can provide estimates of sea ice extent and sea ice concentration. Several different sensors have been utilized to monitor sea ice, starting in 1972 with the Electronically Scanning Microwave Radiometer (ESMR). In 1978, the Scanning Multichannel Microwave Radiometer (SMMR) was introduced, the first in a series of improved microwave instruments including the Special Sensor Microwave Imager (SSM/I) in 1987 and the Advanced Microwave Scanning Radiometer – Earth Observing System (AMSR-E) in 2002 (Comiso et al., 2003). These instruments provide sea ice concentration data on a 25 km by 25 km grid. The archived sea ice concentration data are available from a number of data centers. The NASA Earth Observation System Data Information System (EOSDIS) provides sea ice and other data by type of instrument. The National Centers for Environmental Information (NCEI) and the National Snow and Ice Data Center (NSIDC) provide archived data from each instrument in addition to integrated snow and ice analyses products produced by the National Ice Center (NIC). Internet addresses for these centers are provided in the References section. The fundamental differences between the ESMR and subsequent satellite microwave instruments lead most climate studies to use the microwave sea ice data only from 1978 onward. Merged sea ice concentration datasets formed by linking data from different instruments that take inter-instrument biases into account are available from the National Snow and Ice Data Center (NSIDC) and NCAR.

Active microwave observations made by radars on polar orbiting satellites can provide high resolution, 100-meter or less, images of sea ice. While these data have provided useful detailed information for selected locations since the late 1990s, they have not been extensively used in large-scale climate studies because of the large volume of data, limited geographic coverage, and relatively short period of record. As time goes by these observations are likely to become more useful for climate studies given expected increases in computational capabilities to process and manipulate large datasets.

9.2.2 Sea Ice Thickness

Sea ice thickness is estimated using satellite active and passive microwave data as well as laser altimetry. The sea ice thickness is inferred through the difference in height between adjacent areas of open water and the top of the sea ice surface and assumptions about the percentage of the ice that is below the level of the sea. The difference between water level and top of the floating ice is referred to as the freeboard. Uncertainties in these estimates are introduced by the presence of snow on the sea ice. Laser altimetry cannot differentiate between the top of the snow layer and the sea ice surface. Likewise, the passive microwave estimates are confounded by liquid water and slush at the sea ice-snow boundary as well as melted and refrozen layers in the snow pack (Pope et al., 2017). Active microwave estimates have some of the same difficulties and data are available for limited areas over a relatively short time spans. Microwave snow depth thickness estimates on ice are often confounded where angular shapes and tilted surfaces characterize the sea ice. Limited Arctic sea ice thickness observations taken by submarines during the cold war years and, more recently, from unmanned underwater research vehicles are available for research. These data can provide some climate information but irregular spatial and temporal sampling and short periods of record for any particular location limit their utility for climate studies.

9.3 Land-Based Permanent Ice

During the current era and for at least the past 30 million years, substantial portions of the land surface of the earth have been covered by permanent (as opposed to seasonal) ice. These areas all originate from locations where seasonal snowfall exceeds melting over long enough time periods that the weight of the accumulated snow compresses and recrystallizes the material beneath, initially forming an intermediate stage called firn after a couple of years, and then eventually leading to the formation of glacier ice (see the website listed under NSICD in the References for more details). Such areas of permanent ice are referred to as glaciers, ice caps, or ice sheets, and play an important role in both the physics and the observation of the climate system.

Ice caps include areas of land-based permanent ice with areas of less than 50,000 km^2 (Barry and Gan, 2011) and are constrained by terrain. Ice that flows from higher to lower altitudes, in a manner somewhat similar to rivers of liquid water, is called a glacier. Larger and thicker areas of ice that overwhelm the terrain rather than being constrained by it are called ice sheets. Ice caps and ice sheets often transform at their low altitude/latitude borders into glaciers that either terminate on land or flow into the oceans, generating icebergs. Ice sheets covered much of North America and Northern Europe during the ice ages but the only major ice sheets remaining cover Greenland and Antarctica.

Ice cores taken from ice caps and the Antarctic and Greenland ice sheets are rich sources of data that can be used to infer past climatic conditions (Barry and Gan, 2011). The ice cores contain trapped air bubbles, and analysis of the characteristics of the air in the bubbles, which were trapped at different times in the past, can be used to estimate the time series of a number of climate parameters, including atmospheric temperature and CO_2 concentration. For example, changes in the relative concentration of oxygen isotopes in the trapped bubbles, referred to as "delta-O18," can be used to estimate changes in the large-scale atmospheric temperature (Thompson et al., 2003). Ice cores have been used as proxies for air temperature, precipitation, atmospheric composition, volcanic eruptions (inferred from dust deposits), as well as other quantities, and are discussed further in Section 9.6.4.

9.3.1 Observations

Observations of glaciers and ice caps and sheets are useful as indices of climate variations. In the early nineteenth century scientists deduced that glaciers, then confined to high mountain valleys, were likely the remains of much larger ice sheets that covered a significant portion of the Northern Hemisphere for long periods of time. These periods, the ice ages, have since been well documented in geological and paleoclimatological records. It is estimated that the last great ice age began to wane 20,000 years ago, interrupted by the "Little Ice Age," a slight cooling in parts of the Northern Hemisphere between the fifteenth and mid-nineteenth century (Bradley and Jones, 1993; Mann, 2002). The cause of this relatively recent cooling is not well understood but some researchers believe it may have been related to an interruption of the North Atlantic branch of the oceanic thermohaline circulation (Chapter 8) while others attribute the cooling to a series of volcanic eruptions (Miller et al., 2012). Geological features such as drumlins, spoon or buried egg-shaped hills that are features of the landscape of much of the northern United States and Canada, erratics, large rocks or boulders found in flat terrain far from any mountains, and moraines, which are hills comprising glacial debris dumped at the melting terminus are remnants left behind by the melting ice sheets. Moraines in particular can be extremely prominent features of the post-ice age landscape: Long Island, the largest and longest island in the contiguous United States, is formed largely from the terminal moraines of the two most recent North American ice sheets. Most glaciers have been melting and retreating in recent years.

The movement of glaciers can be monitored *in situ* by placing posts or other markers in the ice and observing their movement relative to the adjacent bare ground. Measurements are generally made every several months or annually. Glacial ice flow can also be estimated remotely using Synthetic Aperture Radar (SAR) data taken several months apart (Alaska Satellite Facility web site). Movements in the near-surface portions of ice caps and sheets can be estimated similarly, but are less useful in understanding overall changes due to their greater thickness.

Changes in the size of a body of ice on land are due to a positive or negative mass balance, which is the balance between the addition of snowfall in the source regions and the melting or discharge into bodies of water. *In situ* or remote estimates of the individual terms in the mass balance of glaciers is possible in a few cases, although with large uncertainties. Recent changes in the total mass of the Antarctic and Greenland ice sheets have been estimated with data from the Gravity Recovery and Climate Experiment (GRACE), discussed in Chapter 3, and ICESat laser altimeter data (Wahr et al., 2000).

9.3.2 Iceberg Observations

Annual and longer-term variations in the number of icebergs have not, as yet, been included as part of routine climate monitoring. We include a brief discussion of them here for completeness. Icebergs are formed by glacial ice collapsing into the sea, a process called calving, at the terminus of coastal glaciers or ice sheets. Recent observations of more rapid movement of Greenland glaciers suggest that the number of icebergs may increase. Icebergs in the Northern Hemisphere are found in Baffin Bay and in the North Atlantic, primarily south of Davis Strait between Canada and Greenland. If the size of the floating ice piece is 5 m or greater it is identified as an iceberg and efforts made to track its movement. In the Northern Hemisphere, icebergs are monitored by the International Ice Patrol (IIP), a consortium of a dozen nations. Icebergs in major shipping lanes are monitored by satellite, airplane, and ships. In the United States, the Coast Guard, in collaboration with the U.S. Navy and NOAA, has formed the National Ice Center (NIC) with the responsibility for monitoring icebergs. Areas near North Atlantic shipping routes are often populated by smaller pieces of ice called "growlers" and "bergy bits" that can also be hazards to navigation. Areas that contain these smaller pieces of floating ice are identified but individual pieces of ice are too small to be tracked individually.

Icebergs generated by the Antarctic ice sheet can often be much larger than those in the Northern Hemisphere. The European Space Agency, U.S. National Ice Center, NASA, the British Antarctic Survey, and the Russian Arctic and Antarctic Research Institute monitor Southern Hemisphere icebergs around the northern boundaries of Antarctica.

9.4 Observations of Ice on Lakes, Ponds, and Rivers

Lakes and ponds modify the climate of the surrounding land areas by acting as sources of moisture and altering the heat and momentum exchanges between the surface and atmosphere. During fall and early winter, lakes will generally be warmer that the surrounding land areas, acting as a source of heat and moisture. In the early spring, the warming of the lake or pond is delayed compared to the surrounding land surface. The interactions between the atmosphere and the water surface change profoundly once ice forms on the lake. The freeze and thaw dates on these bodies of water are indices of climate variability where no other climate records exist.

After the surface of the lake freezes it no longer acts as a moisture or heat source. One well-known demonstration of this phenomenon is evidenced in the extreme snowfall to the lee of the Great Lakes (Petterssen and Calabrese, 1959; Reinking et al., 1993). Snowfalls are experienced to the lee of the Great Lakes throughout the winter but often large snowfall amounts occur early in the season when the lakes are ice free. The largest snowfalls are often displaced a few kilometers from the lakeshore since the water (and ice surfaces) of the lakes are much smoother than the surrounding topography. The downwind change in surface roughness leads to an increase in low-level convergence, resulting in increased snowfall at a distance from the actual lakeshore. Lake effect snows have also been noted near small bodies of water but the effect of these smaller lakes on snowfall is not as dramatic. These lake effect snowfalls are short-term weather events but the snow cover that results may modify the local climate for several weeks after the event occurs, contributing to local climate variability.

Long records of initial freeze and thaw dates, as well as duration of ice cover for mid- and high-latitude lakes, serve as indices of climate variations and change. Historical records of these quantities were usually made from shore stations and the records of some extend to well over a century. Many records show a tendency for shorter frozen lake durations in recent decades, but not all do, nor do all show later freeze and earlier thaw dates. A survey of over a hundred Canadian lakes finds that 30% of these lakes show a trend for longer ice-covered durations, illustrating nonuniformity in this climate index. Ice cover on the Great Lakes has been monitored by satellite since 1973. These satellite-derived ice-cover data show that the interannual variability in lake ice is larger than the mean in each of the Great Lakes (Wang et al., 2012). Given the high variability in ice cover it is difficult to interpret the significance of linear trends over the period of record. Global lake ice and river data can be found at the National Snow and Ice Data Center (NSIDC).

In the discussion so far, we have concentrated on seasonal lake ice and its variability in connection to its influence on local weather and short-term climate. Climate records extending back through several thousands of years have been constructed from sediment data in lakes, especially glacial meltwater-fed lakes. At mid and high latitudes, annual layers in the sediments, called varves, allow differentiation between warm, ice-free parts of the year and cold periods. Warm season sediments contain coarser materials transported by glacial melt in contrast to cold periods when the lake is ice covered and sediments consist of fine particles. Varves, like tree rings and glacial ice cores, are useful in the reconstruction of past climate variations.

Ice often forms on mid-to-high-latitude rivers during the winter, with many of the high-latitude rivers becoming frozen over in the heart of winter. The annual cycle of ice formation and breakup is strongly dependent on surface air temperature. The breakup dates of river ice on rivers in the interior of Alaska are related to the mean surface air temperatures in April-to-May and thus have an association with the El Niño/Southern Oscillation (ENSO). Earlier breakups of river ice are associated with warm ENSO episodes and later breakups with cold episodes (Bieniek et al., 2011). In regions where the surfaces of the rivers freeze over, the spring ice breakup may result in ice jams with initial flooding upstream of a jam followed by downstream flooding once the jam breaks, as occurred with the flooding in the Red River of the North in 2009.

9.5 Permafrost Observations

The term permafrost describes regions where the temperature of the soil remains below 0°C for two or more consecutive years (Barry and Gan, 2011). Areas characterized as permafrost can include regions where the soil thaws at the surface during summer but remains frozen below this thin "active layer," and may also include regions where the frozen layer is spatially discontinuous because of local microclimate or anthropological influences. Permafrost may be monitored through measurements of soil temperatures at various depths (Romanovsky et al., 2002).

Permafrost is generally confined to high latitudes but may include some midlatitude locations at high altitude. *In situ* measurements of permafrost are straightforward using boreholes or probes, but are limited in extent, while remote sensing is of limited utility because permafrost is a subsurface phenomenon. Permafrost is estimated to cover in the order of 24% of the Northern Hemisphere land area (Zhang et al., 1999). The largest continuous regions in North America include northwestern Canada and Alaska, while the largest continuous permafrost regions in Eurasia include the areas east of the Urals, mainly Siberia. Areas of permafrost also include the high latitudes in the Scandinavian countries. The depth of the permafrost is determined by the local geology but may extend to several tens of meters, and in some occasions hundreds of meters. While the name implies permanence, the areas of permafrost have been shrinking since the last ice ages and are expected to continue shrinking at an accelerated rate as the Arctic continues to warm. Among the climate concerns is that permafrost is a significant reservoir of carbon as well as methane gas, which, if released, could exacerbate greenhouse gas warming (NOAA, Arctic Theme Page web site listed in the References).

9.6 Climate Variability in the Cryosphere

The components of the cryosphere are active participants in shaping variability in the climate system, and aggregated observations of their behavior serve as useful indices of climate variability and change at various spatial and temporal scales. For example, snow develops within a few hours as the result of chaotic weather events, but the resulting snow cover can change the albedo over large areas, modifying the seasonal climate, for weeks or months. On the other hand, observations of snow cover and sea ice extent document interannual and short-term variations, while variations in glacier movement and in ice sheet mass balance illustrate long-term changes. Climate variability in the cryosphere is an active area of monitoring and analysis.

9.6.1 Snow Cover Variability

The occurrence of snow and its spatial extent, depth, and liquid water equivalent depend on a very delicate balance of the factors associated with weather events during cold seasons. Once snow has fallen it changes the nature of the land surface in several ways, thus influencing interactions between the atmosphere and land surface as well as influencing the interactions between the land surface and the subsurface. These influences may persist for several days or weeks, modifying the local climate over this period. Alternatively, the snow may melt in a few hours, or a day or two, with the liquid water either running off or adding to the soil moisture. Several empirical and modeling studies have attempted to establish the snow–climate links. At very long time scales, large areas of snow that do not completely melt over the summer may be the harbinger and facilitator of a coming ice age. On shorter time scales, modeling studies have shown that extensive snow cover can modify regional temperatures and wind circulation (Segal et al., 1991) in much the same way as bodies of water alter the local climate. Empirical and modeling studies of snow cover extent over North America have found correlations between snow cover and temperature anomalies in subsequent weeks and months, primarily in zonal bands. In addition, significant monthly snow cover auto-correlations have been documented for the American mid-west. Once a significant snow cover has been established, the likelihood of subsequent snowfall is greater than in the absence of snow cover (Walsh et al., 1982).

The relationship between snow cover and atmospheric circulation patterns, discussed in Chapter 4, continues to be actively studied. The North Atlantic Oscillation (NAO) and Arctic Oscillation (AO) have been linked to seasonal and decadal variability in snow cover extent in Europe, the eastern Mediterranean, and eastern North America. Other studies have attempted to identify the influence of snow cover, particularly Eurasian snow cover and liquid water equivalent, on the NAO (or AO). Climate models, either run retrospectively or in free modes, are often employed in these kinds of studies (see Chapter 11).

9.6.2 Arctic Sea Ice Variability

For the purpose of this discussion we define the Arctic as the region of the world north of the Arctic Circle, roughly 66.5°N. The Arctic encompasses the North Slope of Alaska, the Canadian Archipelago, most of Greenland, the extreme northern portions of the Scandinavian countries (Finland, Norway, Sweden) and northern Eurasia over land. The Arctic Ocean is its largest single geographical feature. Because of long historical national interests, regions of the Arctic Ocean are often discussed in terms of regional seas, including the Chukchi, Beaufort, Barents, Kara, Laptev, and East Siberian Seas. For most of the past century, including the satellite era, the Arctic Ocean,

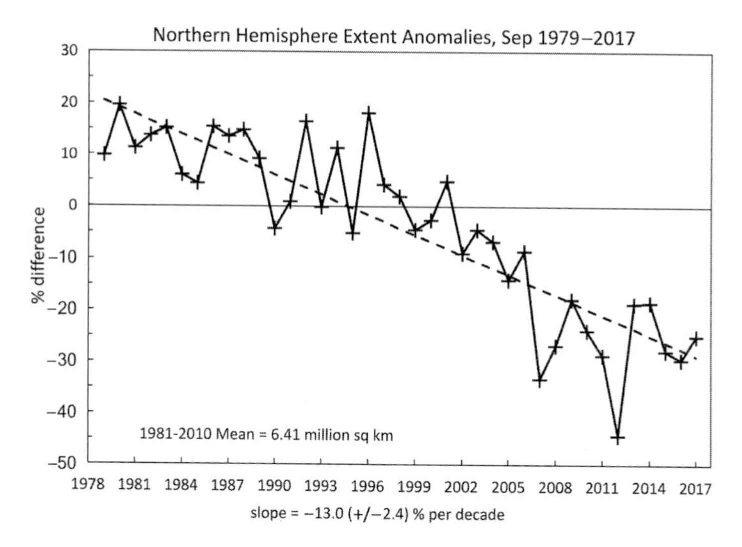

Figure 9.3a Time series of the Arctic sea ice extent anomalies for September, the month with least sea ice in the Northern Hemisphere, expressed as a percentage of the September mean. Source: National Snow and Ice Data Center (NSDIC), University of Colorado, Boulder, Fetterer et al., 2017.

except for some regions of the Barents Sea, was covered by sea ice from November through mid-April. Records show that, in the mean, Arctic sea ice extent starts to decrease in April, reaching a minimum in mid-to-late September. The September minimum in Arctic sea ice extent showed no significant trends from the start of the satellite record until the mid-to-late 1990s, when below average sea ice extent minima begin to occur with greater regularity than expected by chance (Figure 9.3a). There have been no consistent changes in the maximum Arctic sea ice extent observed into the first decade of the twenty-first century.

Until aircraft were capable of flying long enough to routinely venture north of the Arctic Circle and return safely, estimates of Arctic sea ice extent were primarily obtained from ship observations and the records from nineteenth- and twentieth-century Arctic expeditions (Walsh and Johnson, 1979). Of great historic interest was the possibility that ships might navigate through the Arctic Ocean during summer, either north of Eurasia or through the Canadian Archipelago, thus providing a much shorter route from Europe or North America to the Far East. Most early attempts failed. Regular satellite sea ice observations, starting in 1979, confirmed that even during the summer sea ice minimum, there were no consistent, year-after-year, areas of open water in the Arctic in those early years of the record. Since the early 1990s, however, despite large year-to-year variability, satellite observations indicate that the minimum Arctic sea ice extent has been decreasing and a limited amount of commercial shipping has been initiated in the warm season.

The relatively large negative trend in minimum sea ice area is expected to continue in response to global warming and, if this trend does continue, suggests that the Arctic may be completely ice free for at least some portion of the year in the not too distant future. Such a change would result in profound changes to the regional high-latitude climate and likely provide a major perturbation to the global climate system. It is not clear at this time how trends in the minimum Arctic sea ice extent interact with the atmospheric wind patterns at lower latitudes. Some studies suggest that these influences of sea ice on the atmospheric circulation are small (Wallace et al., 2014), while others suggest a more robust relationship (Francis, 2017). Interpretation of contemporary climate data over the Arctic is challenging since it is uncertain whether the past record of short-term climate variability will provide guidance regarding current and future climate variations. This suggests increasing dependence on model-based analysis as well as on proxy records that may identify similar periods of an ice-free or partially ice-free Arctic.

In addition to the trends in minimum sea ice extent, observational evidence suggests that the AO modulates the export of sea ice through the Fram Strait and into the North Atlantic during winter and early spring as well as being associated with an advection of ice away from the shoreline along the Laptev and East Siberian Seas (Rigor et al., 2002) on interannual to multi-year time scales. It is expected that the AO will continue to be a major contributor to interannual and shorter duration high-latitude climate variability.

9.6.3 Antarctic Sea Ice Variability

The Antarctic and Arctic are opposites in a number of ways. Sea ice in the Arctic is largely contained in the Arctic Ocean, which may be thought of as an inland sea or estuary of the Atlantic Ocean (Broecker, 2010), while sea ice in the Antarctic occurs along the northern, equatorward, periphery of a snow and ice-covered landmass. With almost no land to constrain the northward sea ice extent, it has long been known that there is considerable seasonal and interannual variability in both the maximum and minimum extents (Ropelewski, 1983). The areas associated with the Weddell Sea, to the east of the Antarctic Peninsula, and the Ross Sea, south of eastern Australia and New Zealand, account for the largest areas of Antarctic sea ice. Mean Antarctic sea ice extent reaches a maximum in September and a minimum in February. The Antarctic sea ice extent is strongly influenced by ocean currents and the circumpolar westerly winds in the atmosphere. The circumpolar winds and currents are complicated by the shape of the Antarctic continent, especially in the notoriously stormy Drake Passage that separates Antarctica and South America

Antarctic sea ice varies on multiple time scales. Pack ice, ice not grounded to a land mass, can vary rapidly in concentration and extent in response to changes in winds and temperature, in essence reacting to weather events. Changes in Antarctic sea ice concentration and extent are a result not only of atmospheric conditions, but also of interactions with ocean currents. The ocean currents can be responsible for the formation of areas of open water, known as polynyas, surrounded by pack ice. Polynyas can persist for several years. For example, a large polynya in the Weddell Sea ascribed to the upwelling of warm water formed in 1974 and persisted into 1976.

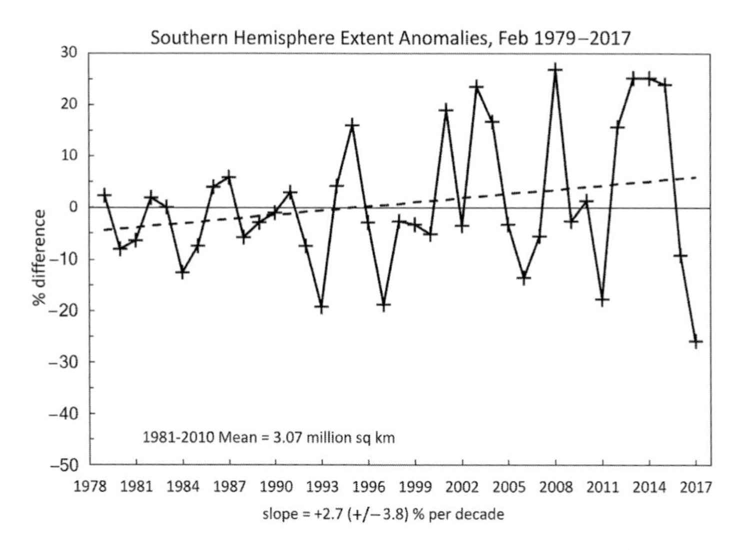

Figure 9.3b Time series of the Antarctic sea ice extent anomalies for February, the month with least sea ice in the Southern Hemisphere, expressed as a percentage of the February mean.
Source: National Snow and Ice Data Center (NSDIC), University of Colorado, Boulder, Fetterer et al., 2017.

The Weddell Sea polynya reappeared in 2017, suggesting climate variability with a time scale of over 40 years.

The total extent of Antarctic sea ice had been increasing slightly up until 2014, with reduced values since then, particularly in 2016. This long-term behavior is in marked contrast to the diminishing minima of sea ice extent in the Arctic (Figure 9.3a). Given the large interannual and spatial variability of Antarctic sea ice, as well as the uncertainties associated with changes in the satellite microwave sensors since 1979, the significance of this linear trend is difficult to interpret (Figure 9.3b). Some researchers have suggested that the observed increase in Antarctic sea ice extent may be related to stronger than average circumpolar westerlies and the resulting northward Ekman transport of sea ice.

9.6.4 Long-Term Climate Variability – Ice Cores

The annual cycle of snowfall creates bands that can provide a chronology in ice cores taken in glaciers and ice sheets. Changes in isotopic concentrations and other characteristics of the ice cores have been used as proxy estimates of interannual, decadal, and longer-term climate variations in temperature and precipitation, as well as other climatic properties. Dust particles in the cores are used to infer occurrence of volcanic eruptions (or to help fix the date of the core data if the eruption date is known) as well as large-scale fires. Extraction of the ice cores from the glaciers and their preservation require considerable effort, as does the subsequent analysis and interpretation of the core data. Uncertainties sometimes arise in fixing the exact dates of the data from the core as well as the extent to which the variations represent large scale, rather than local, conditions.

Comparisons of data from ice cores taken from different locations provide one way to address these uncertainties. Ice cores have been gathered at several high-latitude locations but also from high elevations at low latitudes. Cores from the latter have been used to document variability in monsoons and ENSO for periods spanning thousands of years (Thompson, 2004). Cores from Antarctica have the longest duration records, with some extending for more than 400,000 years; for example, the Vostok ice core record provides estimates of CO_2 concentrations for 420,000 years (Petit et al., 1999) that have been extended to the period between 650,000 and 800,000 years before the present (Luthi et al., 2008). Antarctic research expeditions have been continuing to extract cores with longer and longer records. An archive of ice core data can be searched at the World Data Center for Glaciology maintained by the National Snow and Ice Data Center (NSIDC), the National Ice Core Laboratory (NICL), and ice core repositories maintained by research institutions.

Ice core data have identified periods when Earth was much colder and others when it was much warmer than the current interglacial period. Given the recent warming discussed in Chapter 5, there is considerable concern that melting of land-based ice sheets will contribute to large increases in sea level associated with global warming.

9.6.5 Climate Variability – Northern Hemisphere Icebergs

Icebergs have been known as hazards to navigation since some of the earliest trans-Atlantic voyages, but it wasn't until after the Titanic's fatal collision with an iceberg on April 15, 1912 that the international community organized to monitor North Atlantic icebergs. Early observations were ship-based only but have since evolved into observations by ships, planes, and satellites that are analyzed and archived by the International Ice Patrol. The bulk of North Atlantic icebergs originate from calving of glaciers along Greenland's west coast. The icebergs enter the North Atlantic through the Davis Strait but generally by a circuitous route dictated by the ocean currents (Marko et al., 1994). Ocean currents along western Greenland initially transport icebergs formed in Baffin Bay northward before turning westward at the head of the Bay and then southward along the eastern shores of Baffin Island, entering the North Atlantic east of Labrador and then traveling south past Newfoundland (see the U.S. Coast Guard Navigation Center listed in the website References, Karlson et al., 2001).

The number of icebergs crossing 48°N latitude is taken as an index of iceberg activity. According to this index, iceberg average activity is essentially nil from September through January, rising sharply in February to a peak in April (with mean peak number of 130 icebergs/year), followed by a slow decline from May (with a mean of 120 icebergs/year) through August. The number of icebergs in the western North Atlantic exhibits considerable year-to-year variability. Studies have linked the number of icebergs in the spring to the sea ice extent in Baffin Bay in the previous winter (Marko, 1994). While one might expect that the number of icebergs will increase as global warming accelerates the melting rates of Greenland glaciers, such increases of North Atlantic icebergs have not been observed to this time. Ships are most likely to encounter icebergs near Newfoundland during the spring and early summer. The most profound impact of global warming on icebergs and shipping, however, may not be changes in icebergs in the North Atlantic but changes in commercial shipping lanes, as more ships are likely to move through Davis Strait into Baffin Bay on their way to a northwest passage across the Arctic Ocean. These routes will increase the length of time that ships will potentially encounter icebergs past the current springtime peak in the maximum iceberg season into summer and early fall.

9.7 Summary

The cryosphere is a fascinating part of the complex climate system and will challenge climate researchers for some time to come. A theme that runs through the topics discussed in this chapter is that continental and global scale cryospheric data have been available for climate research only since the late 1960s, when satellites started to provide regular observations. Since this record is marked by changes in the satellite orbital characteristics, instrumentation, spatial resolution, and frequency of observation, great care is needed in the construction of datasets covering the entire time span. A valuable source of information about freely available pre-satellite and satellite-era cryosphere data may be found in Pope et al. (2014). While the cryosphere is not the only element of the climate system to experience these limitations, its complex interactions with other components of the climate system make it especially challenging to understand given the relatively short period for which high-resolution data are available.

Suggested Further Reading

Barry, R. and T. Y. Gan, 2011: *The Global Cryosphere: Past, Present, Future*, Cambridge University Press, 472 pp, ISBN 978-0-521-15685-1. (An in-depth overview of the cryosphere).

The NOAA Arctic Theme Page (A resource for maps, animations and data sources.) www.arctic.noaa.gov/maps.html.

The State of the Cryosphere (SOTC) National Snow & Ice Data Center (This website contains regular updates of the latest cryospheric observations.) https://nsidc.org/cryosphere/sotc/.

Questions for Discussion

9.1 Since the cryosphere as discussed here consists of ice-covered ocean or snow or ice-covered land, what leads us to consider it as a separate component of the climate system?

9.2 Explain the ice-albedo feedback and discuss ways in which this feedback might affect the global climate. What datasets and analysis techniques could be used to document such effects?

9.3 Climate change studies suggest that the global snow cover and sea ice area are likely to decrease significantly in the coming century and longer. What are some of the likely consequences of these decreases to interactions within the climate system?

9.4 How could these changes be monitored?

10 Land Component of the Climate System

10.1 Introduction

Internal variations in the Earth's climate result from interactions among the atmosphere, the oceans, and the land. Most of this book concentrates on the ocean and atmosphere, but in this chapter, we focus on the land, the upper surface of the solid earth. Land covers roughly 29% of the Earth's surface and it is where people spend most of their lives. Its role in the climate, for the most part, is as the lower boundary of the atmosphere, and the lateral boundary of the oceans. The land exchanges heat, moisture, and momentum with the overlying atmosphere, and delivers water and suspended or dissolved materials to the oceans. The land surface is also a significant player in the radiation balance of the climate system through its interactions with radiation from the sun and the impact of its own thermal radiation upward to the atmosphere.

The land component of the climate system, comprising a very diverse collection of solid substances like soil and rock that are often mixed with water in varying amounts, behaves very differently from the atmosphere and oceans, the fluids that form the other two major components. These differences influence how we think about the role of the land surface as well as how we observe, describe, and measure the characteristics of the land in the context of climate. For some climate studies the continents, with their mountains and valleys, prairies and desserts, may be viewed as unchanging over the period of the analysis. On the other hand, variations in atmospheric circulation, cloud cover, and the amount and frequency of precipitation produce short-term variations in the land–atmosphere exchanges of heat, moisture, and momentum. Human activities have also altered, and continue to alter, the land surface. For example, large forested areas are often logged both to harvest the timber and to increase the amount of land suitable for agriculture. In other regions, cropland is paved or built upon to create or expand cities and suburbs. These and other human-produced alterations change the local climate in many parts of the world, and may even influence Earth's mean climate (Pielke et al., 2016). We discuss in this chapter the role of the land surface as a semi-permanent background that limits motions in the oceans and atmosphere, as well as a varying part of the climate system that is influenced by the atmosphere and oceans.

As discussed in Chapter 5, land influences climate on the longest time scales through the configuration of the continents and location of mountains. The land's interactions with the atmosphere and oceans also influence the climate on seasonal through decadal scales since the nature of the Earth's surface profoundly influences the local and regional moisture, heat, and energy exchanges. The intent of this chapter is to introduce the roles of the land surface in the climate system as well to describe land-surface observations and datasets. In the following section, we review how the shape and location of the continents and continental topography influence the mean climate of the Earth. Variable land features and their role in climate variability are discussed in Section 10.3. In Section 10.4 we introduce the land data used to describe topographic features and the general land types. Variable land surface features, such as soil moisture, vegetation, and land use, are presented in Section 10.5. The water and radiation budgets are presented in Section 10.6. The importance of land surface models in climate analysis and Land Data Assimilation Systems (LDAS) in climate models is discussed in Section 10.7, the varieties of drought in 10.8, and a summary is provided in 10.9.

10.2 Mean Land Features and Their Role in Climate

On the planetary scale the orientation and shape of the continents have a commanding influence on the Earth's mean climate by constraining ocean currents and influencing atmospheric flows. In the chapter on climate change we briefly discussed the influence of continental drift on the very early climate of the Earth. Our discussion in this chapter, however, is limited to time scales on which the current large-scale configuration of the continents and their physical geography is unchanging.

10.2.1 The Shape and Orientation of the Continents

The most significant influence of land on the climate system is, arguably, derived from the orientation and shape of the continents that divide the oceans into three major basins, the Atlantic, the Pacific, and the Indian. The continents influence the ocean currents within each of the major basins, as described in Chapter 8, and limit the regions where ocean currents in one basin may directly interact with those in other basins. The continents also determine where direct interactions between the atmosphere and the oceans may take place. At high northern latitudes, North America and Eurasia restrict the Arctic Ocean's direct interactions with the Atlantic and Pacific basins. Antarctica and South America limit direct communication between the Atlantic and Pacific Oceans to the Drake Passage, Africa separates the Indian Ocean from the Atlantic, and Indonesia and Australia isolate the Indian Ocean from the Pacific. The continental constraints on the ocean currents determine the nature of the global heat transport by the oceans. The configuration of the continents has a significant influence on the coherent patterns of climate variability discussed in Chapter 4, and to a large extent determines, for instance, why the climate exhibits ENSO variability in the Pacific Ocean basin but not in the Atlantic.

10.2.2 Topography

In addition to the physical barriers on oceans imposed by the continents, the topography over land has a direct influence on mean local climate. For instance, the decrease of surface air temperature with height, generally about 10°C/km or 5.5°F/1000 ft in the mean, results in relatively cool climates in mountainous regions even at tropical latitudes. The topography can also influence the surface climate by altering the amount of direct sunlight at a location, effectively altering the length of day. In addition, the topography may channel the flow of the winds or support local circulations (See Box 10.1). Topography is well known to have a strong impact on precipitation patterns, as well as in determining whether the precipitation will reach the ground as rain or snow. The influence of the terrain on precipitation is very noticeable in the climate of many locations on Earth, including the western coasts of the Americas. For example, Eugene, Oregon, on the windward side of the Cascade mountains receives an average of 50.9 inches (1,290 mm) of rain per year, while Bend, Oregon, at about the same latitude, but in the lee of the mountains, has an average of 11.2 inches (280 mm) per year.

On the larger scale, the locations and orientation of mountains often have a significant influence on the midlatitude pressure and wind fields. A common example of this is the

Box 10.1 Local Circulations

Mountain/Valley Breezes

It has long been known (Tower, 1903) that depending on the orientation of the mountain and valley with respect to the sun, local thermal effects can influence winds up and down the slopes of a mountain. During the day, the mountain slopes are warmed by the incoming solar radiation, making the air at those surfaces warmer than the air at the same altitude away from the slope. The warmer air rises and is replaced by air moving toward the slope from the surrounding free atmosphere, resulting in a net flow uphill. Conversely, at night, the surface cools more quickly than the free atmosphere, resulting in flow down the mountain slopes, a phenomenon known as "cold air drainage." These mountain/valley circulations are most clearly observed in the absence of any stronger large-scale circulation.

Land/Sea Breezes

Local wind circulations often occur in coastal regions, not due to any large change in the topography, but due to the different heat capacities of land and water. During the day, given the same heating from the sun, the land surface will warm more quickly than the adjacent water. The warmed air rises, and cooler, denser air over water flows onto the shore to replace it, generating a sea breeze. Conversely, at night, the land surface cools more quickly than the adjacent body of water. In this case, the cooler, denser air will be over land and flow seaward, generating a land breeze.

leeside trough. The Glossary of Meteorology (2012) defines the leeside trough as a pressure trough formed on the lee side of a mountain range in situations where the wind is blowing with a substantial component across the mountain ridge; often seen on United States weather maps east of the Rocky Mountains, and sometimes east of the Appalachians, where it is less pronounced. Likewise, large rainfall amounts often occur in tropical areas where the easterly trade winds impinge on steep terrain, as is the case in Central America and many islands of the Western Pacific Ocean. Such influences of topography on the atmosphere can be found on every continent.

10.2.3 Land Types

Other physical characteristics of land surfaces in addition to topography are interactive components of the climate system. For instance, the nature of the surface will influence the mean diurnal cycle of temperature and the portion of precipitation that will be stored locally, in addition to determining whether an area will support vegetation and, if so, what kind. The variety of such characteristics, sometimes referred to as land type or land cover, is enormous, and separating observations into a number of categories of relevance to climate analysis is very challenging. However, as discussed in Section 10.7, climate models and their data assimilation systems must make some estimate or implicit assumptions about the physical nature of the land over the area being modeled.

The term "Land Type" is used in different ways depending on the user community and data sources. Some classifications of Land Type include "Land Use" even though "Use" may change within the time scales of climate analyses. For example, the evolution from farmland to developed suburbs can easily occur within decades. The differences in terminology reflect the complexity in characterizing land surface as well as the variety of disciplines needing to specify Land Type. Land Type generally includes descriptors such as whether the land surface is primarily rock, soil, permafrost, or wetlands. Several research groups have adopted the convention of including the dominant forms of vegetation as part of their description of Land Type. The Integrated Climate Data Center (ICDC) website listed in the References, for instance, includes cultivated vegetation type even though the kinds of vegetation may vary over time, and the global trend toward urbanization may result in changes in Land Type as discussed in the section on Land Use. The term "Land Type" is also sometimes used to characterize the surface by its soil properties, such as soil depth, composition, porosity, and water-holding capacity. As with the descriptions of the topography, these properties are expected to remain unchanged for purposes of most climate analysis.

10.3 Variable Land Features and Climate Variability

While the surface and near-surface geology of the land remains essentially constant on the time scales on which we focus here, the superficial nature of the land surface may vary on shorter time scales. From the standpoint of climate analysis, the most important variables associated with the character of the surface are the moisture content of the soil, its vegetative cover, and its radiative properties.

There are many different definitions of soil moisture, but it is usually thought of as the amount of water held in the soils down to the rooting depth of vegetation. Soils vary considerably in their ability to retain moisture. Sandy soils quickly drain after rainfall, while soils rich in humus retain water for longer periods of time. Some soils, such as clays, are nearly impermeable and form a barrier to movement of water. The actual depth of the soil layer varies from location to location even over small areas but is often taken to be the uppermost 1–2 m. Some analyses refer to the field capacity of the soil as the amount of moisture in that 1–2 m layer, compared to the maximum amount of moisture that type of soil could hold. The difference results because some of the water from precipitation or irrigation percolates[1] through the upper soil layers to lower levels. Changes in soil moisture are determined by the balance of precipitation and irrigation compared to the evaporation, percolation, and runoff. Soils in locations with a high water table[2] tend to be moist, or perhaps even swampy, most of the time. Soil moisture is a critical component of the surface water and energy budget, but is a difficult quantity to measure directly.

Vegetation is another important natural land surface variable. In most parts of the Earth, vegetation exhibits an annual cycle that is tightly coupled to local temperature and precipitation, but vegetation is clearly an active, not passive, part of the climate system. Vegetation conveys moisture into the atmosphere by transpiration,[3] which provides a conduit of moisture from the rooting zone, beneath the surface, into the atmosphere, contributing to the total flux of moisture from the land surface into the atmosphere. The combination of evaporation of moisture from the land surface and the contribution from transpiration from plants is often referred to as evapotranspiration. Vegetation on land and in the ocean is also a central element in the carbon cycle (discussed in Chapter 1) through photosynthesis, in which carbon, in the form of carbon dioxide, is taken from the atmosphere and incorporated into the plant. The carbon cycle is a very important aspect of long-term climate variability, but the current observing system has been characterized as a "sparse, and exploratory framework" (Ciais et al., 2014). Useful large-scale observation-based datasets are not available at present, and we will not discuss carbon cycle datasets in this chapter.

The final element of land surface variability that impacts the Earth's climate system on time scales relevant to humans is the result of animal constructions. Examples include termite mounds, beaver dams, and bird nests, but by far the most conspicuous and significant actors in the regard are humans. Humans pave roads and runways, build houses and buildings, remove native vegetation such as trees and replace it with crops or structures, construct dams and create reservoirs, and dredge waterways or fill in coastal

[1] Percolation refers to the movement of water through a layer of soil, most often from the surface into the water table.

[2] The water table separates the saturated from the unsaturated soils (after the AMS Glossary).

[3] Transpiration is the process through which a plant absorbs water from the surface or soil through its roots and emit it from pores, called stomata, in its leaves.

shallows. All such actions take place on local scales, nearly always on land or coastal regions, but are certainly extensive enough on the whole to exert an impact on the climate. Well-studied examples include deforestation (Nobre et al., 1991; Tolle et al., 2017) and urbanization (Shepherd et al., 2013). Despite the importance of these impacts on the climate, comprehensive observations and datasets are only available to a limited extent, mainly in the form of land use/cover/type categorization, as discussed in this chapter. Time series of such data are scarce, and we will not discuss these aspects of the land surface in the same detail as vegetation and soil moisture.

10.4 Mean Land-Feature Data for Climate Analysis

As discussed in previous sections, the shape and orientation of the continents are taken as constant. One could argue, however, that as global warming continues it is likely that continental shorelines will change over a relatively short time, perhaps as rapidly as over a few years or decades. Sea-level rise will certainly cause significant changes in the areas directly affected and on the people who live within the changed areas. Even though the mean sea-level may be rising, however, we assume that the elevation of the surface above this changing mean sea-level will not be significantly different than those described below for the purposes of most climate studies.

10.4.1 Elevation/Topography

Most official weather observing sites document their exact geographical location including elevation as part of the metadata. Elevation information for other locations can often be obtained, either from local sources or by consulting contour maps. In the United States, for example, the U.S. Geological Survey (USGS) produces high resolution contour maps that can be purchased in hard copy or digital computer format. Many climate analyses cover large areas with varying topography. Detailed Digital Elevation Models (DEMs) based on satellite and aircraft LIDAR observations, Synthetic Aperture Radar interferometry, and stereo imagery provide elevation maps for most of the globe at horizontal resolutions of 1 km and less (Figure 10.1). The DEMs serve as the topography routinely used in global and regional climate analysis. Slope (change of elevation with horizontal distance) and aspect (orientation of the slope) have been computed from these observations and are available commercially. With the notable exception of volcanic eruptions, the DEM elevation estimates are not expected to change substantially for climate analyses spanning interannual, decadal, and centennial temporal scales. These elevation datasets are routinely updated to improve spatial resolution and coverage. The data displayed in DEMs are usually the mean, or representative, elevation over the grid area, and not the elevation for every point within the grid. The raw observations that make up the DEM are often at higher resolution than the final product, so many of the DEM datasets also include statistical estimates of variability, such as standard deviation of height, within the grid. Other statistics may also be available to give indications of the elevation variability within a grid (Hastings and Dunbar, 1999).

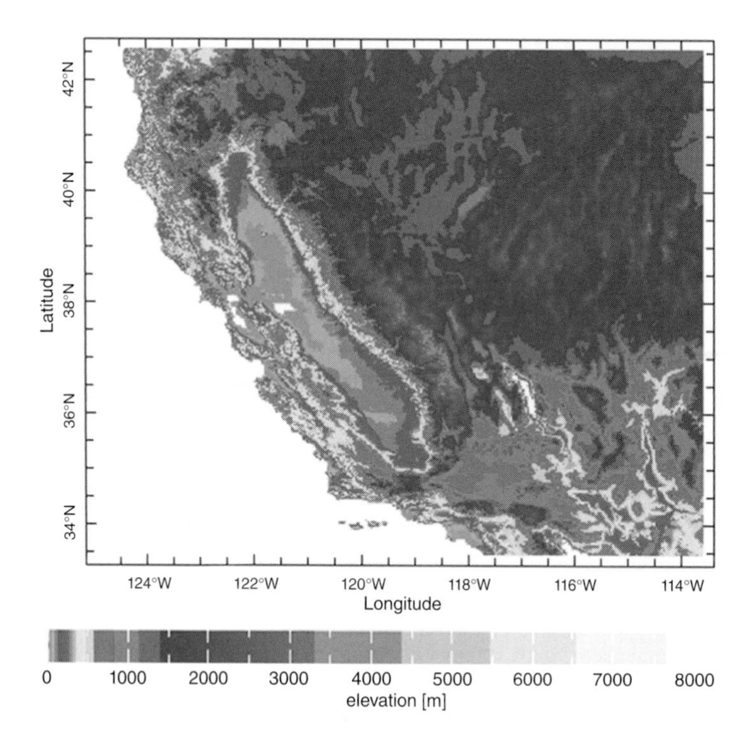

Figure 10.1 An example of a Digital Elevation Map (DEM) at 1 km resolution. This example illustrates the changes in topography from coastal California into the Central Valley and the Sierra Nevada Mountains. A black-and-white version of this figure appears in some formats. For the color version, please refer to the plate section.

Source: Hastings et al., 1999. Data downloaded from and plotted with the IRI/LDEO Data Library by author, CFR.

10.4.2 Mean Vegetation Data

One of the ways in which variations in the land surface can be described is according to the dominant kinds of vegetation growing in a region. Geographic atlases and websites typically display global maps that classify the land as prairies and grasslands, shrubland, tundra, deciduous and coniferous forests, tropical rainforests, and several other classifications (See, for example, the NASA Land Cover Types and the National Map both listed in the Website Reference List). Some atlases refer to these classes, and their subclasses, as the biome.[4] The typical vegetation growing in each biome is determined to a large extent by the climate and, in turn, influences the climate by modulating the heat, energy, and moisture budgets in the atmosphere above. Areas of naturally occurring vegetation may be altered by humans as discussed in Section 10.5.

[4] The term biome generally refers to the dominant type of vegetation in a region, but in some usage will also include all forms of life, including animals.

10.4.3 Mean Land-Type Data

There is no universal agreement on which soil characteristics should be included in an archive of global datasets. Thus, the data in various Land-Type datasets are often not easy to merge, or even compare. The International Geosphere-Biosphere Program (IGBP), however, has led efforts to digitize selected soil characteristics at a resolution of 5 by 5 arc minutes, roughly 1.6 km (Global Soil Data Task Group, 2000). The IGBP dataset includes seven soil characteristics (soil-carbon density, total nitrogen density, field capacity, wilting point, profile of available water capacity, thermal capacity, and bulk capacity). Other soil properties, such as composition, depth, and porosity, are not included in the IGBP global maps. Another source of global soil maps and digital data on soils is the U.S. Department of Agriculture National Resources Conservation Service. This dataset, however, differs from the soils data given above in that it separates soils into 12 categories based on a soil taxonomy (Soil Survey Staff, 1999) developed in conjunction with the United Nation's Food and Agriculture Organization (FAO).

In another dataset, the Integrated Climate Data Center (ICDC), referenced above, archives maps and digital data based on MODIS instruments onboard the NASA Aqua and Terra satellites. The MODIS Land Surface Types (Table 10.1) are based on the International Geosphere Biosphere Program (IGBP) definitions. Additional classifications based on Landsat (De Fries et al., 1998) and NOAA Advanced Very High Resolution Radiometer (AVHRR) data exist as well (Hansen et al., 2000) and are available through the Global Land Cover Facility (GLCF). The European Space Agency (ESA) Climate Change Initiative (CCI) Land Cover project, listed in the References under websites, also provides data and animations of a variety of satellite-based land cover products.

Table 10.1 MODIS Land Cover Types (IGPB convention)

Number	Type
0	Water
1	Evergreen needleleaf forest
2	Evergreen broadleaf forest
3	Deciduous needleleaf forest Nadelwald
4	Deciduous broadleaf forest
5	Mixed forest;
6	Closed shrublands
7	Open shrublands
8	Woody Savanna
9	Savanna
10	Grasslands
11	Permanent wetlands
12	Croplands
13	Urban and developed areas
14	Cropland /natural vegetation mosaic
15	Snow and Ice
16	Barren or sparsely vegetated

Table 10.2 Land Use/Cover categories in the National Land Cover Database (NLCD)

NLCD Class	Description/Type
1	Open Water
2	Perennial Ice/Snow
3	Developed Open Space
4	Developed Low Intensity
5	Developed Medium Intensity
6	Developed High Intensity
7	Barren Land
8	Deciduous Forest
9	Evergreen Forest
10	Mixed Forest
11	Shrub/Scrub
12	Grassland and Herbaceous
13	Hay/Pasture
14	Cultivated Crops
15	Woody Wetlands
16	Herbaceous Wetlands

Data on soil properties such as composition, depth, porosity, chemical composition, and water holding capacity are taken at some weather observing sites, as well as at specific locations, by government agencies and private enterprises with an interest in agriculture, water resources, urban planning, and other practical applications. The soil data taken at weather stations is not generally available through standard data exchange protocols. These and data from other observing sites might be obtained from the agencies involved in making the observations.

The land surface may also be characterized by land use or land cover type. The categories in this sort of classification, which may include urban, cultivated, pasture, and forest, for example, can change due to economic or political reasons that may not be directly related to climate, in addition to varying as a result of long-term climate change. High spatial resolution (30 m) national land use/land cover data for the United States are routinely updated by the U.S. Geological Survey in their National Land Cover Database (NLCD, 2006). These data describe land use/cover in 16 categories and are based on Landsat and other data (Table 10.2). The database also includes the percent of impervious surface and tree canopy cover. There are a number of such products available for different regions of the world.

10.5 Variable Land Data for Climate Analysis

This section briefly outlines the observations and data sources that characterize variability in the land surface component of the climate system. These data help to describe the mean annual cycle as well as interannual and other climate-related variability. While there is near universal agreement that land surface data are important for climate studies, the variables, units of measurement, instrumentation, and observing practices have not yet been standardized. Land surface observations are made for different reasons by different agencies. Thus land surface datasets tend to be heterogeneous, containing records for different variables or, even if for a common set of variables, estimated using different instruments and observing practices. This is true for the limited set of variables, soil moisture, vegetation and land use data that are discussed in the next sections.

10.5.1 Soil Moisture

One reason that soil moisture datasets differ from each other is that, historically, soil moisture has been measured for many different purposes using many different instruments, methods, and practices. The most common soil moisture datasets may be grouped into those based on direct measurements (gravimetric methods and lysimeters), and those based on indirect measurements (changes in soil conductivity or in soil dielectric properties, neutron flux, and gamma radiation measurements) discussed in the next subsections. More recently, there have been several attempts to use space-based instruments to obtain global estimates of soil moisture.

Direct Soil Moisture *In Situ* Measurements

It is difficult, if not impossible, to directly measure soil moisture without digging into the soil when first installing the instruments or as part of a regular soil moisture monitoring process. In the gravimetric technique, cores of soil samples are removed from the ground at representative sites, weighed, dried, and then weighed again to determine the water content initially in the soil. This method has also been used to provide data on the vertical profile of soil moisture by examining small sections of the core (Robock et al., 2000). Successive observations of soil moisture at the same exact observation site are not possible since the soil sample is removed and destroyed by the measurement process. Thus, this technique requires a relatively large area with homogeneous soil properties and drainage if a consistent long-record of soil moisture variations is desired. The gravimetric method has been used extensively in the former Soviet Union and in China, Tibet, India, and some Eastern European countries, and continues to have limited use today.

Lysimeters provide direct estimates of soil moisture without destroying the sample. According to the AMS Glossary of Meteorology (2012), Lysimeters are

> A type of evaporation gauge that consists of a tank, or pan, of soil placed in a field and manipulated so that the soil, water, thermal, and vegetative properties in the tank duplicate as closely as possible the properties of the surrounding area.

The soil container is designed so that water from precipitation or runoff may drain freely from the sample. Changes in soil moisture in the pan are estimated from the changes in weight of the contents in the pan. When used in conjunction with measurements of precipitation and temperature, lysimeters may also be used to estimate evaporation and potential evapotranspiration. Lysimeters are more common at observing sites designed for agricultural studies.

Indirect Methods of Estimating *In Situ* Soil Moisture

A number of instruments have been developed in the latter part of the twentieth century that can probe soil on a regular basis to indirectly estimate soil moisture. Several of these instruments estimate soil moisture by sensing changes in the soil's electrical conductivity or dielectric properties. The instrument sensors are placed into the soil at various depths in the top 1–2 m at the sampling site. The soil at the site is disturbed when the sensors are installed but left unchanged thereafter, unless the instruments cease to function. In the 1990s the U.S. Department of Agriculture Natural Resources Conservation Service (NRCS) implemented the Soil Climate Analysis Network (SCAN), consisting of sites at which several environmental parameters, including soil moisture, are monitored hourly. SCAN soil moisture estimates are made at depths of 5, 10, 20, 50, and 100 cm. SCAN has grown to over 200 agricultural locations with observation sites in most states.

Another common measurement technique uses neutron sensors (Visvalingam and Tandy, 1972). The neutron technique requires a neutron source and a detector. It is based on the fact that hydrogen atoms in soil water moderate the neutron flux through the soil volume observed by the sensor – the greater the percentage of water, the more the neutron flux is reduced. Each of the instruments using neutron sensors must be calibrated at each specific site since the absolute amount of water a soil can hold, its water holding capacity, will vary from soil to soil and location to location for a given soil type.

Data Sources for *In Situ* Soil Moisture

There are very few archives of soil moisture data that span long time periods. *In situ* soil moisture estimates, however, have been made by the Illinois State Water Survey (ISLS) since the early 1980s at 18 locations in the state (Hollinger and Isard, 1994; Scott et al., 2010). These data have often been used as benchmarks in comparisons with empirical estimates based on model output and satellite data.

Recent efforts have resulted in greatly improved access to soil moisture data. Data from SCAN, discussed in the previous subsection, are available online at the SCAN site listed in the Website References. Data from SCAN and 33 other networks in North America have been archived in the North American Soil Moisture Database (NASMD) that provides online access to data from some 1,800 observing sites, some starting as early as the 1990s (Quiring et al., 2016).

Archives have also been developed to provide access to global historical and real-time data from all available soil moisture observing sites. These include the Global Soil Moisture Data Bank (Robock et al., 2000), which has been merged with the International Soil Moisture Network (ISMN) and consists of data from 1,900 sites in 47 networks (Dorigo et al., 2011). This archive contains data from North America, Europe, Asia, and Australia and is available online.

Research quality soil moisture data are available from observations taken as part of the U. S. Department of Energy Atmospheric Radiation Measurement (ARM) program. ARM maintains three permanent instrumented sites located in the U.S. Great Plains, the Alaskan North Slope, and the West Pacific, as well as mobile and airborne facilities that have been deployed in a number of campaigns. Each site makes regular measurements of the components of the energy and water budgets including soil temperature and volumetric water with the Soil Water and Temperature System – SWATS. The ARM project is discussed further in Section 10.7.

Soil Moisture Estimates from Satellites

Each of the observing networks and programs described in the previous sections provide soil moisture data for varying periods of time for some parts of the globe, but there is no network of routine *in situ* observations that is capable of monitoring large-scale soil moisture globally. Given the importance of soil moisture for the water and energy budgets, considerable efforts have been made to develop satellite remote sensing techniques. In general, satellite estimates of soil moisture based on microwave sensors are limited to sampling the top 1–5 cm of soil. An example is the NOAA/NESDIS Soil Moisture Operational Products System (SMOPS). Satellite-derived estimates are limited in vegetated areas and by small-scale variations of land surface features. The European Space Agency (ESA) estimates soil moisture using different satellites and sensors that are likewise limited to directly estimating soil moisture in the top-most layers. Many of the satellite soil moisture estimates are used in conjunction with satellite thermal data and land surface models or with a Land Surface Data Assimilation System (LDAS), discussed in Section 10.7. Estimates of soil moisture obtained from satellites, land surface models, and *in situ* data often disagree (Reichle et al., 2004; Yang et al., 2013), and reconciliation of these data is an active area of research. The differences among these estimates of global soil moisture and *in situ* data, where they exist, suggest that caution should be used in their use in climate analyses.

10.5.2 Vegetation

Vegetation is an indicator of the mean climate of a region, but temporal variation in vegetation is also an indicator and an active participant in climate variations, both through its immediate impacts on the radiation balance, the water cycle, and the exchanges of heat, momentum, and moisture between the surface and the atmosphere, and via its role in the carbon cycle. Vegetation, especially crops and grasslands, has been monitored at the field or farm level since humans first started to cultivate land. Farmers, even in ancient times, had at least a subjective sense of whether their crops matured early or late compared to previous years, as well as a sense of the relative quantity of the yield. As civilization became organized, pharaohs, mandarins, khans, and kings kept records of the amount of land under cultivation and crop yields as a basis for assessing taxes. Records such as these have been used in conjunction with other data to "reconstruct" climate variability for several centuries into the past (Ge et al., 2005).

National governments and private industry continue to monitor the progress of various crops though their growing seasons as well as forecasting estimates of crop yields. Actual yields at the ends of the growing seasons are also recorded. The U.S. Department of Agriculture, in conjunction with U.S. Department of Commerce, publishes weekly reports on the progress of crops for major crop growing regions of the world in the Weekly Weather and Crop Bulletin (USDA and USDoC, 2014).

Since the early 1970s considerable effort has been expended in developing the use of remote sensing to monitor vegetation, primarily for agricultural applications, including estimates of crop yield. The Normalized Difference Vegetation Index (NDVI) and its variants were developed in the late 1970s and produced operationally by the NOAA Environmental Satellite, Data, and Information Service (NESDIS) since April 1982. The NDVI thus provides one of the longest satellite-based records available for climate research. The index is computed from the differences between the observations in the visible and near-infrared portions of the spectrum normalized by their sum (Tucker, 1979; Tarpley et al., 1982). It is well documented that the NDVI is sensitive to the "greenness" of vegetation as it evolves from emergence to maturity, but uncertainty continues to remain as to what the index actually senses. Photosynthetically Active Radiation (PAR),[5] Leaf Area Index (LAI),[6] and plant chlorophyll have all been related to the NDVI in various studies. Attempts to relate the NDVI to net primary production, or crop yield, have led to mixed results in part because of the fundamental differences in crop yields from a particular plot of land at spatial scales of hundreds of meters, at best, and NDVI estimates at 8 km (based on 1 km raw data). The use of the NDVI historical record for developing crop yield estimation methods is also limited by uncertainties in the data.

NDVI data have been used in several climate studies, but analysts have had to deal with artifacts in the historical record of these data arising from many different sources. Contributing to uncertainties in the data are slow drifts in the equator-crossing times of the earlier sun synchronous satellites carrying the Advanced Very-High Resolution Radiometer (AVHRR), data from which is often used to calculate NDVI. Uncertainties arise in the data because the NDVI is sensitive to the geometry of satellite viewing angle and the sun (solar zenith angle). The NDVI value for a location on Earth will change due to satellite drift in way that are not easily quantifiable. The more recent NOAA polar orbiting satellites minimize orbital drifts, resulting in more stable estimates of NDVI. The AVHRR is also sensitive to cloud cover and atmospheric humidity. The NDVI attempts to minimize the effects of cloud cover and moisture by selecting the largest NDVI value observed at a location over a 7- to 10-day period. In the historical record, 8 km NDVI values were sub-sampled at 1 km so that successive NDVI 8 km values may come from different 1 km pixels. Another source of uncertainty, or noise, in the measurement is that the AVHRR instrument scans an area at angles to either side of the sub-satellite point. This leads to differences in NDVI values between the forward and backward scan of the instrument, again because of the sun/satellite geometry.

[5] PAR refers to the part of the spectrum that plants use in photosynthesis.

[6] The leaf area subtended per unit area of land (Glossary of Meteorology, 2012).

The NDVI is also sensitive to variations in atmospheric aerosols, especially large volcanic eruptions. Finally, the AVHRR instrument itself varies from satellite to satellite. Thus, slightly different parts of the visible and near-infrared spectrum go into the NDVI calculations from different satellites. The NOAA/NASA Pathfinder program (Maiden and Greco, 1994) attempted to address all these sources of uncertainty in the production of the AVHRR Pathfinder Land dataset. As with most satellite-derived data, the longest NDVI records are at lower spatial resolutions and coarser spatial sampling than those from more recent satellites.

Extension and enhancement of the NDVI datasets is made possible by estimates from the NASA MODIS instruments, which are available with spatial resolutions of 1 km to 250 m at 16-day intervals beginning in 2000. The Visible Infrared Imaging Radiometer Suite (VIIRS) aboard the Suomi NASA/NOAA National Polar-orbiting Partnership satellite launched in late 2011 and the continuing series of operational NOAA meteorological satellites beginning with NOAA-20, launched in late 2017, will provide high resolution Vegetation Index and estimates of other land surface products. The Land Long-Term Data Record, listed in references under websites, provides a global land surface climate data record that includes NDVI, and other datasets. NDVI datasets are available through the NOAA, USGS, and NASA archive centers listed under Normalized Difference Vegetation Index data in the website references.

10.5.3 Land Surface Radiation Measurements and Data

The climate system is, to a significant degree, the result of the balance between incoming solar radiation and the net outgoing radiation from the Earth. The interactions among the incoming solar radiation, the land surface, subsurface, clouds, and the composition of the atmosphere are complex. Not all of the radiation measurements required for a complete analysis of the radiation balance are routinely made at most operational weather observing sites. The World Meteorological Organization (WMO) has designated 23 meteorological sites as "World, National and Regional Radiation Centers" that are charged with the calibration, maintenance, and operation of surface radiation measurements (WMO, 2008). Instruments used to measure the components of the radiation budgets include pyroheliometers that measure direct solar radiation, pyrometers for the measurement of direct, diffuse, and reflected solar radiation, and pyrgeometers that measure upward and downward longwave (infrared) radiation. In addition, the WMO established the Baseline Surface Radiation Network (BSRN) of about 50 stations as part of the World Climate Research Programme (WCRP) (Ohmura et al., 1998). In the United States, the National Atmospheric and Oceanic Administration (NOAA) maintains the Integrated Surface Irradiance Study (ISIS) network that includes BSRN stations as well as the Surface Radiation Network (SURFRAD). Data from the BSRN are available from the World Radiation Monitoring Center website listed in the References. The ISIS datasets can be accessed through the link listed in the References under websites.

In addition to the international networks discussed earlier, the Atmospheric Radiation Measurement (ARM) Program sponsored by the U.S. Department of Energy maintains three permanent sites, mentioned in the discussion of soil moisture earlier in this chapter, where routine measurements of all relevant components of the surface radiation budget are made (Mather and Voyles, 2013). The ARM Program also maintains the ARM Mobile Facilities (AMF). These are instrumented facilities that remain in place for shorter observing periods at a dozen or more diverse sites around the world. These sites have included locations ranging from the Ganges Valley in India to Graciosa Island in the Azores (Mather and Voyles, 2013). A listing of AMF facilities may be found at the ARM website listed in the references.

In addition to the limited number of surface sites with radiation observations, discussed earlier in the chapter, the number of hours of observed daily sunshine had been recorded at meteorological observing sites since the late nineteenth century. These data have traditionally been used in climate studies that do not require the full surface radiation budget. If there are no clouds, fog, air pollution, atmospheric aerosols, or shading by topography and manmade structures, the potential hours of sunshine at all locations on Earth can be calculated simply from the Earth–sun geometry. Sunshine Tables can be easily found on the Internet. As a practical matter, the atmosphere does include all these factors so that the simple geometric computation can provide only the maximum possible number of hours of sunshine for a given latitude. An evolving variety of instruments have been used to record the observed hours of sunshine since the late nineteenth century, and the United States Weather Bureau (now the National Weather service) has historically used different instrumentation than most other nations (WMO, 2005). The difference in instrumentation has resulted in a mean bias between sunshine duration records in the United States and those from much of the rest of the world. The accumulated bias has been estimated to be as much as 300 hours a year in some studies. Within the United States itself, an analysis of records from 1891 through 1987 indicates that changes in the instruments used by the United States during that period did not introduce significant biases in annual mean sunshine duration (Stanhill and Cohen, 2005). They also found that the sunshine duration records were highly correlated with pyranometer data during a four-year comparison period from 1977 to 1980. The term "hours of sunshine" is ambiguous at dawn and dusk, when the sun angle is low. The WMO has adopted a standard of 120 w/m^2 as the minimum brightness for use in defining sunshine duration (WMO, 2005), and has suggested that recording instruments be calibrated to that standard. Since the introduction of automated weather observations at many meteorological observation sites, sunshine duration is not observed directly but is derived from the observations of net solar radiation.

10.6 Land Surface in Budget Studies

Budget studies in climate, briefly introduced in Chapter 2, refer to analyses that examine the sources and sinks, inflows and outflows, and accumulations and depletions of the mass, energy, moisture, and other properties of the climate system within a geographic area. The area may be as small as a river basin or as large as the Earth itself. Budgets are a powerful tool to provide estimates of quantities that are poorly observed, or not observed at all. They are also useful in helping to assess shortcomings in observations and models. In general, land surface and other data are not sufficient to support budget studies over small areas and short time periods. Routine observations can lend themselves to analyses on spatial scales on the order of 10^3 km and greater and time scales of a month and longer. Budget studies relating to the hydrologic cycle and Earth's radiation balance are discussed in the next sections.

10.6.1 Water Budgets

Water budget studies are useful to further understand the hydrologic cycle by examining the interplay between the moisture in the atmosphere and water on land, including the subsurface. The water budget over a sufficiently large region can be studied by the analysis of the precipitation, evaporation, runoff, and storage (including soil moisture and the depth of the water table). For instance, the difference between precipitation and evaporation, less the amount of water that flows into sewers, rivers and streams, should be reflected in the change of groundwater storage over an area over some period. The average precipitation in an area is often estimated by rain gage or satellite data, while the runoff is estimated from streamflow gage data. If soil moisture and groundwater data are also available, then the evaporation may be calculated (See Appendix D for details). Conversely, if evaporation estimates are available from models, or other sources, then the change in the storage can be estimated. In some studies, the change in storage over an averaging period is assumed to be zero. This assumption is likely to be less valid for long period studies under continuing climate change.

The difference between the precipitation and the evaporation, without measuring either directly, can be calculated independently from radiosonde, or model, estimates of the winds and moisture through the depth of the atmosphere. In this computation, the net moisture convergence or divergence in the atmosphere over some area and averaging time is equal to the difference between precipitation and evaporation (Appendix D). It is common in water budget studies that observations of one of the components of the budget, often evaporation or storage, are not available. In that case, the value of the unmeasured budget component is estimated as a residual in the balance among the measured components in the budget.

The components of the water budgets described in this section are all sensitive to instrument or model biases and uncertainties. In general, the shorter the averaging period and smaller the area, the larger are the relative uncertainties. Water budgets are generally considered useful for averaging periods over a month or longer and for areas 10^6 km^2 or greater. For instance, the mean annual cycle of the water budget for the United States has been estimated from standard meteorological and hydrologic observations (Rasmusson, 1967, 1968) and from numerical models by Roads et al. (1994). Ropelewski and Yarosh (1998) illustrate the annual cycle of the water budget over the Mississippi River drainage basin over the period 1973–1992. Their computations confirm earlier analyses and illustrate some of the fundamental aspects of the mean annual cycle in water budget quantities in the central United States (Figure 10.2). These studies and others document the relationships among key variables in the climate

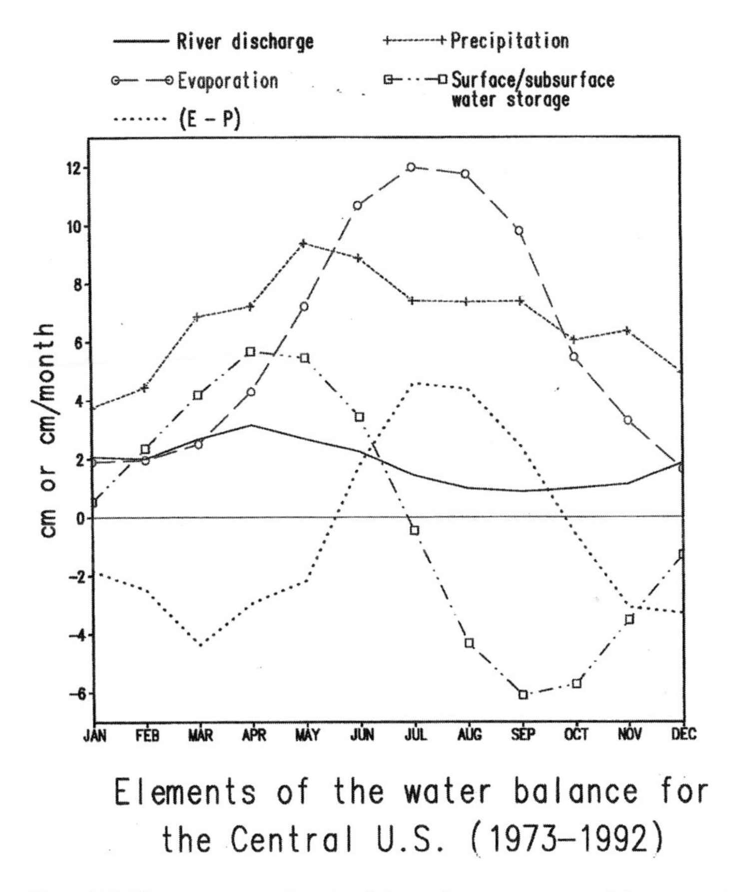

Figure 10.2 The mean annual cycle of the major components of the atmospheric and terrestrial water budgets for the central United States (1973–1992). After Ropelewski and Yarosh, 1998.

system, such as the relative magnitudes of evaporation, precipitation, discharge, and groundwater storage, and serve as benchmarks to support model improvements. The mean annual cycle of the water budget has also been estimated based on numerical model analyses and data for the globe (Trenberth and Guillemot, 1995).

None of the water budget quantities are completely observable everywhere by any one system, due in large part to the extreme heterogeneity of the land surface. Lacking the ability to observe the land surface on the scales on which it varies, climate analysis often treats measurements at specific locations as indices of climate processes on larger scales. From a global analysis standpoint, satellite observations of quantities related to surface albedo, soil moisture, and temperature are used in conjunction with numerical models of land surface parameters and provide the most promising strategy for obtaining realistic routine estimates of budget quantities. Changes in groundwater storage over large geographical areas have been estimated in the Gravity Recovery and Climate Experiment (GRACE) mission (Rodell et al., 2007). Analysis strategies that integrate information from satellites and numerical models that include details associated with the land surface, discussed in Section 10.7, can provide proxies for observations. A major challenge concerning the land surface parameters is that the heterogeneous features of the land surface are on much smaller spatial scales than satellite observations or model grid sizes.

10.6.2 Land Surface Radiation Budgets

The mean climate and its variations at any location, as well as when averaged for the entire Earth, are strongly influenced by the amount of shortwave radiation from sunlight that reaches the surface. Part of that sunlight is reflected by the surface, while the rest is absorbed and partially re-radiated at longer wavelengths back into the atmosphere. As discussed in Chapter 1, the fraction of the incoming visible solar radiation that is reflected into space is called the albedo. The mean albedo of the Earth is estimated to be near 0.30, that is, on average about 30% of the light reaching the Earth from the sun is reflected back into space. Local values of albedo vary considerably in space and time depending on a number of factors, including cloud type and amount, soil type, condition of the soil including wetness, and vegetation, snow, and ice cover. For example, the mean albedo of snow is about 0.8, but the albedo is higher for fresh snow and decreases as the snow ages. Clouds intercept the incoming solar radiation before it reaches the ground and may also have albedos of 0.8 or more (Peixoto and Oort, 1992). (See Chapter 3 for more information on observations and datasets concerning the radiation budget of the Earth measured at the top of the atmosphere.)

Many factors influence the surface radiation budget in addition to the albedo. The solar radiation that is not reflected will heat the surface. While some of that heat will be conducted downward into deeper soil levels, much of it will be re-radiated back into the atmosphere but at longer wavelengths than the incoming solar radiation. The frequency and wavelength of radiation emitted by a body depends on the temperature of the body emitting the radiation. The incoming solar radiation has a spectrum that peaks in visible wavelengths, consistent with the temperature of the sun, while the outgoing radiation from the much cooler Earth peaks in the longer wavelengths of the infrared portion of the spectrum. Some of the upward infrared radiation from the Earth's surface will be absorbed by water vapor in the atmosphere or reflected by clouds back toward the surface, as illustrated in Figure 10.3. The surface radiation budget is extremely complex and its diagnosis requires consideration not only of the land surface but also of the cloud cover, cloud type, and moisture in the atmosphere (Peixoto and Oort, 1992).

It has been long proposed that some elements of the surface radiation budget may be estimated from satellite observations (Pinker and Corio, 1984), and considerable efforts have been expended in the Global Energy and Water Exchanges (GEWEX) Project and related activities. All satellite-based analyses of the components of the surface radiation budget are, of course, limited by the mismatch in spatial scale between measurements at a discrete observation point and spatial averages from satellite observations. The satellite estimates, nonetheless, have provided estimates of large-scale averages of the net surface shortwave and longwave radiation for limited observation periods (Smith et al., 2002) and have been extensively used in model intercomparison studies (Garratt, J. R. et al., 1998).

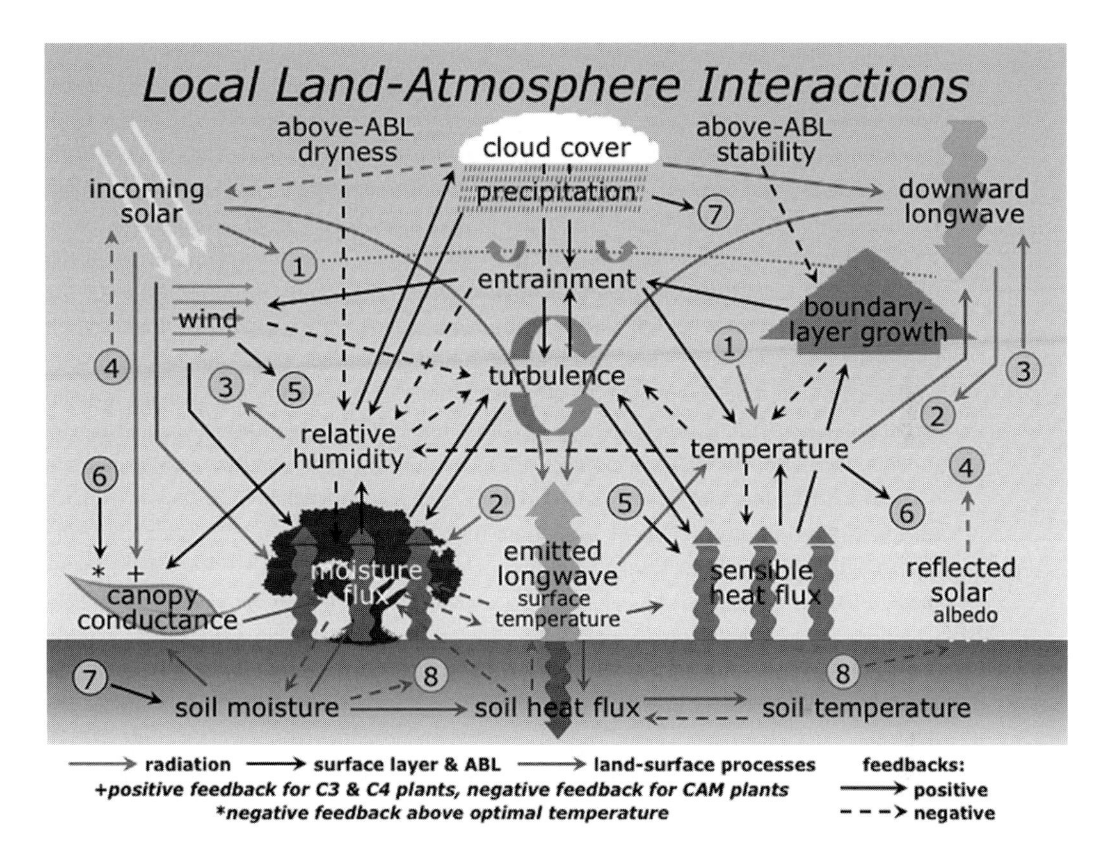

Figure 10.3 This schematic shows the components of the water and radiation budgets typically parameterized or estimated in Land Surface Models (LSMs).) The figure indicates local land-atmosphere interactions for a relatively undisturbed atmosphere, including the soil moisture and precipitation feedbacks. The solid arrows indicate a positive feedback, and large dashed arrows represent a negative feedback, while red arrows indicate radiative processes, black arrows indicate processes near the surface and lower atmosphere (termed the planetary boundary layer – PBL), and brown arrows indicate land surface processes. Thin red and grey dashed lines with arrows represent also represent positive feedbacks. The single horizontal gray-dotted line, with no arrows, indicates the top of the PBL, and the seven small vertical dashed lines, with no arrows, represent precipitation. A black-and-white version of this figure appears in some formats. For the color version, please refer to the plate section.
Source: Mike Ek, adapted from Ek and Holtslag, 2004, *J. Hydrometeorol.*, **5**, 86–99.

10.7 Land Surface Models and Data Assimilation Systems

A complete analysis of the global climate system requires high spatial and temporal resolution estimates of interactions between the atmosphere and land surface. As noted previously, however, there is currently no global observing network that can provide observations and data at the necessary spatial and temporal resolution. One strategy for obtaining such estimates is to use Land Surface Models (LSMs; Figure 10.3) in conjunction with coupled ocean/atmosphere models. The LSMs assimilate data from observing networks and from initial conditions derived from global atmospheric models by Land Data Assimilation Systems (LDAS; Rodell, 2004). In the United States, two land surface models, the Noah[7] LSM and the Variable Infiltration Capacity (VIC) LSM, have been used extensively in weather and climate models and are the subjects of continuing efforts to improve them. The Noah LSM has its roots in modeling efforts first developed at Oregon State University (Marht and Pan, 1984) and was later modified for use in operational weather forecast models (Ek el al., 2003). The Noah model includes several subsurface layers in addition to including incoming solar and outgoing surface radiation, vegetation, evapotranspiration, and all components of the surface water budget. The VIC LSM (Liang et al., 1994) likewise includes the radiation and water budgets, vegetation, evapotranspiration, and has several subsurface layers. Both models attempt to replicate the physical processes influencing the land surface and land-atmosphere interactions but differ in the details and the assumptions going into the models, the parameterizations, numerical methods, and implementation strategy.

Field experiments that have been conducted over recent decades have had as a major goal to support improvements in LDAS and LSMs. These field experiments include the Hydrological-Atmospheric Pilot Experiment (HAPEX), the First ISCCP Regional Experiment (FIRE), the First ISLSCP Field Experiment (FIFE), the Global Land-Atmosphere Coupling Experiment (GLACE), and the Boreal Ecosystem-Atmosphere Study (BOREAS). The Global Energy and Water Exchanges (GEWEX) program also conducted a number of regional observational field programs to better understand and model the hydrologic cycle and energy budgets on regional and continental scales (Lawford, 2004).

When coupled with atmosphere–ocean models driven by observations, the LSMs replicate the major characteristics in the mean annual cycle of the water budget but have not, as yet, shown major improvements over simpler empirical models (Fan et al., 2011). Outputs from the Noah and VIC models are available through the NASA Global Change Master Directory.

10.8 Drought

Throughout human history, civilizations have adapted to a great variety of regional climates. All civilizations depend, however, on some reliable source of water, either from local precipitation or from remote precipitation that is brought where needed by rivers, canals, or aqueducts. The amount of precipitation, either locally or in the source region for imported water, will determine the activities that the civilization can sustain. A drought is said to occur when precipitation falls below the minimum required to support the basic needs of the civilization. We discuss drought in this chapter because droughts are most evident in their impact on the land surface properties, including soil moisture and vegetation as well as other aspects of climate discussed here.

Because of its great economic and social influence, a number of methods have been developed to monitor drought, reflecting the varying ways that it may be defined (see

[7] Noah was named for the four groups that collaborated in its development: the NCEP Environmental Modeling Center, Oregon State University Department of Atmospheric Sciences, the Air Force, and the NWS Hydrology Lab.

Box 10.2 Different Types of Drought

Questions such as "When did the drought begin?" and "How long or severe was it?" are difficult to address without some idea of their context. The answer may vary depending on whether the questioner is focused on knowing the statistics of precipitation, the impact on agriculture, the implications for water resources management, or broader socioeconomic impacts. One way to differentiate among the different kinds of drought is provided in a guidance statement published by the American Meteorological Society (AMS, 2013). The statement suggests four different kinds of drought, (1) meteorological (or climatological) drought, (2) agricultural drought, (3) hydrological drought, and (4) socioeconomic drought. While these drought types are often not mutually exclusive they provide a framework for discussions and studies of drought and drought indices. *Meteorological drought* refers to statistics of rainfall. Descriptors of meteorological drought may be maps of departures of rainfall amounts from a long-term mean, or time series of accumulated rainfall at a location over some time span, or other statistics. *Agricultural drought* is not only concerned with the amount of precipitation but also the influence of evaporation and other losses of moisture to the rooting depth of crops. *Hydrological drought* refers to precipitation shortfalls and increased evaporation, as well as increases in other losses to surface and ground water. For instance, hydrological drought may be indicated by reduced water levels in reservoirs, falling water tables, or other measures of water availability, and is related to agricultural drought through increased demands for irrigation. *Socioeconomic drought* is less concerned with the physical definitions of precipitation and increased water loss than it is with how meteorological, agricultural, and hydrologic droughts influence the economy and social activities. There are no simple metrics that link droughts to their socioeconomic impacts. These impacts are often influenced by factors outside of the physical characteristics of a particular drought.

Box 10.2). Very often droughts are described by an index that is based on selected aspects of the phenomenon. Some of the indices widely used in climate analysis are discussed here. No individual index, by itself, can characterize all the different kinds of drought. Given the lack of global high-resolution soil moisture and other land surface observations, many climate analyses have relied on drought indices based on empirical estimates, numerical models, or combinations of both. Consensus on an appropriate index for monitoring drought has often been difficult to achieve because different users have different concerns. It is often not clear to the users which of the types of drought an index was designed to monitor, or what the criteria are for declaring drought conditions.

Several indices are used to monitor droughts in the United States (Heim, 2002). Meteorological drought is defined solely by precipitation deficits, irrespective of other conditions. One common index of meteorological drought is the weekly or monthly departure of rainfall *amount* from average, or median, precipitation over the same period. A related index might be the *percentage* of the mean, or median, weekly or monthly rainfall for the same period. The first is an example of a quantitative drought measure, the second an example of a relative measure. Each has an advantage but both are needed to get a full assessment of precipitation over the period of interest. For instance, the significance of a 10 mm precipitation deficit for a region where the average rainfall over the period of interest is 25 mm, or 40% of the average, is much greater than a 10 mm precipitation deficit in a region where the average rainfall is 100 mm, since that is only 10% of the average. In general, characterization of meteorological drought requires both quantitative and relative measures of precipitation (see Figures 12.5a and 12.5b). There are other relative measures of precipitation deficit that may be used to provide context. Many drought indices assume that the overall distribution of precipitation is Gaussian even though this is not usually the case for precipitation or precipitation anomalies. These include indices that scale the precipitation anomaly by the standard deviation of precipitation, for example, the Standardized Precipitation Index (SPI; McKee et al., 1993; Hayes et al., 1999), the Weighted Anomaly of Standardized Precipitation (WASP) index (Lyon, 2004; Lyon and Barnston 2005), as well as a host of others.

At most locations, it does not rain every day and extremely large precipitation amounts are rare so that the frequency distributions of daily precipitation usually exhibit the largest values at low amounts. Such a distribution is skewed (see Appendix A), and this behavior is observed in the distributions of weekly, monthly, and seasonal precipitation. Some precipitation analyses assume skewed distributions to routinely monitor precipitation and meteorological drought. For example, the NOAA Climate Prediction Center often uses percentiles of a two-parameter gamma distribution (illustrated earlier in Figure 2.3). Other skewed distributions used in practice include the three-parameter gamma, mixed exponential, Weibull, Gumble, log-normal, and others described in statistical texts (for example, Wilks, 2011). Each of these distributions has advantages and disadvantages but have in common that each requires a sufficiently long historical record of rainfall data, generally 30 years or longer, to obtain stable estimates of the parameters of the distributions. None of the theoretical distributions fit the histograms of observed precipitation for all locations, observation periods, or climates.

Statistical characterizations of precipitation and meteorological drought are particularly difficult in extremely dry and seasonally dry regions such as the African Sahel. Mediterranean climates, having an annual cycle with pronounced wet and dry seasons, pose difficulties because the median precipitation value of daily rainfall during the dry season is often zero or near zero. In these regions and seasons, the statistics of rainfall are often computed using data only from days with rain, and subsequent interpretation must be made in that context. If a fitted distribution is necessary, it is advisable to plot the empirical frequency distributions of the precipitation observations before deciding which theoretical distribution to use.

Agricultural drought is more difficult to monitor than meteorological drought since it requires estimates of evaporation and of soil moisture down to the rooting depth of crops. In the United States, the Palmer Drought Severity Index (PDSI; Palmer, 1965) has been widely used to monitor the status of agricultural droughts. The PDSI attempts to characterize drought through a simple empirical water balance model, driven by meteorological observations of precipitation and temperature and with assumptions about the character of the soil and potential evapotranspiration. The limitations and weaknesses of the PDSI have been well documented (Alley, 1984). However, despite its assumptions and empirical estimates, the PDSI is relatively easy to compute given standard meteorological observations and continues to enjoy wide, if sometimes grudging, use. The PDSI, sometimes with variations, has been used to characterize drought in the United States for half a century and has been adapted to characterize historical droughts for the entire globe (Dai et al., 2004). The PDSI values are dimensionless and generally range between plus and minus 4. Weekly updates of the PDSI are produced by NOAA's Climate Prediction Center and are archived there and with other drought indices at the NOAA National Centers for Environmental Information (NCEI).

Among the limitations of the PDSI is that it is relatively slow to identify changes in agricultural drought status. This is a serious limitation for monitoring drought conditions during the growing season. The Crop Moisture Index (CMI) was developed to augment the PDSI (Palmer, 1968). It is based on precipitation, weekly mean temperature, and the previous value of the CMI. It is more responsive to short-term climate variability than the PDSI and thus can monitor short-term conditions to which crops may respond even though they may occur within a period of longer term drought or excessive rainfall. In addition to the drought indices discussed in this section, attempts to provide estimates of soil moisture have been made using simple empirical "leaky bucket" models (Huang et al., 1996; Fan and van den Dool, 2004). These models are analogous to the PDSI in that they attempt to make estimates of soil moisture through empirical modeling of various components of the water budget based on assumptions developed and calibrated at a small number of instrumented locations (sometimes only one site) and driven by observed precipitation and temperature. The analyses based on these empirical models are particularly useful in predictability studies of monthly and seasonal drought, for estimating the role of soil moisture in formation of climate anomalies, and for comparison with more sophisticated numerical models (Fan et al., 2011).

Hydrologic drought focuses on the water supply and thus the status of groundwater and reservoir levels. Hydrologic drought is monitored through observations of streamflow, lake and reservoir levels, and variations in groundwater including soil moisture. *In situ* observations of soil moisture are challenging as discussed earlier. Observations of streamflow and lake and reservoir levels are easier to make, but consistency of methodology and availability across national borders can be challenging. Within the past few decades, data derived from satellite-based altimeter observations have been used to monitor lake and reservoir levels with some success. While the datasets derived from these efforts (Birkett et al., 2010) are limited in their length of record, they provide a unique global perspective on some aspects of the surface water cycle and hydrologic drought (see the Global Reservoir and Lake Monitor website listed in the References for examples).

Variations in these quantities generally occur more slowly than variations in precipitation and evaporation so that hydrologic drought has an inherently longer time scale than either meteorological or agricultural drought. This leads to situations in which an area may be experiencing one kind of drought but not the others. Hydrologic drought is often discussed for areas defined by the drainage basins of rivers or combinations of several drainage basins where streamflows are managed by several reservoirs. In the United States, hydrologic drought is monitored by several agencies including the U.S. Geological Survey (USGS).

Socioeconomic drought is the most difficult to characterize objectively. Discussions of socioeconomic drought may be found in the US Drought Monitor. Socioeconomic drought may be defined as the "Impact of meteorological, agricultural, or hydrologic drought on the supply and demand of economic goods." (After the US Drought Monitor definition, NOAA, 2012). Estimates of the impact of socioeconomic drought require not only data describing the physical climate system, but also socioeconomic data and analyses. One common socioeconomic impact of droughts occurs when municipalities or other entities declare drought emergencies, restricting or completely banning water use for certain purposes. Socioeconomic aspects of drought may also influence, and be influenced, by policy decisions (Kohl and Knox, 2016).

10.9 Summary

In this chapter, we have seen that the land surface directly influences many aspects of the climate system. The nature of the underlying surface strongly modulates the transfer of heat, moisture, and momentum at the interface between the atmosphere and the land. In addition, all components of the surface radiation budget are strongly influenced by the nature of the land surface. Even as we become able to observe the land surface and climate on finer and finer spatial scales, it is unlikely that global and regional climate analysis systems and models will ever match the smallest of these scales. Thus, we are limited to the observations, analysis, and modeling of the spatially and temporally aggregated land surface quantities in the analysis of the role of the land component of the climate system.

The diverse topics discussed in this chapter reflect the complexity of land surface climate processes and point to the number of related, but different, disciplines that are involved in various aspects of the study of these land surface processes. In searches for data and in scanning the literature we find a variety of different variable names and nomenclature for the same or related phenomena as well as different physical units of measurement. These differences are bound to diminish as the roles of land surface in climate variability continue to become more integrated into studies of the climate system.

Suggested Further Reading

Heim, R. R., 2002: A Review of Twentieth-Century Drought Indices Used in the United States. *Bulletin Of the American Meteorological Society*, **83**, 1149–1166.

van den Hurk, B., M. Best, P. Dirmeyer, A. Pitman, J. Polcher, and J. Santanello, 2011: Acceleration of Land Surface Model Development over a Decade of Glass. *Bulletin of the American Meteorological Society*, 92, 1593–1600. (This technical article provides insight into the complexity of land surface modeling and attempts to address outstanding issues.)

Questions for Discussion

10.1 Explain what microclimates are and explain the role of land surface properties in creating and sustaining them.

10.2 What are the principal roles of the land surface in the regional and global climate system? What physical quantities must be measured to create datasets and analyses that can be used to characterize and understand that role?

10.3 Areal average values for some climate parameters can be obtained in more than one way. For example, precipitation averaged over an area can be calculated from the values at stations within the area, from satellite-derived analyses, or from model forecasts. Discuss the differences that might arise from using these different methods.

10.4 Discuss which land surface characteristics may require special consideration for climate studies in (a) Rocky Mountain Park, (b) Mohave Desert, (c) Central Park, NYC.

11 Climate Models as Information Sources and Analysis Tools

11.1 Introduction

For most of history the word "data," when used in relation to climate studies, has referred to direct measurements taken by instruments located at a specific observing site. In recent decades, the meaning of "data" has been expanded to include remotely sensed observations, and inferences from those observations, from satellites, radars, and other instruments. In recent years, climate scientists have begun to refer to the output of mathematical models of the atmospheric and oceanic components of the climate system, data assimilation[1] schemes, and reanalyses (as discussed in Chapter 3) as "climate data," a usage that bothers many traditional climate analysts, who feel that model results and observations should not be equated. Model output differs fundamentally from traditional observations in that the models estimate the value of a parameter by constraining it to be consistent with laws of physics as embodied in the model. Even when observations are used as additional constraints, as in data assimilation systems, the model calculations play a vital role.

As discussed in Chapter 2, observational data often have discontinuities in space and time as well as being subject to non-climatic influences. Standard mathematical and statistical climate analysis tools are most easily applied and interpreted if the datasets being analyzed are spatially and temporally complete. Thus, observational climate datasets often include some estimates to fill in for missing or erroneous values. Since the output from numerical climate models is spatially and temporally complete over the geographical domain of the model, this output is ideally suited to analyses using standard mathematical tools. It is crucial to remember, however, that even though the model output is spatially and temporally complete, the data assimilated by the models probably are not. The advantage provided by using the model is that the output values are constrained by the model physics and will provide reasonable values even in the absence of observed input data. On the other hand, it can be quite challenging to determine the degree to which model output benefits from observed data assimilated in the model.

[1] The combining of diverse data, possibly sampled at different times and intervals and different locations, into a unified and consistent description of a physical system, such as the state of the atmosphere. From the AMS Glossary of Meteorology, 2012.

In this chapter, we discuss the different ways that mathematical models of the climate or its component systems are generated, and how they are compared to observations. We then discuss the characteristics of climate model output used as data, including parameters, grids, and units. We conclude with discussions of how model-derived data are used in analyses of the climate system.

11.2 Generating Numerical Model Data for Climate Analysis

Scientists use models of the individual components of the climate system, particularly the atmosphere and the ocean, and often combine these models through a process called coupling that permits the outputs from each component model to influence the evolution of the others (see Section 11.2.3). Climate models are executed, or "run," from some initial values and a set of defined boundary conditions. The initial values are needed at every point in space for which the model will calculate values at future times, while the boundary conditions are needed for parameters that the model does not calculate but which are needed in its calculations. The models can be run in a number of different ways. Here we discuss two of the most common: the unconstrained, or free, model runs, and models constrained by observations through data assimilation. In both cases, boundary conditions must be specified using observations, sometimes augmented by the results of specialized models. In addition, model runs require an initial period of time (sometimes called the "spin-up") in order to eliminate any inconsistencies between the initial state and the climate of the model so that that stabilization process does not contaminate the experimental results.

11.2.1 Free Running or Unconstrained Models

Climate model runs unconstrained by the input of observed data are useful in evaluating the characteristics of the model and in testing hypotheses regarding climate variability and change. Simulations can proceed for model years, decades, or centuries, allowing the model to reach its own mean climate and exhibit climate variations determined solely by the physical interactions embodied in the equations that make up the model. Changing boundary conditions can then be imposed to test their impact on climate variability.

Models can be run to simulate conditions during a particular season, or through a long series of annual cycles. For example, an atmospheric model experiment representing perpetual boreal, or Northern Hemisphere, summer conditions would use specified boundary forcing with oceanic, land surface, and solar radiation values appropriate to that season. The outcome of such a "perpetual boreal summer" might be useful in identifying the equilibrium climate response to variations in any of the specified forcings. More realistic but more computationally demanding experiments use boundary forcings that vary through the year to cause the atmospheric model to exhibit a realistic annual cycle, permitting experiments that illuminate a wider range of climatic variations and changes.

For studies of global climate change the models may be run with no changes in constituent gases or with observed or assumed changes in the greenhouse gases as discussed in Chapter 5. The effect of volcanoes on the model climate is sometimes taken into account by introducing increases in aerosols and greenhouse gases randomly in time and geographical location.

These kinds of runs can be very helpful in understanding the characteristics of models, identifying biases and areas that need improvement, and comparing different models. However, output from such unconstrained model simulations is rarely used as if it represents observations of the real climate, primarily because even the most realistic of such models are not capable of correctly and accurately simulating the real climate. The differences between the model climate and the observed climate, without the inclusion of actual observations to provide corrections, lead to unrealistic depictions of climate states. On the other hand, the use of such simulations as forecasts or projections of future climate, for which we have no actual observations, is certainly an important goal of climate model research and development.

11.2.2 Models Constrained by Observations

When a climate model state is updated during the period of its run by ingesting, or assimilating, observed data, the process is called data assimilation. The mean climate of a model differs from the observed climate, and a long model run will always tend to drift away from the real climate toward its own climate. Data assimilation assures that this drift is minimized and the model's fidelity to the observed climate is maximized. However, individual observations include the effects of small-scale variations that the models do not include (this is one aspect of representativeness error; Hoffman et al., 2017), and so it is important not to constrain the model exactly to every observation.

Climate models that assimilate data (as in reanalysis, discussed in Chapter 3) are often used to analyze past climate events in greater detail than was possible when the event occurred and, in fact, the use of numerical models to analyze past events has helped to revolutionize climate analysis. In order to examine the details of specific events, the climate model must assimilate observations made at the time the event occurred. Simulations of past extreme events, such as the 1982–1983 and 1997–1998 ENSO events, using modern climate models that assimilate all available observations, can facilitate far more detailed analyses of the similarities and differences between the two events that are possible with the observations alone.

Earlier data assimilation approaches required that some satellite observations be transformed into pseudo-observations of some type already used by the assimilation system. For example, satellite observations of radiances at wavelengths near atmospheric absorption bands provide information about the temperature and humidity at different heights in the atmosphere. Methods were developed to use that satellite-based information to estimate vertical profiles of temperature and moisture at the same levels as radiosonde observations so that they could be assimilated into the models in much the same way. It became clear that such satellite-derived pseudo-radiosonde observations were subject to biases that made them difficult to use. Contemporary numerical models compute radiances based on the physics in the model in conjunction with a radiative transfer model and thus are capable of assimilating observed satellite radiances directly. All modern numerical weather prediction centers use some form of data assimilation to create the initial conditions for their forecast models. These systems continue to evolve over time.

11.2.3 Uncoupled and Coupled Climate Models

The earliest climate models concentrated on the atmosphere, treating the other components of the climate system as external boundary conditions for the model. In these early uncoupled models the atmospheric variations were not influenced by feedbacks, for instance in moisture flux or radiation, with the ocean, cryosphere, or land surface. While these atmospheric models provided valuable experience in how to model the mean circulation and climate variations of the atmosphere, they were not able to cope with the full complexity of the climate system. The realization that ENSO is a coupled ocean–atmosphere phenomenon spurred the development of coupled ocean–atmosphere models. Initially, such models concentrated on the equatorial Pacific, with each component of the coupled system producing independent values for the fluxes of moisture and radiation at the water–air interface. The values of the fluxes from each component of the models generally did not agree, requiring adjustments (referred to as "flux corrections") in order to ensure successful simulations. Improvements in these models, greatly aided by improvements in both observations of ENSO and the relevant theoretical understanding, have resulted in coupled climate models that do not require such ad hoc adjustments.

Models for the cryosphere (Chapter 9) and land surface (Chapter 10) components of the climate system have been developed. These have been traditionally run uncoupled, using outputs from the global and regional atmospheric climate models as external forcing with no feedback. The current generation of earth system models, such as the NCAR's Community Earth System Model (CESM), have coupled versions that incorporate these components of the Earth's climate system in an interactive fashion.

11.3 Characteristics of Model Generated Data

In this section, we describe some of the characteristics of model data, including what variables are available, units of measurement, and the typical spatial and temporal resolutions.

11.3.1 Model Grids

In principal, model values are available for each calculated variable at each time step for each point in the model's representation of physical space. In practice, there is most often no need for such a large volume of data, and substantial culling of the model

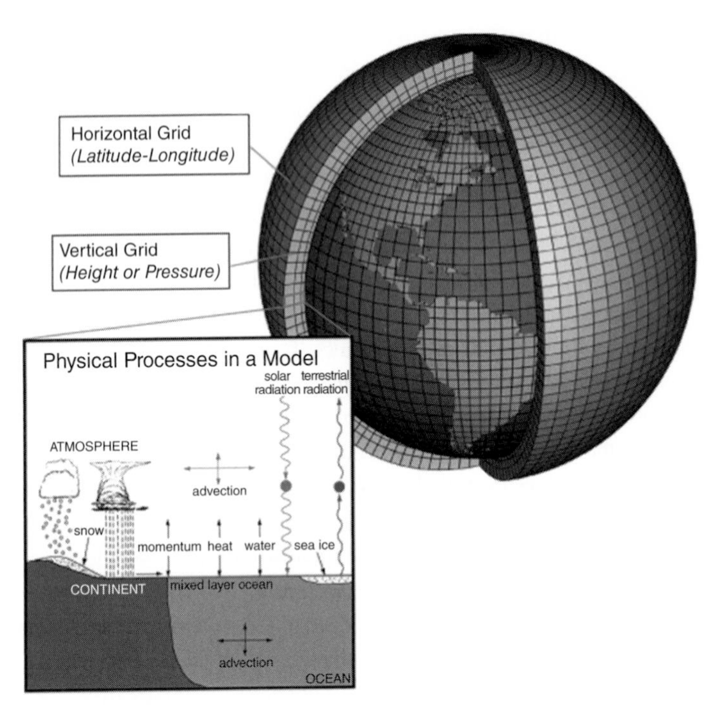

Figure 11.1 Schematic showing typical grids in the atmospheric component of climate models. Early reanalyses had horizontal grids with dimensions of a few latitude/longitude degrees and 20 or so levels. Several contemporary reanalyses have dimensions less than one latitude/longitude degree at 60 or more vertical levels. A black-and-white version of this figure appears in some formats. For the color version, please refer to the plate section.
Source: NOAA Climate.gov, courtesy Frank Niepold and Rebecca Lindsey.

values is usually done for the model dataset archives. In weather and climate models the physical descriptions of the atmosphere and ocean are often presented as datasets using a regular lattice of points, generally referred to as grid points, at uniformly spaced latitude and longitude locations. For example, a 2.5° latitude/longitude grid could have values located at 0° latitude and 0° longitude and at all points whose latitude and longitude are multiples of 2.5° (Figure 11.1). Such grids are convenient for a number of mapping and analysis techniques. Other grid arrangements are often used, depending on the goals of the analysis.

Each model has a specific number of levels in the vertical at which values are computed and available. Long-standing practice in the analysis of atmospheric data has established a set of pressure values for which data are typically available, and, while the model levels are rarely the same as those, output values are generally interpolated to the conventional levels. The time intervals at which model values are available is handled similarly: The model values are generally computed at time intervals of less than one hour, but output values are usually averaged or sampled to longer periods.

One common method for using station observations in conjunction with gridded model values is to average the station observations in areas that are geographically coincident with the model grid areas.[2] This allows for easier comparisons between observations and models in historical studies in addition to providing model-based estimates of climate parameters where observations are sparse or nonexistent. Several observation-based gridded datasets exist (for example, Jones et al., 1982, 1986a, b). It is not uncommon for the observational and model grids to have different spatial extents and different geo-referencing, as discussed later, and thus either the model outputs or the observational data must be interpolated or averaged to make them directly comparable. For example, observational data averaged onto 5° latitude by 5° longitude grid areas, centered on the equator, are not directly comparable to model output with a 2° resolution grid centered at 1° north and south. There is no universal agreement among modeling centers and among producers of gridded observational datasets on how these data are stored. For example, some models will start with the first value centered on the prime meridian and South Pole with subsequent data values to the east and then north, while for others the data are stored so that subsequent data values are to the west. Special care is needed when attempting to use gridded data for different components of the climate system. For example, oceanic, atmospheric, sea-ice, and land surface data often have different geo-referencing conventions for data storage. The international research community has begun to address this issue in a systematic way (see, for example, Teixeira et al., 2014), but inconsistencies are likely to continue to be found for some time.

Some of the observational gridded datasets with the longest records are available on fairly broad spatial scales, with resolution of 5° latitude by 5° longitude or coarser. This arises from the fact that the spatial density of observational data for many areas decreases dramatically previous to the mid-twentieth century. Grid areas spanning several hundred km include data averaged over a large number of microclimates. For example, a 5° by 5° grid centered on Richmond, Virginia, includes an area that reaches from North Carolina to Pennsylvania in latitude and from the Allegheny Mountains in West Virginia to Chesapeake Bay and the coastal Atlantic Ocean in longitude. An average of a climate parameter over such a large heterogeneous area may be better thought of as an index of that parameter over the grid rather than representative of any location within the grid.

As discussed earlier, site-specific data are often averaged to the same grid as used by a model for ease in comparisons. By definition, an average value over a grid, by itself, masks climate variations within the grid even if the grid covers a relatively homogeneous area. The averaging of station data over an area acts as a spatial filter and thus influences the statistics of gridded values so that they will differ from the observations within the grid. In particular, the character of the statistics of weather events within the grid becomes distorted, as evidenced in the probability distribution of gridded weather phenomena compared to any location within the grid. For example, the number of days with rainfall for a grid area over a season will always be equal to or larger than the

[2] A sometimes-confusing issue arises when trying to distinguish between grid points, which are the individual points where model calculations are carried out, and grid areas, which are the areas bounded by the lines that are equidistant from adjacent grid points. The example discussed in the text uses grid areas.

Table 11.1 Examples of units for atmospheric variables in climate models

Variable	Unit
Zonal (u) wind component	meters/sec
Meridional (v) wind component	meters/sec
Vertical velocity	Pascals/sec
Temperature	Kelvin
Stream function	meters^2/sec
Precipitation rate	kg/meter^{2}sec
Pressure	Pascals

Box 11.1 Example – Rainfall Frequency

Many climate models tend to produce more days with precipitation than are observed at individual stations. This may occur due to the mismatch between a model's climate and the initial conditions used for a given run: In locations where the observation-based initial conditions are moister than the model prefers, the model produces rainfall to reduce atmospheric moisture to a level consistent with its own climate, thus producing more precipitation days than appear in observations. There are also occasions when the observed atmosphere is too dry compared to the model climate, but the corresponding balancing process is excess evaporation, which still leaves the model run exhibiting more rainy days than are observed in reality. This phenomenon is more common during warm seasons and in showery conditions. Given that the gridded observation-based precipitation datasets also produce more days with rainfall than occur at individual stations, a comparison may suggest closer agreement between the model and observations than is actually the case for any location within the grid. The modeling community will eventually correct this aspect of models, but this example illustrates that comparisons between a model quantity and some index derived from spatial averages of observations may not always accurately define model bias at specific locations.

number of days with rainfall for any particular location within the grid, since the former includes rainfall occurring at any location within the grid. (See Box 11.1 – Example – Rainfall Frequency.) Likewise, the maximum daily rainfall or maximum daily temperature averaged over the grid will be less (and the minimum greater) than at some locations within the grid and the entire probability density function is distorted.

11.3.2 Model Units of Measurement

The units of measure assigned to variables in models are not necessarily the same as the units conventionally used for observations and in other climate datasets. In order to facilitate comparisons among models most producers have agreed to store their model output in MKS (Meter, Kilogram, Second) or SI (International System of Units) units. Table 11.1 lists the units used for some common meteorological variables. The use of standard units helps to ensure that the physical constants and parameterizations that make up the components of climate system models are internally consistent as well as facilitating comparisons among different models.

11.4 Model Validation

The development and application of models depends on confidence in the fidelity of their outputs. Thus, an essential step before using any climate model is the validation of its products by comparing them to observations of the climate system. Such comparisons can reveal biases and errors and identify needs for changes or improvements, and might help determine which of the attempted improvement options is best. An improved model is then put in place and used, and further validation performed in a cycle of continuing improvement.

An accepted definition of validation in the field of simulation modeling is "substantiation that a computerized model within its domain of applicability possesses a satisfactory range of accuracy consistent with the intended application of the model" (Schlesinger et al. 1979; see Sargent, 2013 for a more detailed discussion). Most validation is performed using inspection and simple statistical comparisons between analyses based on observations and model products representing the same quantity. For example, maps of the global distribution of precipitation from a combination of rain gages and estimates derived from satellite data (see Chapter 7) may be compared to the precipitation forecast produced by the model. Pattern correlations over the spatial domain, along with mean biases, temporal correlations, and biases at each point, may be calculated for the globe or for a variety of subareas. Patterns of climate variability can be extracted from observations and model simulations and compared in various ways, and prominent features of the climate system, such as the mean annual cycle or mean diurnal cycle, can be examined. This section and the following one will describe model validation and use of models for examining climate as closely linked activities. Both are necessary in order for the use of models to improve our understanding of the climate system. It is important to note that the validation of climate models is a highly subjective and intuitive process, since these models are so enormously complex that even verification (the process of ensuring that the computer program is producing the results it should) is quite challenging, and in fact is in certain ways impossible (Oreskes et al., 1994).

Model validation has a number of definitions and the meaning intended often depends upon the context of its use (see Box 11.2 – Validation and Verification). Here, we use "model validation" to refer to the process of comparing specific aspects of the model- and observation-based climates in order to document differences between them. The differences identified provide a basis for the interpretation of model-derived products and provide model developers with some guidance on aspects of the model that require improvements. Of course, simply finding that some aspect of the model, such as the diurnal cycle of precipitation over the oceans, does not agree with certain observations may be useful information, but further validation studies are likely to be

Box 11.2 Validation and Verification

There is a large body of literature on the topics of model validation and verification (for example, see Schlesinger et al., 1979; Oreskes et al., 1994; Sargent, 2013), with no universal consensus on the meaning of these terms across, and to some extent within, disciplines. These terms have been applied to economic models, investment models, disease models, traffic control models, and, indeed, to just about anything that can be modeled. Verification generally refers to the process of ensuring that the algorithms in the model have been implemented as intended (without "bugs," in the programming sense), while validation is the process of ensuring that the results of the model are correct. This usage of the term validation is more relevant to the model developers, the people that actually build the models, than to the people who are attempting to use the model's output to aid in describing and understanding the climate system. Most users of the output from climate models tend to assume that the model developers have performed whatever validation is necessary, an assumption that is not always warranted.

required to more clearly identify what aspects of the model require adjustment. Model developers and users are much more concerned with fixing the models; climate analysts use such information primarily to identify areas of uncertainty in studies of observed climate variability and change, including uncertainties in the observations.

Comparisons of observation-based and model-based analyses can provide guidance on where, and under what circumstances, the model performs well by identifying where and when the model and observations agree within the uncertainties associated with the observation-based analysis. Close agreement of some aspects of the climate system, however, does not necessarily imply that all aspects of the model faithfully replicate those occurring in nature. It is possible for the models to arrive at the "right" answer for the "wrong" reasons. Sorting this out is beyond the expertise and resources of most analysts, who, therefore, must rely on the model developers for guidance. For example, questions such as "Since there is good agreement in the seasonal temperature patterns in the model and observations, can we assume that the parameterizations in the model radiation budget algorithms faithfully replicate nature?" need to be addressed. If this analysis has not been performed or is not included in the documentation accompanying the model, then there is little that an analyst can conclude, in our example, about the relationships of the observed temperature and components of the model radiation budget.

11.5 Model-Based Climate Diagnostics

In the context of this book, "model diagnostics" refers to the use of climate models, or their components, in climate analysis. Such usage is predicated on the assumption that the biases between the models and observations being used have been documented or can be computed. As discussed earlier in the chapter, before embarking on an analysis of some aspect of climate based on this type of model-derived information we need to have some idea of how well the model climate replicates the observed climate.

When a climate model is run unconstrained by observations, we do not expect it to replicate specific past events, but we expect that a skillful model will accurately reproduce the statistics of the observed climate. The extent to which that expectation is correct serves as a guideline for whether output from a particular model is appropriate to the problem being studied. Once again, the statistics of the model climate should be compared to those of the observed climate and any biases and significant differences between the two taken into account when using the model as a diagnostic tool. If the model statistics replicate observations within the limits imposed by the study, then the outputs from the model may be used to provide insight into climate variability in the real world.

When run without data assimilation, the models are often used to test hypotheses relating to the evolution of climate anomalies, such as "Are droughts in a certain area of the world related to sea surface temperature patterns?" In long simulations, the model output is often thought of as a surrogate for historical data that are not available, and is inspected to test the hypothesis. Alternatively, anomalous sea surface temperatures might be imposed as boundary conditions on a climate model's oceans to investigate whether the atmospheric component of the model behaves in a way consistent with reduced precipitation in the geographical area of interest. In summary, the unconstrained or artificially constrained models provide the climate analyst with the closest thing to controlled experiments in the model world that might help provide insight into the workings of climate in the real world. It should always be kept in mind, however, that even if a model replicates the statistics of the past observed climate there is no guarantee that the model correctly replicates the relevant processes in the real world or that it is capable of forecasting climate variability in a changing climate.

11.5.1 CMIP/AMIP and Other Long Model Runs

Spurred primarily by the recurring Intergovernmental Panel on Climate Change (IPCC) assessment reports, discussed in Chapter 5, major climate modeling centers regularly generate a number of climate simulations. These simulations follow a specific set of protocols, and the results are compared to each other and to observations. The physical parameterizations and algorithms (how the physical processes of the climate system are rendered by mathematical expressions and how the mathematical expressions are rendered in the model computer program) vary from model to model. In addition, the spatial and temporal scales and a host of other details differ from model to model. Major modeling centers have agreed on a number of guidelines in order to facilitate the intercomparison of these climate simulations.

It has become routine for climate modeling centers to cooperate in a series of Coupled Model Intercomparison Projects (CMIP; Meehl et al., 2005). The CMIPs provide a framework that allows comparisons of climate models to each other and to observations by agreeing to sets of experimental protocols with defined inputs and outputs. The intercomparisons often include Atmospheric Model Intercomparison Projects

(AMIPs; see Gates, 1992) to examine the characteristics of the atmospheric component of the climate system independent of other components. All of the various intercomparison projects go to great lengths to ensure that scientists from both the modeling and data analysis communities have access to the output datasets. CMIP and AMIP data are distributed in a common format for all of the models involved in these projects by the Earth System Grid Federation (ESGF). (See: Cinquini, L. et al., 2014). Studies using datasets from these models must include careful consideration of what model outputs are necessary for the study of the phenomenon of interest. Different models perform differently, and their relative strengths and weaknesses must be assessed in the context of the study at hand.

11.5.2 Climate Analysis Based on Reanalyses

In the previous section, we discussed the analysis of climates generated by models under the assumption that model climate and model climate variability can provide insight into the real-world climate. Here we take the view that models using observations as constraints through data assimilation can provide detailed estimates of the observed past climate. In this discussion it is useful to keep in mind that the output from a reanalysis (see Chapter 3) is quite sensitive to both the physics of the model and to changes in the observations provided to the data assimilation system.

Since the mid-1990s a number of reanalyses have been performed, as discussed in Chapter 3. Aside from the differences in the ways the models parameterize the basic physics of the climate system, the various reanalyses differ primarily in their spatial resolution and period of record. Both of these factors need to be considered when contemplating the use of a particular reanalysis in a climate study. If a study is focused on the climate for a particular location, for example at an agricultural observing site, then the analysis needs to establish the relationship between the observations at the site and the estimates represented by the nearest grid point of the model used in the reanalysis. If there are no observed data at the specific location of interest, then the nearest grid point values are used to estimate the parameter required, or some way to downscale the climate parameters of interest for a location within the grid is needed (see Box 11.3 – Downscaling).

11.6 Advantages and Limitations of Models in Climate Analysis

As discussed in Chapter 2, observational data are often rife with discontinuities in space and time and exhibit variations due to non–climate-related influences. Standard mathematical/statistical analysis tools most often require that the data being analyzed be spatially and temporally complete. Before being subjected to analysis, the observational data must often be adjusted to account for non-climatic influences and to include estimates for missing or erroneous values. Since the output from climate models is spatially and temporally complete over the geographical domain of the model, it is well suited to analysis using standard mathematical tools. However, even though the model

Box 11.3 Downscaling

The process of estimating the value of a parameter at a specific location from the values contained in a reanalysis or forecast model grid is termed downscaling. Downscaling has traditionally been performed by developing statistical relationships between observations at specific locations within a grid area and the model grid values. The statistics are developed by comparing the historical time series of the model values to those at observing sites. Another form of downscaling utilizes regional climate models that have higher spatial resolution than global climate models. The global climate models provide atmospheric lateral boundary conditions to the regional models that are "nested" within them. The nested regional models assimilate observations within their domain, while using the values from global models at their boundaries. While regional models can provide higher spatial resolution than global models, they too are represented by their grid values and will often differ from observations at a specific location. For example, an estimate of temperature from a regional model that has a spatial resolution of 10 km is likely to be more locally representative than a global model with a 100-km grid, but neither of them are identical to the temperature observation at an airport or other observing site. As available computing power continues to increase, the spatial resolution of global climate models increases as well. Downscaling then requires finer resolution regional models in turn.

outputs are spatially and temporally complete, the data assimilated by the models in reanalyses probably is not. The model's advantage is that the model outputs are constrained by the model physics and will provide values in the absence of observed input data. A disadvantage is that the analyst generally is unaware of which model outputs represent information derived from the assimilation of observations, and which represent mainly the constraints of the model.

Gridded model output datasets, nonetheless, have proven invaluable in analyses of atmospheric circulation. A number of diagnostic studies (Barnston and Livezey, 1987; Wallace and Gutzler,1981, for example) have identified atmospheric circulation features, or patterns, that appear often, some at quasi-regular intervals, and account for a significant percentage of the variance that is associated with the atmospheric circulation. These patterns, as discussed in Chapter 4, are often referred to as "modes" of climate variability and include ENSO, NAO, PNA, MJO, QBO, and others. Climate models replicate the modes of observed climate variability to a lesser or greater degree depending on the model and whether the model is run with data assimilation, as in a reanalysis, or without the benefit of systematic assimilation of data. None of the existing climate models faithfully mimic all modes of observed climate variability, nor do they exhibit the same relative intensity in the model as in observations.

Most climate models can claim to do a decent job of replicating the mean annual cycle of temperature and atmospheric circulation and, to a lesser extent, precipitation, especially for reanalyses, but less so when the models are run unconstrained by the data. Whether this is good enough depends on the requirements of the problem at hand. If the mean annual cycle is an important aspect of a climate study, the differences between the observed and modeled annual cycle need to be evaluated. It is not uncommon to find that even a model that has some deficiencies in replicating the mean annual cycle may do a reasonable job of replicating observed anomaly patterns based on differences between model means and modes of climate variability in the model (Ropelewski and Bell, 2008). As climate models have become more sophisticated in how they include land surface, oceanic, and cryospheric processes, a complete analysis of the climate system must include evaluation of these parts of the climate system in addition to the atmospheric component. Integration of all components of the climate system is a major challenge for climate models but offers great potential for performing climate analysis that furthers fuller understanding of how the complete system works.

11.6.1 Biases

The climates produced by contemporary models, whether in reanalysis mode or unconstrained, generally replicate the major features of the observed atmospheric climate. These models have cold poles, warm equators, jet streams, Hadley circulations, and many other realistic features of our climate, but some aspects of all of the model climates differ consistently from our best observation-based estimates of the same parameter. Documenting these differences is essential to interpretation of the climates and climate variability of the models. The simplest form of documentation is through visualization of the mean differences, or biases, between the observed and modeled climates. Many studies require, in addition to the annual mean bias, a description of the biases through the annual cycle and sometimes on shorter time scales. For example, computation of mean monthly biases of surface temperature tend to show larger values during cold seasons, and similarly precipitation biases in wet seasons are generally larger than in dry seasons. The biases will often also have a well-defined and informative spatial structure and annual cycle.

It is often necessary in addition to examine the biases in variability between observations and model outputs. Such biases can be identified by examining the differences between the modeled and observed variances at various time scales. The biases in variance need to be examined for the period appropriate to the specific study to determine the time scales associated with the largest percentage of variance. The model variances in frequency bands associated with observed phenomena should be compared, since many model datasets have far different intraseasonal (30–70 day) and interannual (2–7 year) variance than is found in observation-based datasets.

Model outputs often exhibit quite different dynamic range, or variance, compared to observations, but these models may still provide useful insight for climate analysis. One way to do this is through examination of the standardized biases (the differences between modeled and observed means divided by their respective standard deviations). In this type of analysis, the analyst acknowledges that the actual model values differ significantly from observations but assumes, or hopes, that the behavior of the model relative to its own climate may provide some insight with respect to the climate variability observed in the real world. Examination of the standardized biases can provide some indication of the validity of that assumption. In addition, some models may replicate the observed climate at one time scale more accurately than at others; for example, some models may have stronger or weaker variability associated with intraseasonal variability, the annual cycle, or ENSO than is found in observations. If the analysis is focused on a particular time scale, then both the observations and the model output datasets can be time filtered to focus the analysis on that time scale.

11.6.2 Teleconnections

Spatial fields of correlations between the time series of a climate variable at a point and variables at other geographical locations in the area studied are called teleconnections. One of the earliest and best known of teleconnection studies was performed by Sir Gilbert Walker, who, in attempts to find methods to predict the onset of the summer monsoon rainfall in India, correlated sea level pressure at locations in India and pressure at all available locations over the globe. A number of other teleconnection-based studies followed during the twentieth century, leading to the identification of the Southern Oscillation (SO) as well as a host of other modes of climate variability (as discussed in Chapter 4). Teleconnections continue to be useful tools in climate analysis, both for exploratory research and as a tool to aid in seasonal climate prediction.

Models provide excellent datasets for teleconnection studies. The data from various reanalyses, as well as from AMIPs and CMIPs, can span many decades and are relatively easy to access from a number of sources. Many teleconnection studies focus on correlations between precipitation or temperature in a region and SST. As we have stressed repeatedly, however, a high or low correlation does not prove cause and effect. We need also to bear in mind that relatively high correlations may arise by chance and statistical significance must be tested to minimize the probability that the observed teleconnection pattern may have arisen by chance (Appendix A).

Scientists' interpretation of teleconnection patterns may give rise to hypotheses that attempt to explain the physical reasons for the patterns. Model output data can be helpful in examining whether consistent features of the climate system, such as changes in the atmospheric circulation, are present in association with a teleconnection pattern. Model-derived datasets may also be used to help understand features of the teleconnection pattern such as seasonality or longer-term variations in the strength of the teleconnection.

11.6.3 Pattern Correlations

Simple side-by-side comparisons of modeled and observed fields give a qualitative indication of how well the model replicates the observed climate. For example, do the locations of the major precipitation zones correspond well to those observed by satellite? Pattern correlations provide one way of quantifying how well the model and observed fields correspond (or how one model compares to other models). In the computation of pattern correlation, the values of a climate parameter at model grid points are correlated with the values at the corresponding grid points in the observed field. The resulting correlation value provides a quantitative estimate of how well the two compare and can be useful in comparing modeled and observed mean climate (Chelliah and Ropelewski, 2000).

Time series of pattern correlations between observations and model-derived datasets from several years of monthly or seasonal mean map pairs can provide estimates of how well a model or reanalysis performs through the annual cycle and during the occurrence of other modes of climate variability. Pattern correlations can also be used in retrospective evaluation of climate forecasts. The mean annual cycle accounts for the largest percentage of observed and modeled climate variance, and the mean annual cycle should be removed from datasets in any analysis of the time series of pattern correlations. Pattern correlations provide estimates of how well the *patterns* of a field compare but do not quantify the agreement of the *magnitude* of the quantities being correlated. Thus, even though there may be a relatively high correlation between two fields, the two may explain far different fractions of the variance within their own climate.

Pattern correlations can also be used to compare the climates of models unconstrained by data assimilation. It is not unusual to find that pattern correlations between model climates tend to be larger than pattern correlations between any model's climate and observations; model climates tend to resemble each other more than they do observations. Aside from the annual cycle and other phenomena forced by variations in boundary conditions, we do not expect the pattern correlations between unconstrained model runs to be meaningful since even if the models faithfully reproduce the variations associated with a phenomenon there is no mechanism that will synchronize their timing. The set of pattern correlations between individual unconstrained model runs can be useful for providing an estimate of significance for the pattern correlations between model climates.

11.7 Summary

Mathematical models of the climate and its component systems have revolutionized our ability to analyze the past and monitor current climate variations. They have contributed to the identification of the modes of climate variability discussed in Chapter 4 and to the documentation of climate change over the past century and projections into the future (Chapter 5). Their continued routine use in monitoring the evolution of climate in real time is discussed in the following chapter. Climate models also provide a powerful tool that can be used to test hypotheses, design observational networks, plan field experiments, and examine interactions among components of the climate system. Improved climate models promise to provide better understanding and forecasts, but must be accompanied by robust observing networks.

Suggested Further Reading

Nebeker, F., 1995: *Calculating the Weather: Meteorology in the 20th Century*, Academic Press, 265 pp. (A history of the development of numerical models for weather analysis and prediction.)

Taylor, K. E., R. J. Stouffer, and G. A. Meehl, 2012: An Overview of CMIP5 and the Experiment Design. *Bulletin of the American Meteorological Society*, **93**, 485–498. This is an overview of Coupled Model Intercomparison Project's (CMIP) fifth phase.

Questions for Discussion

11.1 Can the output of climate models be used as forecasts of the changes in climate over the next several decades? If so, what caveats must be considered? If not, why not?

11.2 What distinguishes a coupled climate model from a model of the atmosphere alone? What are the strengths and weaknesses of coupled models for assessing past climate variability and change? Future variability and change?

11.3 Observed changes in global and regional temperature during the twentieth century have been reproduced by coupled climate models. Does this constitute validation of these models? Verification? Discuss the differences.

11.4 In many scientific fields, studies that attempt to evaluate the impact of some imposed change, such as a change in medication or the use of a different fertilizer, rely on a pair of parallel studies in which the change is made in one and not the other. Explain the ways in which climate models can be used in such controlled studies of climate variability and change.

12 Operational Climate Monitoring and Prediction

12.1 Introduction

Climate analysis, the title and topic of this book, is the science and art of creating and using environmental datasets. Many of the uses for such datasets are in the realm of research, and research will be the focus for many of our readers. Their interest will be in achieving a better understanding of the phenomena involved in climate variability and change and advancing knowledge in general, the underlying goal and motivation for research scientists. Society indulges research (to varying degrees) since it has proven over and over that knowing more occasionally turns out to be useful in future practical applications.

In addition to research, basic awareness of current climate conditions, and what they might portend, are of interest and potential use in many other endeavors. In the first chapter of this book we introduced the climate system and its various components. This was followed by the chapters that described the principles and tools of climate analysis, the known patterns of climate variability, and long-term trends that characterize Climate Change. The remaining chapters, previous to this one, identified the variables, observational practices, instruments, and models that have been developed to enable and perform climate analysis. In this chapter we discuss how all of these activities have provided the means to monitor the climate as it evolves, a process generally labeled real-time climate monitoring.

Real-time climate monitoring is made possible by our increased understanding of the climate system and the data now available from a host of observing systems. Real-time climate monitoring is essential because it allows us to identify known patterns of climate variability when they appear and, since their evolution is relatively slow, provides a basis for empirical climate prediction. In addition, climate monitoring contributes to providing the initial conditions needed by numerical models of the atmosphere and ocean for climate prediction.

12.2 History

It seems likely that the earliest systematic attempts to collect and use information on climate came about in conjunction with agriculture and animal husbandry. Agriculture in particular depends significantly on an appreciation of the local sequence of warm and cold, wet and dry, over the course of a year. Adequate and accessible supplies of water are essential for any large human settlements, and most early civilizations – in the Nile River valley, the Fertile Triangle in the Middle East, the Indus River valley, and the Huang Ho River valley in China – depended on an annual cycle of river flooding to replenish the soil and water crops. The dates and amplitude of the flooding were crucial to scheduling cropping decisions, and records of both, initially oral but eventually written, were certainly kept.

As agricultural civilization expanded into other regions, other factors became important as well. In areas with cold season temperatures below freezing, or other thresholds limiting to various plants, records of latest and earliest freezes, and other measures of growing season length, became important. In most areas, local rainfall was of great significance, and so records of wet and dry season timing and rainfall amounts were needed.

These early efforts, predating modern science, most often consisted of qualitative descriptions of the weather at specific locations and were not aggregated over large areas until sometime after the fact, if at all. It wasn't until the seventeenth century and into the eighteenth century that Halley and Hadley made the first attempts to describe the global-scale climate, as discussed in Chapters 1 and 4. Climate was initially viewed as the mean state of the weather, and until the end of the nineteenth century nearly all activity that we would now refer to as climate analysis was directed toward better defining that mean state. The notion that climate can vary evolved in concert with the geological discoveries of ice ages, and by the latter half of the nineteenth century it was generally accepted that at some time in the past glaciers and ice caps had extended much further than during the present era. While this obviously implied that the global climate had changed significantly, there was little consideration at the time of climate change or variability on the shorter time scales of relevance to society.

The Indian Monsoon provided one of the first concrete examples of climate variability on human time scales. The drought and famine of 1877 led Blanford, the first director of the India Meteorological Department (IMD), to investigate interannual variations in monsoon precipitation and to attempt predictions. As discussed in Chapter 4, a successor, Sir Gilbert Walker (1924, 1928), used atmospheric data from around the world to identify coherent variations in atmospheric pressure, temperature, and precipitation in an effort to uncover predictive equations. Routine real-time global climate monitoring, however, did not start to emerge until the 1970s.

12.2.1 The Rise of Contemporary Climate Monitoring

The growth in international trade and travel following World War II provided incentives for expanded regular weather observations that included the entire globe. (Chapter 3 includes more details about the expansion of atmospheric observations.) This expansion was supported by the rapid development of improved instrumentation, telecommunication systems, computers, and satellite-based observations. The development of sophisticated computer models for weather prediction required improved initial conditions derived from data assimilation systems and provided another strong incentive for expanding observing systems. By the 1980s, these systems, primarily designed to support weather forecasting, began to be used to monitor the global climate. By aggregating weather data, climate scientists had, for the first time in history, accurate and nearly real-time global estimates of monthly and seasonal temperature, precipitation, and other basic climatic variables.

These data provided a great potential resource for climate analysis but also presented great challenges. Among these challenges were the fundamental differences between the historical records and the contemporary data, much of which was in the form of gridded datasets either from satellite observations or from assimilation systems associated with numerical models. In the late 1970s and early 1980s monthly and seasonal anomalies derived from satellite observations or numerical models were often identified and discussed in terms of year-to-year differences. As time went by, seasonal and monthly anomalies were also defined by comparison to short (three- to five-year) base period.[1]

The modern climate monitoring activities that began in the 1980s have evolved and flourished since. These monitoring activities support: (1) research into the description, understanding, and prediction of the climate system; (2) improved analyses of climate datasets; and (3) the extension of weather prediction to longer lead times. These efforts initially focused on the monitoring and prediction of seasonal-to-interannual climate anomalies, and in particular those associated with the El Niño/Southern Oscillation (ENSO). Within a decade, however, climate variations on both shorter, intraseasonal, and longer, decadal, time scales became topics of interest, and of course anthropogenic climate change, global warming, did as well. Nearly all of these activities are carried out by national, or sometimes multinational, organizations, but international coordination has also been an important component.

In the United States, the recognition that knowledge of current climate conditions provides a basic nowcast[2] and the potential for climate forecasts led to the establishment of the National Climate Program (NCP) in the late 1970s. The NCP combined the Long Range Prediction Group (LRPG) in the National Weather Service with other existing NOAA elements to create the Climate Analysis Center[3] (CAC) to utilize research results to inform climate monitoring, nowcasting, and climate prediction.

The CAC combined three mission thrusts: climate prediction, climate monitoring, and climate analysis. The prediction mission was a natural extension of the preexisting role of the LRPG, which had provided empirically based 30-day outlooks for the

[1] Base period is the time period over which an average was taken to represent the background climate. Long-standing practice, codified by the WMO, defines 30 years as the standard base period, but in the early years of model- and satellite-derived datasets, when the total period of available observations was much shorter than 30 years, the length of data used was explicitly identified as the base period used in an analysis or monitoring product.

[2] Nowcast is a term used in meteorology for a description of current weather conditions and those expected in the very near future. A "climate nowcast" is a description of current climate conditions in the context of the background mean climate. As in short-range weather situations, a climate nowcast can be used as a very short-range climate forecast by simply persisting the existing situation.

[3] Now the Climate Prediction Center, located within the National Centers for Environmental Prediction in the National Weather Service.

coterminous United States (CONUS) since the early 1940s and experimental seasonal forecasts starting in 1958 (Namias, 1968; Rasmusson, 1998).

Climate monitoring obviously requires good climate datasets, and in the 1980s in particular a great deal of effort went into creating the systems needed to compile these data and turn them into useful products. A number of organizations in the United States and elsewhere contributed to the efforts in assembling and updating datasets as well as quality control, archiving, and data distribution. Many of these efforts were discussed in the previous chapters on temperature, ocean variables, precipitation, and land surface variables. The evolution of some of the most frequently used datasets for climate monitoring is summarized next.

Temperature and Precipitation from Land Station Observations

Temperature and precipitation datasets derived from station observations were needed both to validate the CAC/CPC forecasts and to support efforts to detect long-term trends due to anthropogenic influences. The CAC created the Climate Anomaly Database (CADB) based on daily weather observations and the Climate Anomaly Monitoring System (CAMS; Ropelewski et al., 1985) based on WMO monthly data exchanges to monitor temperature and precipitation anomalies. The National Climatic Data Center (NCDC), now part of the National Centers for Environmental Information (NCEI), created a station temperature observation dataset[4] that could be used to create long time series of global land temperature to be used together with ocean temperature observations to examine long time scale variability in global mean temperature. Gridded station temperature data are also produced and routinely updated by the Climatic Research Unit (CRU) at the University of East Anglia (Jones et al., 2012), and by the NASA Goddard Institute for Space Studies (GISS).

Sea Surface Temperature Datasets

Sea surface temperature (SST) datasets were required to complete global temperature datasets, but they were also needed for monitoring of shorter time scale variations and to serve as lower boundary conditions for atmospheric forecast models, both weather and climate. NCEI, in order to complete its global temperature time series, worked with other NOAA components to create a comprehensive dataset[5] of ocean conditions based on ship observations. Since the ship observations were quite sparse in most of the world's oceans, estimates of surface temperature derived from infrared satellite observations were used as well. The first near-global SST analysis based on a combination of ship and satellite observations, together with high quality data from research buoys as calibration, was produced by the CAC in the early 1980s (Reynolds, 1988). This analysis was significantly more complete than any available before, and enabled

[4] The NCDC (NCEI) dataset, the Global Historical Climatology Network (GHCN), discussed in Chapter 6, contains temperature and precipitation data from 5 to 10 thousand observing stations worldwide. It was updated most recently in 2011. Further updates are expected.

[5] The Comprehensive Ocean-Atmosphere Dataset (COADS) was also discussed in Chapter 8. The current version is referred to as International-COADS (ICOADS), and contains a number of other atmospheric and ocean surface parameters in addition to SST.

the creation of reliable global temperature time series by both NCEI and GISS. The Reynolds SST analysis was used by CAC/CPC to monitor ENSO and by NMC/NCEP as the surface boundary condition for weather forecast models. The initial analysis eventually evolved into two versions, one produced at NCEI and used mainly in global temperature analyses, and the other produced at NCEP and used in weather and climate forecasting. Paralleling these efforts in the United States, the Hadley Center in the United Kingdom Meteorological Office (UKMO) constructed global SST datasets (Parker et al., 1995) that have also been used extensively in climate monitoring.

These global temperature data are used in two main ways: to monitor monthly interannual fluctuations in regional temperatures in order to identify heat waves and cold spells, and to create global mean temperature time series and extend them in real time to monitor the evolution of long-term trends. Both NCEI and GISS issue these time series to the public monthly, shortly after the end of the preceding month. The maps of global temperature anomalies can be very effective in illustrating how large local anomalies, such as the cold northeastern US winter of 2014–2015, can offer a misleading impression of global temperature changes: globally, the winter of 2014–2015 was the warmest during the period (since 1880) for which reliable instrumental records are available.

While mean temperature has been used most, both the GHCN and ICOADS datasets contain additional temperature data as well as other parameters. Over the United States, minimum and maximum temperature data have proven useful in demonstrating that nighttime temperatures have increased more than daytime highs in recent years (Karl et al., 1993).

Precipitation Datasets

Precipitation is measured at more locations than temperature, but since it varies more on short space and time scales it is actually less adequately sampled. Both CAC/CPC and NCEI created datasets of precipitation observations in their CADB and GHCN products. The CADB is continually updated in real time to permit prompt monitoring of precipitation anomalies around the world, but is not as useful for historical and diagnostic studies as the GHCN. The GHCN contains about 8,000 different precipitation time series covering different periods of record, and is quality controlled to reduce erroneous data. It has been used extensively in the production of precipitation analyses for land areas.

Many land areas, and all ocean areas, are too poorly sampled with rain gages to permit the creation of useful precipitation datasets without additional information. Since about 1980, this supplemental information has been provided by precipitation estimates from satellite data, as discussed in Chapter 7. In the United States, both NASA and NOAA produce precipitation analyses based on these estimates. The primary application for these satellite-only analyses is in monitoring weather variations, such as precipitation associated with tropical storms. For climate monitoring purposes, the satellite and gage data are combined into precipitation analyses that cover much, but not all, of the globe. The most widely used of these are the GPCP analyses (see Chapter 7), produced by a group centered at the NASA Goddard Space Flight Center and including other US scientists and organizations in Germany, Japan, and the

European Union. CPC produces other precipitation analyses (CMAP, CMORPH) using very similar data and different methodology (discussed in Chapter 7).

The use of satellite data for precipitation estimates and other variables has stimulated the development of Climate Data Records[6] (CDRs), first recommended by the National Research Council, and adopted by NOAA, NASA, USGS, and a growing number of institutions in the international climate community. The CDRs are subject to the highest standards of quality control. The NOAA CDR website states:

Satellite instruments are often affected by problems, such as changes in a sensor's sensitivity to light, over the lifetime of the satellite. These problems can impact our ability to accurately see changes in Earth's environment. The CDR Program applies proven and accepted scientific techniques to correct these problems, resulting in a more useful and reliable record of Earth's behavior.

Within NOAA, CDRs have been developed for a number of satellite-derived products covering all components of the climate system.

These precipitation analyses are used extensively in research but quite a bit less than are the temperature analyses in climate monitoring due to several factors. First, long-term trends in global mean precipitation do not attract the same degree of public attention as those in temperature, presumably because the possible signal due to anthropogenic climate warming is smaller and because the datasets used to estimate the trends are less trusted. For shorter time scales and regional variability, where the precipitation datasets are quite good, datasets using gages alone are easier to understand and are adequate. This is because precipitation variations on such scales matter far more when they lead to drought or flooding where people live and work, and in such regions the density of gage observations is typically quite high. The large-scale precipitation analyses have proven useful in monitoring precipitation in regions where for any reason gage observations are not sufficient or are not available.

Sea Level Pressure Datasets

While temperature and precipitation datasets are necessary for climate monitoring and prediction, they are by no means sufficient. Variations in both are related to variability in the large-scale atmospheric circulation, and understanding and predicting the anomalies in temperature and precipitation requires good datasets of atmospheric circulation parameters. Such data were even more essential for weather prediction, and the first large-scale atmospheric circulation analyses were created for weather forecasters using surface pressure observations. Sea level pressure[7] (SLP) observations have been collected since the 1800s at many locations, including over oceans. Some of the earliest

[6] The National Research Council defines a CDR as a time series of measurements of sufficient length, consistency, and continuity to determine climate variability and change. The full report, Climate Data Records from Environmental Satellites (2004), can be downloaded from the web site listed in the References.

[7] Surface pressure has been observed at thousands of locations with varying elevations. In order to make analyses created from the data more useful, the standard practice is to convert observed station surface pressure to sea-level pressure using a consistent procedure.

climate diagnostic and prediction efforts (e.g. Walker, as discussed in Chapter 4 and earlier in this chapter) made use of these data to uncover statistical relationships among variations around the world.

Until recently, historical surface pressure observations were collected in a variety of forms and held in a number of databases. Regular observations made for weather forecasting were, and still are, archived by national meteorological and hydrological services. A subset of those observations, as with other surface observations, is exchanged under WMO auspices, and many national archives have been contributed to the WMO World Data Centres. Within the past decade, an international group of scientists has led a concerted effort to collect all available surface pressure archives into a single quality controlled database called the International Surface Pressure Databank (ISPD). Several versions of the ISPD are in various stages of processing; information about the latest versions can be found at the ISPD website listed in the References. The ISPD is maintained by a group of scientists who accept data from any interested source, with a particular interest in the digitization of historical observations from physical media such as ships' logs.

In the past, analyses of surface pressure observations were used, as for instance by Walker (1924, 1928), to identify and characterize large-scale atmospheric oscillations from the historical record. Climate diagnostic studies of this sort continued, and continue to the present, but the reliance on collections of surface pressure observations has lessened as comprehensive analyses and reanalyses of parameters at multiple levels have become more robust and available. A series of papers written from the 1960s to 1980s, for example those by van Loon (1965, 1979; Kiladis and van Loon, 1988), illustrate the nature of the transition: together with many collaborators he greatly advanced the understanding of the oscillations in surface pressure originally discovered by Walker and discovered more interesting features, but his later analyses generally combined surface pressure data with analyses of observations through the depth of the troposphere and into the stratosphere.

More recently, surface pressure observations, and in particular the ISPD, have been used as input to model-driven analyses and reanalyses of the global atmosphere (as discussed in Chapter 3). The NOAA-CIRES 20th Century reanalysis for example, which uses no atmospheric observations other than surface pressure, provides a very long time series of atmospheric analyses that is more independent of changes in observing systems than are the other available reanalyses.

Upper Air Climate Monitoring

Atmospheric scientists recognized quite early (Meisinger, 1922) that the characteristics of the mass and motion fields above the surface were as important as, or more important than, surface values. The US Weather Bureau (predecessor to the NWS) began regular observations of upper air conditions using balloons in the 1930s. Initially, the analyses of these upper air observations were performed at predetermined heights, 5,000 ft for example. By the 1940s, however, these analyses were supplanted by analyses of geopotential height for given pressures, that is the height above sea level of standard pressure levels, 500 hPa for example (see Chapter 3 for further discussion of Standard

Pressure Levels). During World War II, the Weather Bureau created a Central Analysis Center to produce master upper air analyses for use by forecasters across the country. These centrally produced and distributed analyses signaled the beginning of the transition from using individual observations, which generally involved making an analysis locally, to the eventual general practice of working with analyses produced elsewhere and with the grids of values interpolated from those analyses.[8]

Beginning in the 1950s, computer models of the atmosphere were used, initially experimentally but operationally before long, to make forecasts. Automated analysis procedures were developed to produce initial conditions for these models, and the analyses produced were archived and eventually used in diagnostic studies. The convergence between the manual and digital analyses and their archives took place over about 15 years from the late 1950s into the early 1970s, and their use in climate diagnostic and prediction studies began in the 1970s.

Upper air observations made using radiosondes began in the late 1930s, and datasets dense enough to support climate studies were available by the early 1960s. The MIT General Circulation Project made use of these data to produce the first large-scale climatologies of atmospheric circulation. For this project, analyses were produced from averages of quality-controlled radiosonde observations. This approach continued in use into the early 1980s (Angell and Korshover, 1964), but was eventually completely replaced by the use of averages of analyses of conditions at specific times produced for forecast model initialization. Such analyses were used beginning in the 1970s for a very large number of diagnostic studies (for example, Blackmon et al., 1977; Wallace and Gutzler, 1981; Arkin, 1982) that served to describe both the details of the global climate and its seasonal evolution and the seasonal-to-interannual variability associated with phenomena like ENSO.

Monitoring of atmospheric climate variability was a high priority for CAC when it was established. Some large-scale atmospheric analysis datasets based on operational initial analyses from the National Meteorological Center were available for research and climate monitoring: geopotential height at 700 hPa and 500 hPa for the Northern Hemisphere (Barnston and Livezey, 1987; Wallace and Gutzler, 1981), and the 200 hPa and 850 hPa winds for the tropical belt between 23° S and 23° N (Arkin, 1982). CAC created the Climate Diagnostics Database (CDDB) in the early 1980s, using the operational analysis produced for the NMC global forecast model. This analysis used an optimal interpolation approach with a first guess derived from the model forecast, and was the first global analysis to provide somewhat realistic depictions of the zonally averaged meridional circulation. Data began to be collected in real time in 1982, and retrospective processing of analyses for the period beginning in 1979 that had been archived at the National Center for Atmospheric Research was completed by 1984, providing a five-year climatology that was extended into the future.

The CDDB provided CAC/CPC with the capability to calculate monthly and seasonal mean atmospheric fields and anomalies both historically and in near real-time, and

[8] As a reminder, analysis is used here to mean the contour map of some property, usually geopotential height, created using irregularly distributed observations.

proved invaluable in both monitoring and diagnostic studies. A number of investigations into the dynamics of the 1982–1983 ENSO warm episode, the largest in a century, were based on data from the CDDB (Arkin et al., 1983). Beginning in 1984, CAC/CPC began production of the monthly Climate Diagnostics Bulletin (CDB), a publication that provided the climate monitoring community with regular updates of monthly and seasonal maps, time series, and other graphical depictions of recent climate variations. The CDB continues to be produced, now in online form, to the present, and may be accessed on the CDB website address listed in the References.

Reanalysis

By the 1980s it had been recognized that the analyses created for operational weather prediction were not ideally suited for climate purposes, leading to the initiation of the NCEP/NCAR reanalyses. As discussed in Chapter 3, the NCAR/NCEP reanalysis became available in 1996 (Kalnay et al., 1996) and has been extended in real-time since then. It is used both as the historical database for the calculation of anomaly fields and as the real-time data for CPC's climate monitoring efforts.

The other main US atmospheric climate monitoring effort was originated by the NOAA/CIRES Climate Diagnostics Center, now the Physical Sciences Division (PSD) of the NOAA Earth System Research Laboratory (ESRL). Scientists at the CDC have created and maintained the twentieth century reanalysis project (see Chapter 3), and have used its archive as well as other available datasets to monitor and diagnose current climate variability. While the main mission of PSD is research to support NOAA's climate monitoring, assessment, and prediction mission, PSD scientists also provide diagnostic assessments of recent climate variations (Dole et al., 2011).

Other Sources of Climate Monitoring Datasets

Climate monitoring and prediction rely on ocean, cryosphere, land surface, and other land surface datasets in addition to those described in this chapter. In the United States, most operational data are acquired, organized, and archived by the NOAA National Centers for Environmental Information (NCEI), which was formed in 2015 by combining the National Climatic Data Center (NCDC), the National Oceanographic Data Center (NODC), and the National Geophysical Data Center (NGDC) as well as other relevant elements. The NCEI combines the climate and weather data archive and diagnostic analyses previously offered by NCDC with the oceanographic datasets and analyses of the former NODC and the cryosphere and solid earth products of the NGDC.

A great deal of data relevant to monitoring the ocean and land surface climate is acquired through research programs supported by the National Science Foundation (NSF) in addition to NOAA, NASA, and the Department of Energy (DoE). These data are not always readily available for real-time climate analyses and monitoring, but like the satellite datasets discussed later in this chapter can be extremely helpful. Projects like the NSF-sponsored Ocean Observatories Initiative (OOI) and the DoE-sponsored Atmospheric Research Measurement (ARM) Program's Climate Research Facility collect information about the current and past state of the climate system and make it available to researchers and to climate monitoring efforts (see References for website addresses).

Such data can be used in two ways: in the construction of single variable analyses or time series suitable for tracking some critical climate variable; and as input to a reanalysis or as initial conditions for a climate forecast model. Examples of the former are observations of atmospheric carbon dioxide from many sites around the world, and observations of polar sea ice cover, used to create time series of total coverage. Just about any data that can be used to create single variable analyses can also serve as input to reanalyses and climate model forecasts, but in addition observations that are difficult to use for climate monitoring and prediction on their own can be valuable input to reanalyses and model initial conditions. For example, moored buoys in the tropical ocean are too widely separated to permit detailed spatial analyses, but their information on equatorial surface and subsurface conditions is vital to defining conditions in those regions in ocean analyses.

Satellite Data for Climate Monitoring

Similarly, observations from operational meteorological and research satellites are used both to create standalone analyses and as input to broader analyses and reanalyses. In the United States, NOAA manages operational satellites, and the standalone analyses of individual variables are provided by the Center for Satellite Applications and Research (STAR). STAR scientists create satellite-based datasets that describe the current state of the atmosphere, ocean, and land surface and that can be used to monitor the state of the climate and predict its future evolution (the STAR website is listed in the References). NOAA satellite observations also provide input to the analyses of initial conditions for NCEP's Climate Forecast System, and contribute to improving those analyses for climate monitoring.

NASA research satellite programs continue to collect enormous amounts of specialized observations that are used to analyze climatic parameters that are observed poorly or not at all by other systems. While NASA's original mission was outward looking, concerns about the potential for climate change due to greenhouse gas emissions as well as a desire to see NASA contribute to a better understanding of the Earth System led Congress to expand that mission, first to study stratospheric ozone and then to the complete system (see the Earth Data Centers at the website listed in the References for a Summary of NASA's role). The Earth Observing System (EOS), later designated Mission to Planet Earth (MTPE), was NASA's contribution to the US Global Change Research Program. MTPE comprised a constellation of large research satellites carrying a number of instruments designed to observe climatically relevant characteristics of the Earth's atmosphere, ocean, and land surface. As often happens with such programs, budgetary and political circumstances led to considerable modification of the original configuration, but by 2007 NASA was operating more than 15 space missions collecting information on Earth's climate system. NASA's Earth Science Mission continues to develop and implement new missions as of this writing (mid-2018), with at least 20 satellites currently in orbit, and will undoubtedly undergo further evolution in the future.

Transforming the enormous amount of data (an estimate 350 petabytes for climate change data alone) collected by NASA research satellites into coherent analyses of climate conditions is a huge challenge. Operation of each instrument is guided by an Instrument Team of scientists that works with data systems experts to create single and multi-parameter datasets using the quality controlled observations. The data products generated have facilitated climate research on time scales from seasonal to decadal and have served as the basis for thousands of scientific publications.

These data have also been used to develop initial conditions for ocean–atmosphere climate models, both within NASA and elsewhere. The Modern Era Reanalysis for Research and Applications (MERRA; see Chapter 3) has used NASA satellite observations to improve its initial analyses, and has investigated the information added by various data sources. The MERRA analyses have the advantage of incorporating more satellite information than other reanalyses in general, but their suitability for climate monitoring is limited due to the continuing large changes in the makeup of the observing system. These changes lead to changes in the analyses that are difficult to distinguish from variability in the climate system.

Operational Climate Monitoring in the United States

At present (2018), real-time operational subseasonal-to-seasonal climate prediction for the United States is carried out by the CPC. Climate monitoring also takes place in a number of US Government laboratories, including the CPC, the NCEI, and ESRL in NOAA, NASA's GSFC and GISS, and in generally more limited fashion in many university centers. The IRI Maprooms provide real-time climate analyses in the context of specific applications (website listed in the References). The large amount of data and monitoring products available allow for useful intercomparisons that have led to considerable progress in the past decade.

These US investments in climate monitoring and prediction as a national mission were accompanied by international efforts led by the WMO with its several climate programs. In the past 30 years, other nations and multinational consortia have ramped up their efforts significantly as well.

Other Climate Monitoring Activities

European nations have had a long history of leadership in the development of weather prediction and the sponsorship of meteorological and oceanographic research programs. During the 1970s, the European Community (EC) began to plan for a multinational weather prediction center that would have the resources required to be a global leader in extended range prediction, which was defined at that point as out to 10-day lead time. The European Centre for Medium Range Weather Forecasts (ECMWF) was created in 1975 and began to issue operational forecasts in August 1979. During the 1980s, ECMWF scientists began the development of a global coupled ocean–atmosphere climate model that now supports extended range (out to one month) and long-range (out to seven-month) forecasts. Support for climate monitoring is also a part of the ECMWF mission, and is carried out through the conduct of ongoing atmospheric reanalyses as described in Chapter 3. ECMWF itself does not publish climate-monitoring products operationally, although it does conduct and publish relevant diagnostic research.

The European Space Agency (ESA), like ECMWF a cooperative effort sponsored by nations of the European Community (now European Union), develops, launches, and operates Earth-observing satellite missions. ESA's first geostationary meteorological satellite program, Meteosat, was the European Community contribution to the international effort to achieve global geostationary satellite coverage in time for the First GARP Global Experiment in 1979.[9] Meteosat Second Generation (MSG) continued the Meteosat sequence in 2004, with an updated and enhanced instrument complement. At about the same time, the first satellite in the Meteorological Operational (METOP) satellite program was launched. METOP is a collaborative venture with the United States intended to ensure twice-daily coverage of operational low Earth orbit meteorological satellites for observations critical to weather prediction models. Both MSG and METOP are planned to continue indefinitely to assure Earth observation capability for both weather prediction and climate monitoring.

In the 1980s, the EC established the European Organisation for the Exploitation of Meteorological Satellites (EUMETSAT) to manage the operational meteorological satellites launched by ESA. EUMETSAT provides data and derived products to its user communities through a set of eight Satellite Application Facilities (SAFs). One of these, the Climate Monitoring SAF, is specifically charged with the development, delivery, and archiving of continuing climate datasets derived from operational meteorological satellites. Other SAFs, including the Land Surface Analysis (LSA) and Oceans and Sea Ice (OSI), also provide products and datasets that are useful for climate monitoring.

In addition to its role in operational meteorological satellite missions, ESA has a continuing mission to develop, build, launch, and operate satellites that observe other aspects of the Earth. The earliest of these, ERS-1 and -2, focused on observations of land and ocean surface characteristics, while the third, Envisat, carried 10 instruments that provided observations of the land and ocean surface, atmosphere, and ice caps. While these satellites could not be considered operational, since there were no plans for continuing the time series of observations, they did provide extensive datasets that, especially in combination with similar data from the United States, greatly advanced our ability to describe the global climate system and monitor its evolution. ESA Earth Observation Data may be found at the website listed in the References.

The success of these initial satellite programs led ESA to establish the Living Planet and Copernicus programs for Earth observation missions to monitor the climate and environment. These programs include a large number of satellite missions, some already

[9] The Global Atmospheric Research Program sponsored several international meteorological research field programs, including GATE and the Alpine Experiment, and the First GARP Global Experiment (FGGE) was planned as a grand culmination. In the event, FGGE, also called the Global Weather Experiment, took place during the year 1979 and incorporated the winter and summer Asian Monsoon Experiments (MONEX) and the West African Monsoon Experiment (WAMEX). During the FGGE year, complete global geostationary coverage was achieved, with two operational US satellites, GOES-EAST and GOES-WEST, Meteosat, GMS (the Japanese Geostationary Meteorological Satellite), and the GOES-Indian Ocean, which was a temporarily relocated operational US satellite.

launched and more planned for the coming two decades, which will together constitute an Earth observation network suitable for climate monitoring on a wide range of time scales. The Living Planet Programme, which includes the European meteorological satellites as its Earth Watch element, has already launched four Earth Explorer missions with four more in various stages of planning and preparation. The Earth Explorer missions are research satellites with instrument complements aimed at key scientific questions involving the climate system. While their lack of continuity prevents them from yielding long-term climate monitoring datasets, these missions will facilitate diagnostic studies that should improve our understanding of the physics of the climate and guide the development of future monitoring efforts.

Copernicus is an ambitious and unique program intended to coordinate and integrate information derived from at least 30 operational and research satellites to support "a range of thematic information services designed to benefit the environment, the way we live, humanitarian needs and support effective policy-making for a more sustainable future." The Copernicus space component consists of six Sentinel missions, each with instrument packages optimized for specific categories of Earth observations. Some will be accommodated on other ESA satellites while the others will be individual missions. Copernicus is designed to provide a continuing stream of observations that will permit long-term monitoring of the environment for a number of goals. One of the explicit goals is the monitoring of climate change, while others, such as the atmosphere, land management, and marine environment, will enhance monitoring of climate variability on shorter time scales.

Copernicus will operate through services devoted to each thematic area. The Copernicus Climate Change Service (listed in website section of the References) is intended to support European Union efforts to monitor and respond to climate variability and anthropogenic climate change. At present, this service consists of a set of projects aimed at climate modeling and observational analyses, including the ECMWF 20th Century reanalysis. The plan is to incorporate new observations, from the Sentinel missions and others, as they become available, and to create a complex of analyses and integrated products that will enable the supporting nations to monitor, predict, and attribute climate change and to improve adaptation and mitigation policies.

Most of the nations of the European Community participate in the ESA, EUMETSAT, and ECMWF climate observation, monitoring, and prediction activities through their financial support and by using their datasets and products. The United Kingdom (UK) is the only one to develop its own global climate analysis and prediction activity, starting from the UK Meteorological Office (UKMO) weather prediction effort. The organization that became the Met Office was founded in 1854 and, like the US NWS, is focused on the protection of life and property through the observation and prediction of significant weather events. In 1990, the Met Office created the Hadley Centre to conduct research and coordinate policy on anthropogenic climate change.

The focus of UK climate monitoring and prediction has been on long-term climate change, through modeling and diagnostic research together with observational analyses. The Met Office, even prior to the establishment of the Hadley Centre, was a world leader in the analysis of ocean temperature (see Chapter 8) and upper air observations

(See Chapter 3). At the present time, the Met Office provides a great deal of regional climate information, both historic and current, for the United Kingdom, as well as forecasts for the coming month. The UKMO uses a Unified Model approach (listed under websites in the Reference list) that permits the use of a single model for a variety of weather and climate products. Climate forecasts use a version with a coupled ocean model. The same model is used to simulate long-term climate change resulting from various social scenarios to provide guidance to policy makers and other potential users of climate services.

Other national climate monitoring and prediction efforts resemble those of the United States and UK, although generally smaller in scale. We will briefly describe a few of these here, but with the caveat that all such efforts continue to evolve and advance and the details we offer will quickly be outdated.

Japan may have the largest such enterprise because of its Earth-observing satellite program: a series of operational geostationary satellites has been in place since the late 1970s, and a number of research satellites have produced important datasets. The Japan Meteorological Agency issues one and three month forecasts weekly and a variety of climate monitoring products through the Tokyo Climate Center (website listed in the Reference list).

Australia is strongly influenced by ENSO-related climate variability (Chapter 4), and some have speculated that regional-scale long-term climate change related to greenhouse gas emissions is already being observed there (IPCC 4th Assessment Report, 2007). While lacking the resources to create large national observing programs such as satellites, the Australian Bureau of Meteorology (BoM) and Commonwealth Scientific and Industrial Research Organization (CSIRO), together with several universities, were early and significant contributors to climate diagnostic and modeling research. The BoM currently issues operational monthly and seasonal outlooks for the country as well as various diagnostic and monitoring discussions (listed under BoM in the website Reference list).

India has a longstanding interest in a very specialized climate forecast due to the dominant influence of the southwest monsoon on rainfall over the area. India, aside from high elevation regions, is a tropical and subtropical climate, and water is the critical factor for agriculture and an important constraint on many other aspects of Indian society. The India Meteorological Department (IMD) was created in part to monitor monsoon rainfall, and began to develop statistical prediction methods almost immediately. Operational monsoon forecasts are issued several times each year for various parts of the country using a number of statistical models (the IMD website is listed in the References). IMD has recently begun to use a dynamical climate model in monsoon prediction. In addition to *in situ* observations, India also operates a series of operational geostationary combined-application satellites with meteorological instruments and has launched several research satellites to observe the regional land and ocean surface. All of these observations can contribute to improved climate datasets.

In addition, several other nations monitor climate in their region using satellite and other data from the activities in the previous paragraphs augmented by local observations and local expertise. In Asia, the Beijing Climate Center (BCC) was designated a

WMO Regional Climate Center (RCC) whose services include support for climate monitoring in China and surrounding nations. In South America, Brazil obtains data from automated observation sites on land and ocean using satellites designed by the National Institute of Space Research (INPE) to provide real-time monitoring of weather and climate. The Brazilian Weather Service (INMET) routinely monitors the climate within the nation. CIIFIN[10] (the International Center for the Investigation of El Niño), a consortium of South American nations in collaboration with the WMO, holds an annual Regional Climate Outlook Forum (RCOF) to interpret and share information on the current state of ENSO and its likely evolution. Similar RCOFs are regularly held in many parts of the world (see RCOF link in the References). Climate monitoring is also a major activity of the International Research Institute for climate and society (IRI). The IRI, part of the Climate Institute at Columbia University, monitors climate with a focus on seasonal to interannual climate variability through a series of maprooms. "The climate and society maproom is a collection of maps and other figures that monitor climate and societal conditions at present and in the recent past. The maps and figures can be manipulated and are linked to the original data." (The IRI maprooms may be accessed at the website listed in the References.)

The current state of climate monitoring and prediction around the world can be described as promising, with many national and international efforts underway. However, it might also be described as chaotic, with numerous overlapping and relatively poorly coordinated national efforts, and with a startlingly large number of international efforts all with similar goals and overlapping activities.

WMO Climate Monitoring Activities

The World Meteorological Organization (WMO) recognized the significance of both climate research and operational climate monitoring and prediction and sponsored and coordinated large research programs to help to advance the efforts of their members. The World Climate Research Programme initiated observing programs such as the International Satellite Cloud Climatology Project and the Global Precipitation Climatology Project, along with broad experiments that combined field work with theoretical modeling, including WOCE (the World Ocean Circulation Experiment), TOGA (the Tropical Ocean – Global Atmosphere program), GEWEX (the Global Energy and Water Cycle Experiment), now renamed the Global Energy and Water Exchanges Project, CLIVAR (Climate and Ocean: Variability, Predictability, and Change), and others. Other organizations arose alongside WCRP[11] to help coordinate other activities relevant to this mission. The Global Climate Observing System defines and coordinates critical observations, both ongoing and new. The Intergovernmental Panel on Climate Change (IPCC) coordinates ongoing scientific assessment of research into the

[10] CIIFEN includes the nations of Argentina, Bolivia, Chile, Columbia, Ecuador, Peru, Uruguay, and Venezuela.

[11] The World Climate Research Program (WCRP) was one element of the World Climate Programme, but the other elements, aimed at observations, applications, and users, remained relatively small and were to a considerable degree overtaken by similarly motivated international efforts.

theoretical understanding of climate along with its past and future behavior. The IPCC assessments, published every five to seven years, provide global decision-makers with "the most clear and up to date" synthesis of scientific consensus on the current state and future course of climate change.

In recognition of the fact that transitioning research results into the various communities that can use them is quite challenging, the Climate Services Partnership (CSP) was formed. The CSP:

> *is an informal, interdisciplinary network of climate information users, providers, donors and researchers who share an interest in climate services and are actively involved in the climate services community. Members of the CSP recognize that their collaborative efforts have the potential to exceed those of any single institution acting alone.*

Other important international and multinational coordinating groups include the CGMS (Coordinating Group for Meteorological Satellites), the CEOS (Committee on Earth Observation Satellites), and GEO (Group on Earth Observations), all of which work to improve the functioning of the enormous constellation of observing systems that are necessary to monitor the climate of the Earth.

12.2.2 The Rise of Contemporary Climate Prediction

Until the 1970s, meteorologists interested in prediction did not clearly distinguish between weather and climate. For example, the US Weather Bureau began making "long-range" forecasts in the late 1930s, initially for just a few days but eventually extended to one month in the early 1950s. A great deal of theoretical research into the behavior of the atmosphere took place during this time, and one of the most relevant results (Lorenz, 1963) was the demonstration that deterministic prediction of atmospheric features is limited by the growth of small errors in the analysis of the initial state. Lorenz's estimate of the limit of predictability of "weather," in which weather is defined as the variability of the circulation about the background "climate," was 8–16 days, and further research has supported that conclusion. While the limit is both location and scale dependent, most commonly the practical limit on weather prediction in midlatitudes is thought to be about two weeks, as discussed in Chapter 1.

Initially, US Weather Bureau climate forecasts were based on statistical relationships between current observations and future temperature and precipitation, modified by subjective input from forecasters. It had long been understood that both temperature and precipitation anomalies were strongly influenced by anomalies in atmospheric circulation. By the time the Climate Analysis Center was founded, the Long-Range Prediction group at the National Weather Service routinely used statistical relationships between current and subsequent 700 hPa heights developed from several decades of geopotential height data as an important forecasting tool. Predicted 700 hPa heights were then used to "specify" temperature and precipitation for the United States. These purely statistical forecasts were augmented by 10-day numerical model forecasts.

As time went on, other elements were factored into the forecasts. The forecasts were validated[12] against 30-year normals or against persistence. By the 1990s it became clear that a "climatological" forecast based on fewer than 30 years was often better than one based on 30-year normals. For example, if temperatures at a station exhibited an upward trend, a forecast based on some recent period would likely be more skillful than one based on either 30-year normals or other methods. This finding was used to create a method called Optimal Climate Normals (Huang et al., 1996), variations of which are still in use. Persistence is a forecast made by assuming that conditions will remain unchanged. In a climate context, this can be interpreted in a couple of ways: that the anomaly will be the same as the previous month, or that the anomaly will be the same as for the same month one year previously.

Until the coupling between the ocean and atmosphere was recognized as a fundamental part of ENSO, monthly and seasonal climate prediction had no strong theoretical underpinning. The realization that the slowly evolving ocean temperatures exerted an influence on the atmosphere that might potentially prove predictable motivated prediction efforts using coupled ocean–atmosphere models. Early efforts to predict the evolution of ENSO were based, primarily, on statistical/empirical models (Wyrtki, 1975; Barnston and Ropelewski, 1992; Penland and Magorian, 1993). ENSO prediction efforts using numerical models were spurred by the successful prediction of the onset of the 1986/87 ENSO warm episode using a relatively simple coupled ocean/atmosphere model (Cane et al., 1986; Zebiak and Cane, 1987).

Coupled ocean–atmosphere models were developed at several national modeling and research centers. Improving and expanding such models to include the land surface and other aspects of the climate system continues to be a major activity. The numerical models were found to be sensitive to the initial conditions, and most numerical modeling centers have found that the overall forecast performance is improved if the models are run several times with slightly different initial conditions, producing several forecasts, generally termed an ensemble of forecasts (discussed in Chapter 11). It turns out that different numerical models have different strengths and that, in the long run, the (statistically) best forecasts are those produced when the ensembles for several numerical models are averaged together to form Multi-Model Ensemble (MMEs).

Despite the success of the early ENSO forecasts, subsequent efforts at numerical modeling centers have produced only modest improvements over ENSO predictions based on statistical/empirical methods (Barnston et al., 2012). Perhaps the most significant impact of the early success in the numerical model-based prediction of the 1986/87 warm episode was to encourage further research (Sarachik and Cane, 2010), and to encourage operational prediction centers to develop and implement coupled ocean/atmosphere/land models for use in both weather and climate forecasts. Real-time ENSO

[12] Here, validation refers to comparison between the forecasted values and actual observed values. Initially, CAC forecasts were validated at NWS observing stations with relatively long records, usually at airports in major cities. Validation is often done by comparing the error in a forecast to errors in a forecast made with some simple method. Two such methods are persistence and climatology. More typically in climate forecasting, a climatological forecast is used, in which the forecast value is the arithmetic mean of the previous 30 years at the same location.

predictions can be found at the Climate Prediction Center and Climate.Gov websites, in regular statements from the World Meteorological Organization (WMO), and on the websites of major operational climate modeling centers.

12.3 Climate Monitoring and Forecast Products

Several national and international organizations now routinely monitor the climate and make regular climate forecasts either for particular regions or for the entire globe. Climate monitoring and forecast products are produced monthly and seasonally by a number of the organizations discussed above. In addition, NCEI publishes annual climate reviews in the AMS Bulletin and IPCC publishes extended summaries of the state of the climate system at longer intervals. These publications and websites discuss climate for periods ranging from two weeks, monthly, and seasonal, through to the century scale. In the following sections, we provide examples of many of the regularly produced climate monitoring and forecast products.

12.3.1 Climate Monitoring Examples

The purpose of this section is to provide examples of the typical kinds of available climate monitoring information and to explain how to interpret them. There are several sources of this information. The CPC and NCEI produce maps and time series of monthly and seasonal surface temperature and precipitation anomalies for the United States as well as for the globe. Many of the CPC monitoring products concentrate on subseasonal through interannual climate variability. The NCEI also updates monthly and seasonal information but has a stronger focus on time series and other products designed to monitor long-term climate trends. These and various other aspects of the climate system are also routinely monitored by a host of government and academic research organizations cited in previous chapters and identified in the website Reference List. Many of the satellite-derived climate monitoring products are available through the NASA Distributed Active Archive Centers (DAACs). Drought is routinely monitored in the North American Drought Monitor discussed in Chapter 10. Droughts worldwide are monitored at the International Activities web page of the National Drought Mitigation Center (NDMC).

The American Meteorological Society (AMS) Bulletin publishes an annual State of the Climate that reviews the major climate anomalies occurring in the previous year. These reports were initiated in 1989 as annual Climate Assessments published by the US Government Printing Office and coordinated and edited by the CPC (then called the Climate Analysis Center – CAC) until 1995 when publication was shifted to the AMS Bulletin as a supplement (Halpert et al., 1996). Coordination of the Assessments was transferred to NCEI (then NCDC) in 2002 and the Annual Assessments renamed the State of the Climate in 2006. These reports have grown in size from a couple of dozen pages to over 200 pages and have expanded in scope to include all aspects of climate variability, from the ocean subsurface to land surface and the atmosphere into the stratosphere. Other climate monitoring publications include WMO annual statements on the Status of the Global Climate that have been issued since 1993. Comprehensive assessments of climate change, published by Cambridge University Press, have been issued at approximately five-year intervals by the IPCC since 1990.

The following examples illustrate some of the products used to monitor the climate as it evolves. Our first example is the monitoring of ENSO. The emerging understanding in the 1960s that the ENSO phenomenon depended on coupling between the ocean and the atmosphere motivated the development of improved observations in the Equatorial Pacific Ocean Basin in the 1970s and 1980s. It also motivated the development and, eventually, routine monitoring of several climate indices related to the various components of ENSO. Due to the relatively long lifetime of an episode, simply monitoring ENSO conditions provides some guidance on the character of precipitation and temperature conditions in the coming seasons for many parts of the world. Examples of the most commonly used ENSO monitoring products are illustrated with examples produced by the CPC in collaboration with several climate and data centers. Real-time updates of these and others can be found at the ENSO monitoring websites listed in the References.

Time Series of Climate Indices

Time series of climate indices are used to monitor the various climate modes discussed in Chapter 4, Climate Variability. For example, ENSO has traditionally been monitored with time series of the SOI, Equatorial Pacific OLR, SST, and others. NINO SST indices are designed to monitor the SST in three equatorial regions and one along the west coast of South America (Figure 12.1).

The NINO1&2 index is indicative of local climate variability in the traditional El Niño areas along the west coast of South America, while the NINO3, NINO4, and particularly the NINO3.4 indices tend to be more closely related to the global impact of ENSO. Many contemporary discussions, concerned with the global ENSO impact, refer to the NINO3.4 index as the Oceanic NINO Index (ONI). The magnitudes of the SST anomalies as well as the amplitude of the mean annual cycle are largest in the easternmost index areas and decrease westward (Figure 12.2).

ENSO is not the only climate variation monitored using time series in this manner. The CPC and others also monitor indices from many of the patterns of climate variability discussed in Chapter 4 and publish monthly tables and updated time series graphs of the North Atlantic Oscillation (NAO), Pacific North American (PNA) pattern, Arctic Oscillation (AO), Antarctic Oscillation (AAO), and other indices. In the United States, NCEI monitors state, national, and global surface air temperatures monthly.

Time-Longitude Diagrams

Propagating climate features are often tracked through the use of time-longitude graphs that are often referred to as Hovmöller diagrams, named after the Danish meteorologist who devised them to track propagating features (ridges and troughs) in the upper atmosphere (Persson, 2017). These diagrams have time as the vertical axis and longitude on the horizontal axis, and use data from a single latitude or an average over a latitude band. In the example shown (Figure 12.3) Outgoing Longwave Radiation

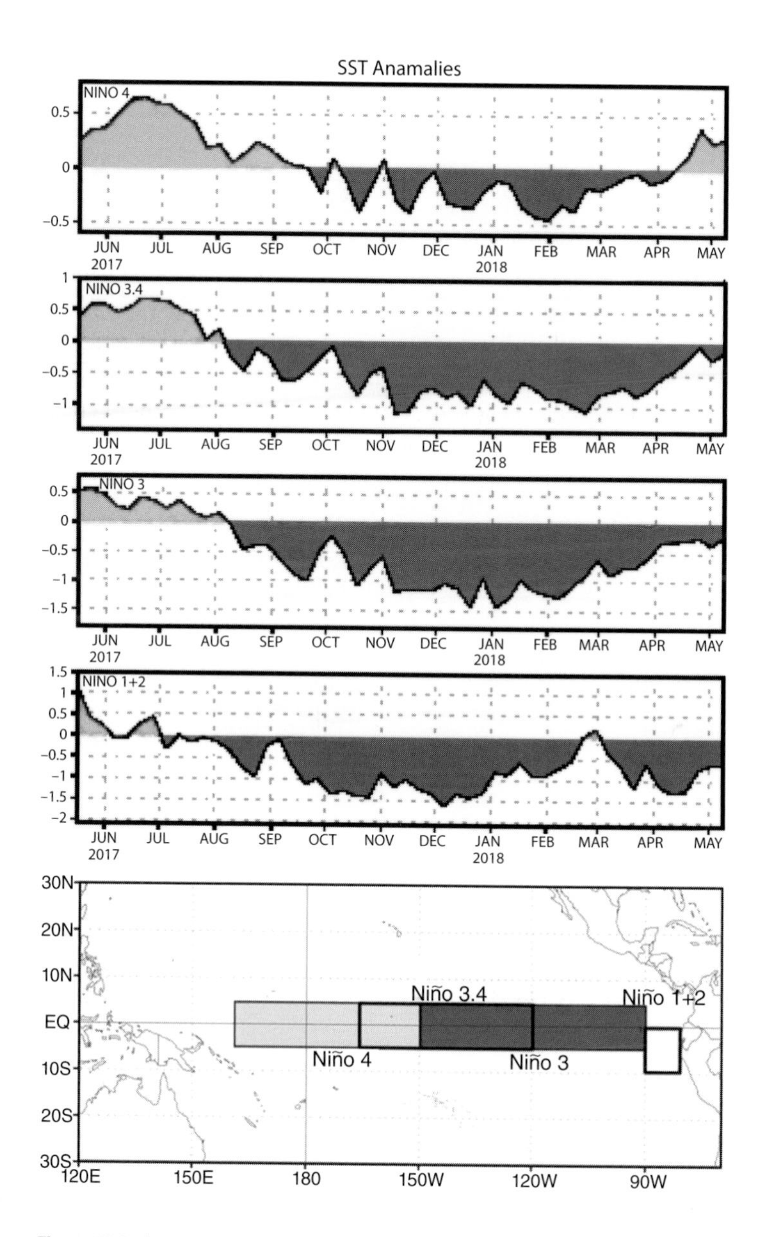

Figure 12.1 An example of time series of SST in the NINO areas. Schematic shows the locations for the NINO SST indices. NINO3.4 is a useful index for monitoring global ENSO impacts. NINO1&2 is a useful index for monitoring El Niño and La Niña and their local effects along the west coast of South America.
Source: NOAA Climate Prediction Center.

(OLR), discussed in Chapter 7, is used as a proxy for convective precipitation along the equator.

This figure illustrates the eastward progression of convective precipitation anomalies across the equatorial Pacific in association with the 1997–1998 El Niño episode. Such diagrams are also used to track features of the SST, 850 hPa, and 200 hPa winds,

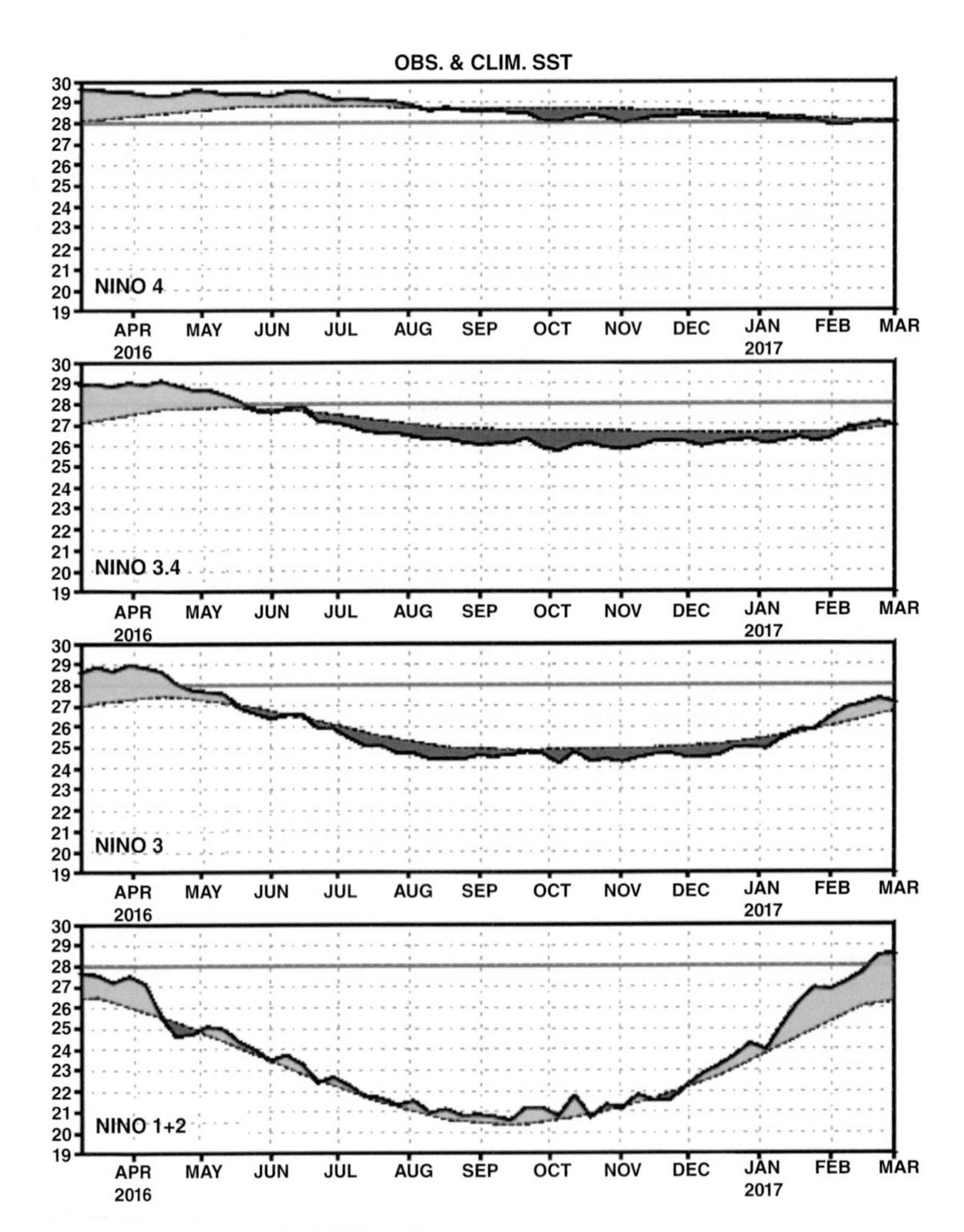

Figure 12.2 An example of a time series of the SST anomalies in the NINO SST index areas superposed on the amplitude of the mean annual cycle going from the Central Equatorial Pacific (NINO4) to the eastern part of the basin (NINO1&2). This illustrates the large differences in the magnitudes of the anomalies and the mean annual cycle. The anomalies in NINO4, the westernmost area, typically have amplitudes comparable to the amplitude of the mean annual cycle. The mean annual cycle in the NINO1&2 area has larger amplitude and illustrates the annual warming associated with the traditional area off the coasts of Ecuador and Peru as well as larger amplitudes in the anomalies. NINO3.4, or ONI, with modest annual cycle and anomalies is typically used to monitor ENSO.
Source: based on data from the NOAA Climate Prediction Center.

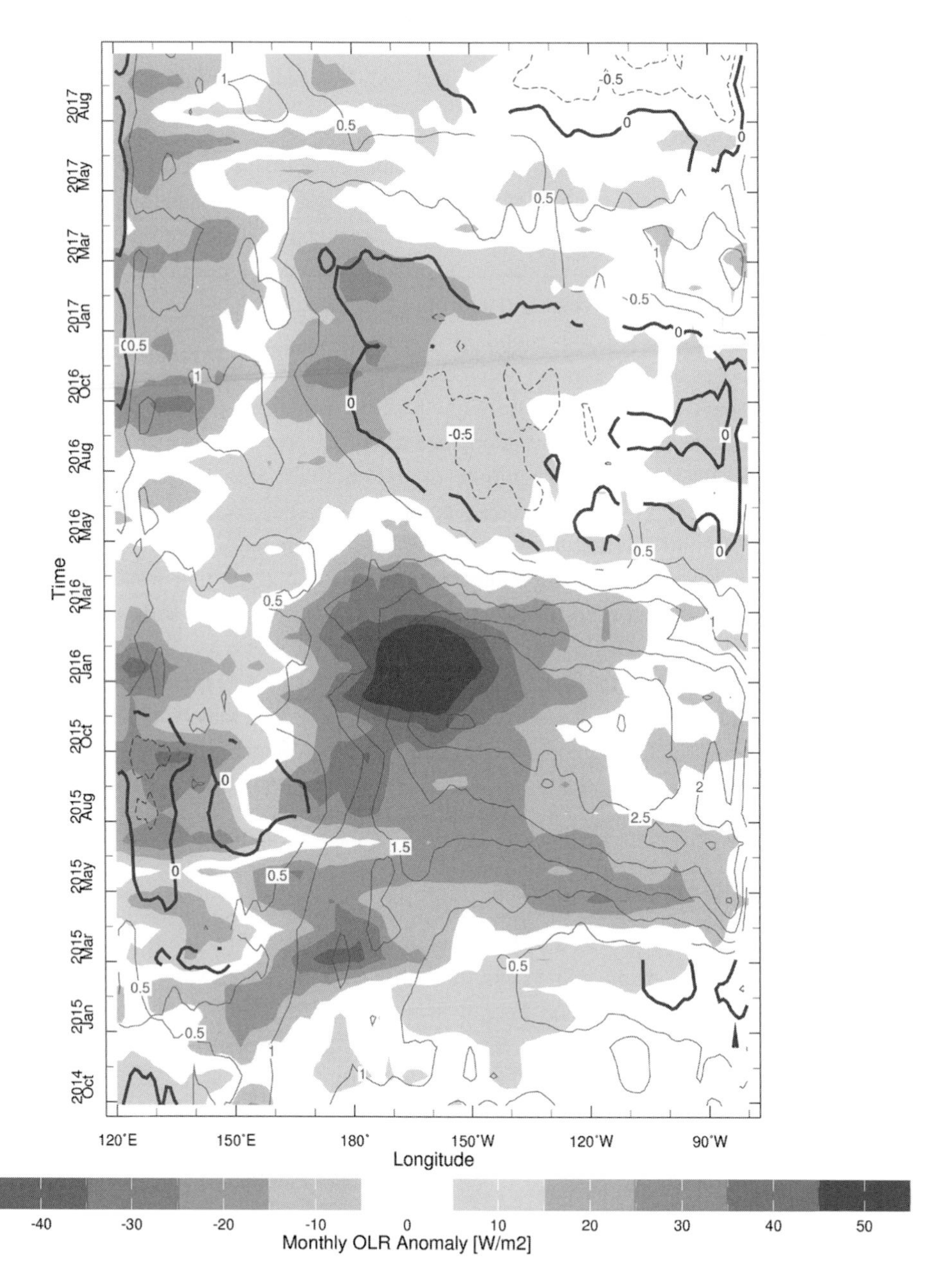

Figure 12.3 Time-longitude (Hovmöller) diagram of sea surface temperature (SST) anomalies (positive anomalies in red contours, negative anomalies in blue contours – °C) and outgoing longwave radiation (OLR) anomalies (negative anomalies indicating enhanced convection in green, positive anomalies indicating suppressed convection in brown). The data are averaged between 5°N and 5°S, from 120°E to 80°W. The earliest data are at the bottom of the diagram. Contour interval is 0.5°C. The OLR color scale is given below the diagram. Anomalies are departures from the 1981–2010 base period monthly means. A black-and-white version of this figure appears in some formats. For the color version, please refer to the plate section. Source: data are from CPC/NOAA and figure plotted by M. Bell using the LDEO/IRI Data Library.

and the depth of the ocean mixed layer (estimated from the 20°C isotherm) in the equatorial Pacific to monitor ENSO and to track intraseasonal variability associated with the MJO.

Anomaly Maps

Many climate-monitoring discussions center on maps of temperature and precipitation anomalies in the context of anomalies in atmospheric circulation and SST. The Climate Prediction Center, the NCEI, and other centers produce preliminary maps of monthly temperature and precipitation anomalies based on preliminary records derived by aggregating daily weather data from hydrometeorological stations in the first week after the end of the previous month to support their real-time monitoring efforts. For research purposes the data that go into these monitoring maps are subsequently replaced by data that have been more carefully quality controlled.

Monthly and seasonal anomalies are often presented in pairs of maps: one map to represent the actual magnitude of the anomaly and the other showing a statistical measure of how unusual the anomaly might be. Examples of these are shown for temperature (Figures 12.4a and 12.4b) and precipitation (Figures 12.5a and 12.5b).

In these maps the anomalies are shown as contours that result from performing an analysis on the point observation data. As an alternative, some analyses at the NCEI show the data from individual observation locations as circles whose diameters are proportional to the magnitude of the anomaly in average temperature for longitude/ latitude boxes. There are a number of ways these data may be displayed. Anomaly maps of winds including zonal and meridional components as well as velocity potential and stream function are also produced by the centers mentioned here as well as others referenced in earlier chapters.

12.3.2 Climate Forecast Examples

As discussed previously, forecasts of climate variability are made for a wide range of time scales, from two weeks through a year or more. Monthly and seasonal climate forecasts are illustrated here with examples for the United States from the NOAA Climate Prediction Center. Forecasts of climate anomalies of United States temperature, precipitation, and winds at 850 hPa and 200 hPa as well as 500 hPa heights, based on the NCEP Coupled Forecast System model out to nine months into the future are issued by NCEP. These forecasts are routinely used as key input to construct CPC experimental outlooks for Week 3–4, and operational outlooks for the following month, and overlapping seasons out to a year.

Initially, following the standard practice in weather forecasting, climate forecasts were issued with as little lead time[13] as possible. Since weather prediction models often provided some skill out to 10 days or more, both monthly, and eventually seasonal, forecasts made use of that information, which made the monthly and seasonal forecast skill appear higher. Eventually, forecast lead time for seasonal forecasts was increased to two weeks, which removed the artificial skill in the monthly and seasonal forecasts

[13] Lead time is the time between forecast issuance and the start of the period for which the forecast applies. For example, if a forecast for December were issued on November 27, the lead time would be three days.

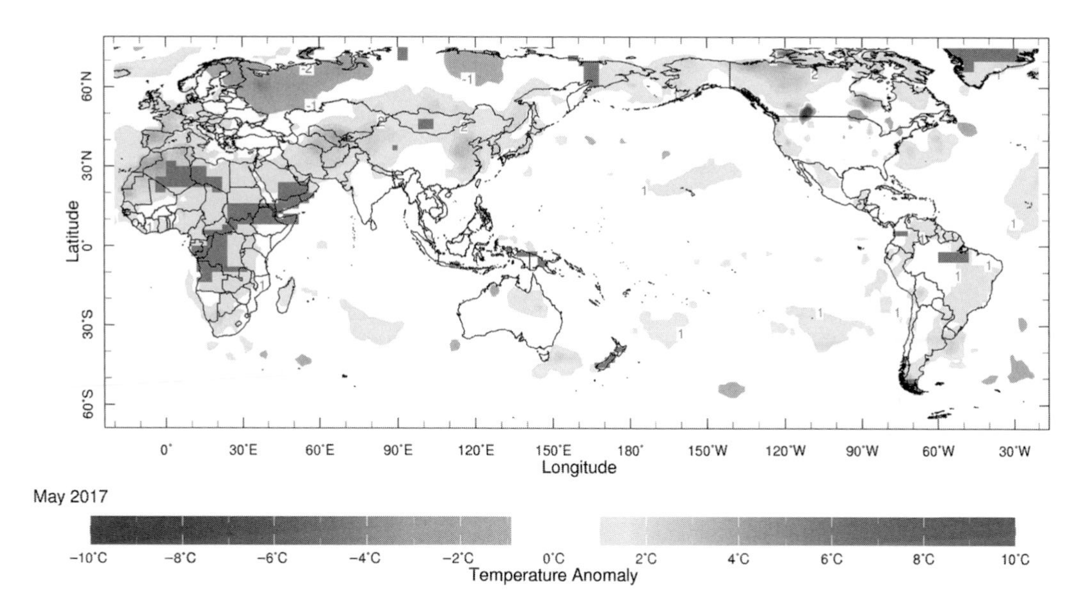

Figure 12.4a An example of surface temperature anomalies (°C). The analysis is based on station data over land and on SST data over the oceans. Anomalies for station data are departures from the 1981–2010 base period means, while SST anomalies are departures from the 1981–2010 adjusted OI climatology (Smith and Reynolds 1998, *J. Climate*, **11**, 3320–3323). Regions with insufficient data for analysis are indicated by gray shading. A black-and-white version of this figure appears in some formats. For the color version, please refer to the plate section.
Source: data are from the CPC/NOAA; figure courtesy of M. Bell, LDEO/IRI Data Library.

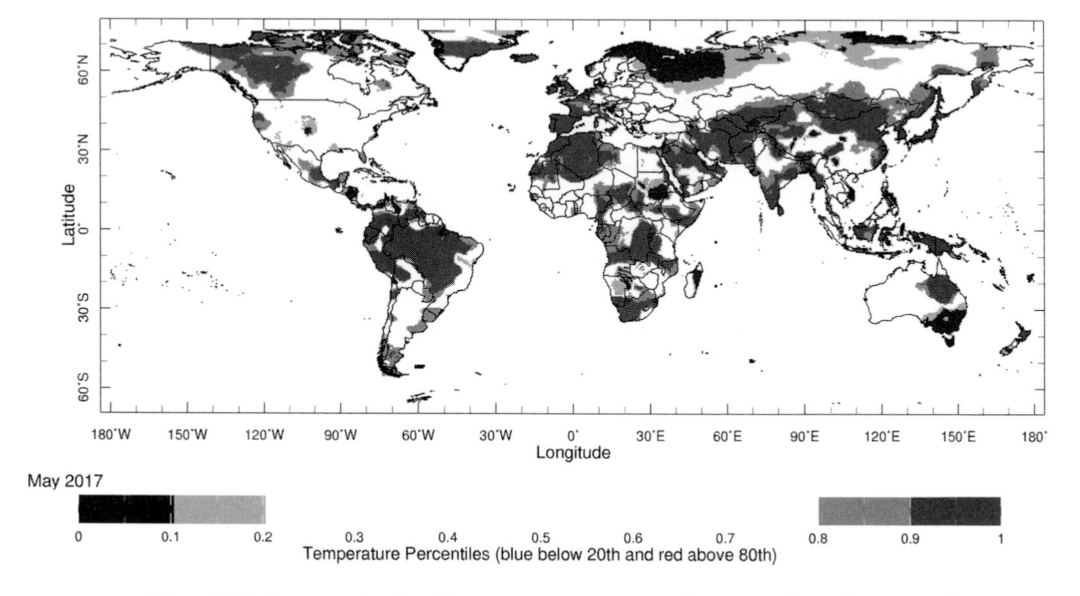

Figure 12.4b An example of surface temperature expressed as percentiles of the normal (Gaussian) distribution fit to the 1981–2010 base period data. A black-and-white version of this figure appears in some formats. For the color version, please refer to the plate section.
Source: data are from the CPC/NOAA; figure courtesy of M. Bell, LDEO/IRI Data Library.

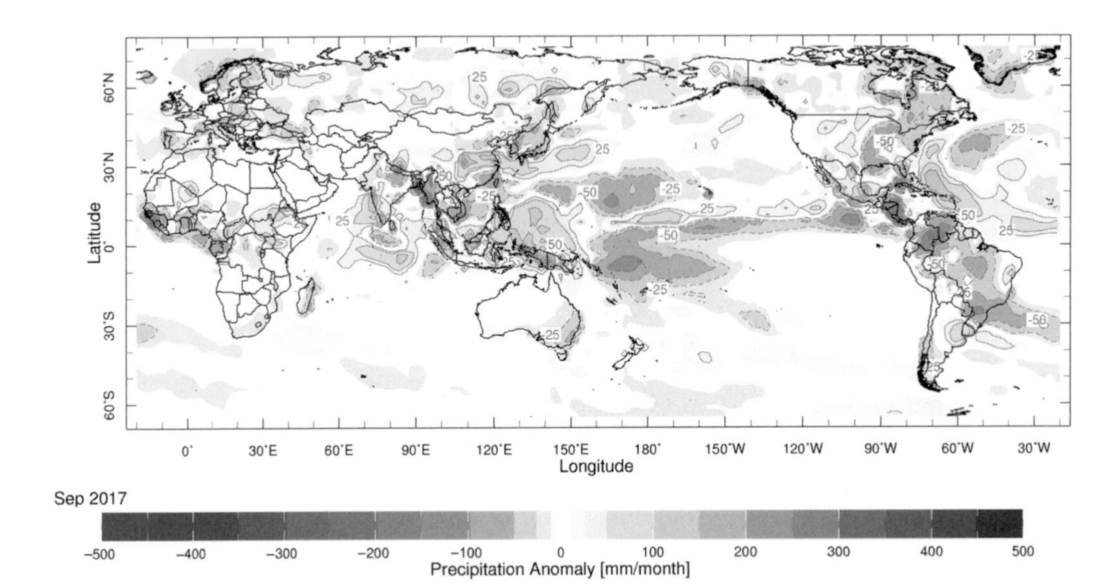

Figure 12.5a An example of anomalous precipitation (mm). Precipitation data are obtained by merging rain gage observations and satellite-derived precipitation estimates (Janowiak and Xie 1999, *J. Climate*, **12**, 3335–3342). Contours are drawn at 200, 100, 50, 25, −25, −50, −100, and −200 m. A black-and-white version of this figure appears in some formats. For the color version, please refer to the plate section.

Source: data are from the CPC/NOAA; figure courtesy of M. Bell, LDEO/IRI Data Library.

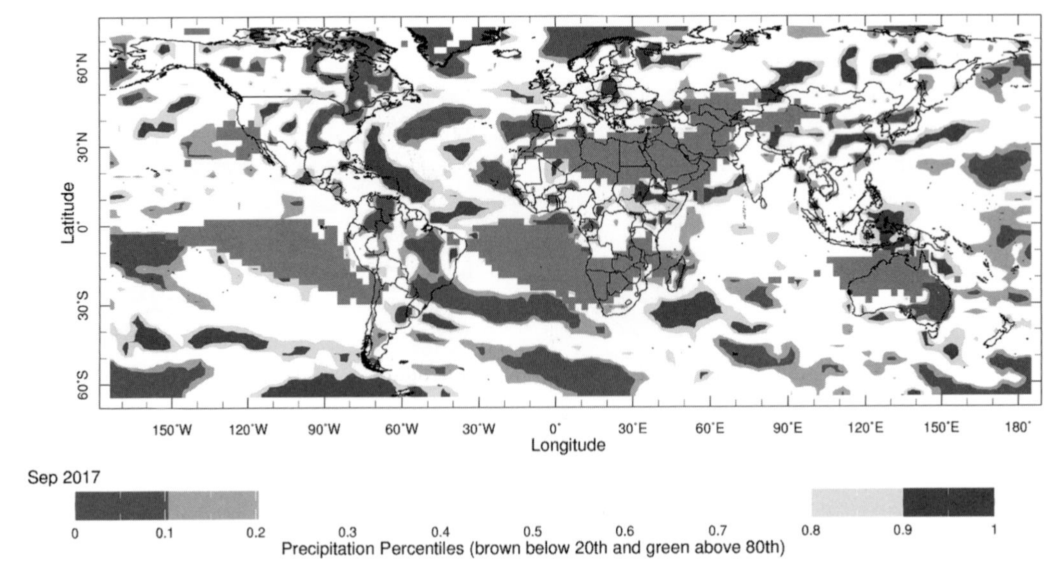

Figure 12.5b An example of anomalous precipitation expressed as precipitation percentiles based on a Gamma distribution fit to the 1981–2010 base period data. Gray shades indicate regions where mean monthly precipitation is <5 mm/month. A black-and-white version of this figure appears in some formats. For the color version, please refer to the plate section.

Source: data are from the CPC/NOAA; figure courtesy of M. Bell, LDEO/IRI Data Library.

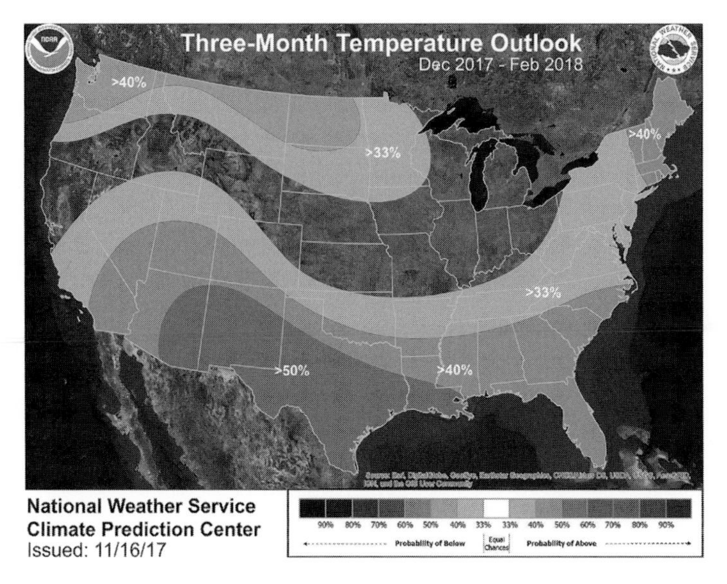

Figure 12.6 An example of a seasonal forecast from the CPC/NOAA using a tercile, or three-class (above, normal, below) system. The class limits are determined from the data in the 1981–2010 period. A black-and-white version of this figure appears in some formats. For the color version, please refer to the plate section.
Figure courtesy of the CPC/NOAA.

that came from the use of deterministic weather prediction model results. The increased lead time also improved the forecast utility for applications that could take advantage of greater advance notice.

The CPC issues probabilistic forecasts for the United States of monthly and seasonal climate anomalies each month. The CPC forecasts are presented on maps as probabilities for three classes (terciles[14] – below, near, and above average or median, Figure 12.6). These probabilities are determined by subjective analysis of all of the forecast tools available to the forecaster, including the forecasts from the NCEP Coupled Model System, a National Multi-Model Ensemble (NMME) and several empirical prediction tools as well as the current state of ENSO. The NMME includes ensemble forecasts from NOAA/NCEP, NOAA's Geophysical Fluid Dynamics Laboratory (GFDL), NCAR, NASA, and the Canadian Meteorological Center. The CPC produces forecasts of temperature and precipitation for the initial month and the 13 following three-month overlapping seasons.

The International Research Institute for climate and society (IRI) issues probabilistic seasonal forecasts of temperature and precipitation for the globe on a monthly basis (Figure 12.7). The probabilistic forecasts for the next four overlapping seasons are

[14] Tercile – for a set of data arranged in order, terciles are values that partition the data into three groups, each containing one-third of the total number of observations. For climate forecasts the historical data are partitioned into three subintervals such that climatological probability of the data lying in any interval is 0.333. (After the AMS Glossary of Meteorology, 2002.)

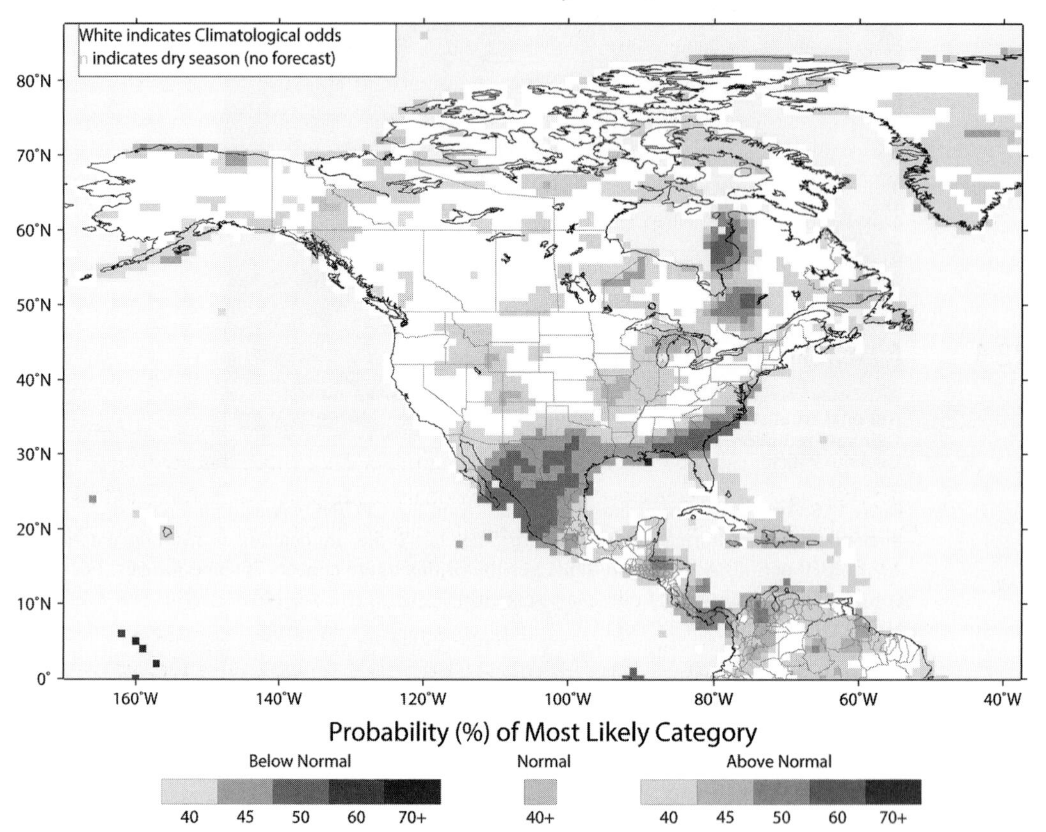

Figure 12.7 An example of a seasonal forecast from the IRI in a tercile, or three-class (above, normal, below), system. The class limits are determined from the data in the 1981 to 2010 period. A black-and-white version of this figure appears in some formats. For the color version, please refer to the plate section.
Figure courtesy of the IRI.

based on a technique that statistically corrects NMME model data. The IRI seasonal forecasts are based solely on NMME averages with no subjective adjustments by human forecasters. Global seasonal probabilistic forecasts of precipitation, temperature, and sea surface temperature are issued by the ECMWF to their member states, but publically available products are restricted to within 30° of the equator.

Since ENSO has such a large influence on seasonal climate there are several websites that provide guidance on forecasts of ENSO. CPC ENSO discussions are updated monthly. Monthly ENSO probability forecasts based on numerical and statistical model products from several operational and research centers are issued by the CPC (ENSO Forecast web addresses are listed in the Reference list.) El Niño/La Niña Updates are issued by the WMO when episodes are occurring or expected to occur. Several hydro-meteorological services, for example the Bureau of Meteorology in Australia and the

Center for Weather Forecasting and Climate Studies (CPTEC) in Brazil, issue precipitation forecasts based on ENSO state.

Regional Climate Outlook Forums (RCOFs), are held in several regions of the world in anticipation of upcoming ENSO episodes. The RCOFs are often held at WMO Regional Climate Centers, and other regional centers. The RCOFs often include educational sessions that target user communities who are likely to be affected by an upcoming ENSO episode. All of the ENSO forecasts, discussions, and forums discussed here are generally available on the respective institutions' web sites. CPC issues the only officially sanctioned seasonal climate forecasts for the United States, and for lead times out to about one year.

12.4 A Look Forward

Predicting the future of human behavior is challenging at best, and often impossible. Climate monitoring and prediction efforts are grounded in science, but are fundamentally dependent on social enterprise – while individuals can hope to make some predictions of upcoming weather, no one person can monitor climate variability on any except the most local scale, and meaningful predictions are impossible without near global observations at a minimum. At present, climate monitoring and prediction, along with the necessary supporting research, observation, and dataset creation, are well supported and vigorous, with lots of promising activities underway. Obviously, there are areas where improvement is possible, even essential, and there is no guarantee that the resources required to make those improvements will become available, nor that the current efforts will continue.

In the next section, we describe what we believe to be the national and international plans and prospects for advancing climate monitoring and prediction. We will finish by pointing out some gaps and presenting our thoughts on the shape of an ideal climate monitoring and prediction enterprise.

12.4.1 Current Plans and Prospects

First of all, we must re-emphasize that any successful monitoring and prediction effort, whether national, multinational, or on any other level of organization, will depend on a vigorous and well-supported research program combined with ongoing and constantly improving observations. Current levels of research and observations are sufficient to permit a modest monitoring and prediction capability, but sustaining the current level is vital, and expansion in some areas is essential.

The current US climate monitoring and prediction effort is diverse and extensive, but for the most part is not one that could be called unified or coherent, and that situation seems unlikely to change. On short climate scales, from a few weeks to a year, climate predictions for the United States are likely to continue to be issued by the NOAA Climate Prediction Center, with a significant body of forecasting also coming, as it does now, from the research community and private sector. The NWS plans to continually improve all of its operational products, and presumably is committed to improved versions of the Climate Forecast System analysis and model, but resource constraints are always a potential stumbling block. For longer time scales, the research community will continue to be the main source of projections especially through the research done to support the Assessment Reports issued by the Intergovernmental Panel on Climate Change. Climate monitoring reports will be produced by the various elements in NOAA, NASA, and the research community that do so at present.

New and improved observing systems, particularly satellite systems, are planned by both NOAA and NASA. Some of those observations are explicitly aimed at monitoring climate, while other observations are likely to improve climate datasets, monitoring, and prediction through improved weather analysis and prediction. The continuation of recently implemented ocean observing systems such as Argo (discussed in Chapter 8) is planned and should lead to improved datasets and analyses. Similarly, continued digitization of historical observations will improve atmospheric datasets and analyses by extending their temporal coverage farther into the past and improving their accuracy.

Reanalyses (gridded datasets incorporating all available observations and a model or theoretically based background) are essential for climate diagnostic studies, monitoring, and prediction. Several current US reanalyses, as discussed in Chapter 3, are quasi-operational, and continued production is planned. However, the financial and institutional support for these and other efforts in climate monitoring and prediction is fragile, and no specific plans for improved future reanalyses by US organizations have been put forward.

A venue where robust multinational climate monitoring and prediction activities, with strong supporting research and observational efforts, are planned is the European Union (EU). The European Commission Copernicus Program has begun, with an expected lifespan that will provide a wide range of Earth monitoring and forecasting benefits to contributing nations past 2030. Climate monitoring and prediction are a part of these benefits.

The EU appears to have a strong commitment to Earth environment, including climate, monitoring, and prediction. However, the significant costs of such systems and the relatively unwieldy nature of the European polity means that the resources required may not always meet the requirements. Those planned activities that are closely associated with robust and well-accepted continuing activities such as the meteorological satellites managed by the European Space Agency (ESA) and the medium range weather and climate forecasts provided by ECMWF are the ones most likely to continue.

The World Meteorological Organization (WMO) and its continuing coordination of national and multinational weather observing, forecasting, and research activities is an example of how an international organization can facilitate and foster continuing monitoring and prediction efforts at the national level. The WMO provides a forum for international interaction on weather-related activities that, while sometimes challenging and slow moving, is far superior to attempting to rely on piecemeal bilateral arrangements.

The WMO has made repeated attempts to create organizing structures that will coordinate and facilitate climate observing, monitoring, prediction, and research activities at the national level, as we have described in the preceding section. While progress has been made, particularly in the World Climate Research Program (WCRP), much remains to be done in the operational arena. One challenge in international programs is that the termination of any program is rarely seen, often leading to duplication of responsibilities and confusion on the part of member nations. The most recent restructuring of international climate programs at WMO has brought a number of activities such as the Global Climate Observing System under the organizational umbrella of the World Climate Program, but others, such as the Global Earth Observing System of Systems (GEOSS) have similar goals but are thoroughly distinct.

Different countries have various ways of discriminating between governmental and nongovernmental, whether private sector, nonprofit, or academic, responsibilities. One useful concept is that of *public good,*[15] something that is generally available and usable by anyone interested. Clean air and water are sometimes considered to be public goods, although certainly there are circumstances in which neither is generally available. In the United States, a common approach to defining government responsibility is to argue that something "should be" a public good, and therefore that the Federal government should take action to make that the case. The establishment of the Environmental Protection Agency was at least partially justified by such an argument.

Climate services and the supporting observations and research can be viewed through this lens: some basic level could be provided by the Federal government as public goods, with value added services provided by nongovernmental agents, or by state or local governments. A model of this sort is used for weather, in which many basic observations and a certain level of forecasting is provided as a public good, and enhanced services, such as detailed forecasts for specific applications, are provided by the private sector.

Climate services consist of analysis, monitoring, and prediction products. Climate applications are the use of any of the products or observations to support some other activity. Climate monitoring and prediction products in the United States are in reasonably good shape, with a satisfactory range of products and support, and applications are being developed. Climate research in the United States, historically, has been well supported but the future is less certain; although there are topics where further support would be helpful, including the integration of observations into analyses and applications, in general.

12.4.2 A Comprehensive Climate Enterprise

An idealized comprehensive climate enterprise consists of an observing program, an analysis effort, monitoring and prediction operations, and applications, with all components supported by a continuing research and development program. As we shall see, the various national, multinational, and international[16] activities that exist take a variety

[15] In economic theory, a public good is a *good*, a tangible object or a service, the consumption of which by an individual does not reduce the availability to any other individual. Sometimes public goods are also said to be non-excludable, meaning that if it is available to any individual, it is available to all. Forecasts of imminent dangerous weather, such as tornadoes, can be thought of as public goods.

[16] By multinational we mean activities that are supported by consortia of countries, such as the ECMWF. International means organizations that are open to, and include nearly all of, all nations.

of forms. Climate observations are the basis for all climate monitoring and prediction products. In the United States the Federal government supports most climate observations, either because of weather forecasting requirements or specifically for their utility in climate. An optimal climate observing system would reduce duplication of efforts and ensure that gaps are filled by some level of coordination among the various observing systems that have grown out of research programs. This coordination would include some level of analysis, since the current situation has resulted in roughly one analysis system for each observing system without much cooperation and intercomparison among them.

Much of the data used for climate monitoring is derived from observations made to support weather analysis and forecasting. There is a basic tension between the requirement that observing systems remain consistent over time to facilitate support for climate services and the constant efforts to introduce new instrumentation and technologies for improved and more cost-effective observations in support of weather analysis and prediction. An idealized climate monitoring system is one that could systematically incorporate the data from new technologies into the climate record without introducing significant biases relative to data in the historical record. Such a monitoring system requires dedicated efforts to compare the data from older instruments to the data from the new systems over several annual cycles sampled in different climate regions.

In an attempt to provide a benchmark for continuing stable climate observations within the United States, NOAA has established the US Climate Reference Network (USCRN). The USCRN, established in 2008, has a 50-year design lifetime with 114 observation sites (See the USCRN web address in the References). The system was designed primarily to observe temperature, precipitation, soil moisture, and soil temperature, but solar radiation, wind speed, relative humidity, and skin temperature and wetness are also recorded.

The USCRN is a logical attempt to establish benchmark climate observations, but history has not been kind to the long-term support for such efforts. Before the establishment of the USCRN there were a small number of locations designated to document the climate record. Many of them were established well before that advent of modern instrumentation, computers and communication systems. For example, the Blue Hill Observatory (BHO) in Milton Massachusetts was established in 1885 and has a long history (Conover, 1985). Meteorological observations continue to this day. In its long history, however, there have been many changes, not only in instrumentation, perhaps compromising the climate record, but also in financial support. The BHO was built privately and was then supported by Harvard University through a grant from the observatory's founder. Support then moved to the Boston Metropolitan District Commission, and currently the BHO is supported by contributions to a non-profit volunteer organization (see the Blue Hill Observatory website listed in the References).

The history of support to the BHO illustrates the difficulties in maintaining support for continued climate observations over long time periods. Only time will tell whether the USCRN will continue to be federally supported through changes in administrations and national priorities between now and the mid-twenty-first century. An ideal climate monitoring system then, needs to be robust to changes in any particular network of observations by adopting procedures for maintaining continuity. In the case of the USCRN, particular attention must be paid to any signs of slow but steady attrition in the number of observation sites, instrumentation degradation, or instrument change.

The USCRN provides a benchmark for climate observations, but other observational networks, such at the National Weather Service (NWS) Cooperative Observing Program (COOP, discussed in Chapter 6), Automated Surface Observation System (ASOS), operated primarily by the NWS, and Automated Weather Observation System (AWOS), operated by the Federal Aviation Administration at airport sites, are also important resources. Data from over 8,000 observation sites in the COOP, binned into 344 climate divisions, extending back to 1931 with estimates derived from statistical models reaching back to 1895, have long been a major source of data for climate studies. Data from the ASOS, AWOS, and other networks discussed here may be incorporated into the climate records after being corrected for systematic bias. The future climate monitoring system should build on these networks to increase the number of observation sites and the kinds of data that might be gathered.

Traditionally, climate monitoring has focused on monthly and seasonal data based on summaries of daily data. Many contemporary and emerging observing systems are capable of sampling the environment at much higher frequencies and storing and transmitting these data. Higher frequency observations will allow for the definition of climate monitoring to evolve into analyses of the statistics of daily weather over time scales ranging from weeks to several months.

Other future sources of surface climate data could come from sensors incorporated into cell phones, tablets, and other portable platforms. The instrumentation that needs to be developed to support driverless vehicles may also serve as a data source. As satellite observations become able to provide data at higher spatial resolutions, *in situ* observations from the sources listed above could be used to help anchor those satellite estimates. The challenge lies in developing methods to gather, archive, and merge these data with more traditional climate observations. We will need to learn how to relate the data from new sources to the traditional observations and how to make the data available for use in model data assimilation and other applications, including climate.

Short-term climate prediction, like climate monitoring, has traditionally focused on the monthly and seasonal time scales. The enhanced capabilities in climate monitoring, discussed earlier, as well as societal demands, are likely to lead to the production of forecasts tailored to user requirements. For instance, climate forecasts might include outlooks for the planting season of a particular crop at some region of the country, or the forecasts might be tailored for construction schedules.

We expect that climate research and model improvements will lead to further improvements in short-term climate predictions. Despite continued progress, however, climate outlooks are inherently probabilistic. Very often, the user is faced with making yes or no decisions based on only a slight difference in the probabilities between contrasting climate outcomes, say a 58% versus 42% chance of a wet period spanning some critical activity such as planting a crop. In order to meet its full potential, the climate enterprise will need to provide the end users with improved tools to interpret current climate conditions and the forecasts in the context of their applications. The development and application of appropriate tools will require a new class of applied scientist, who is well grounded in climate analysis and prediction, but who also has expertise in the fields that wish to apply climate monitoring information and forecasts. It is our hope that this book will be helpful to any who pursue that profession.

Suggested Further Reading

Samuelson P. A., 1954: The Pure Theory of Public Expenditure, *Review of Economics And Statistics*, **36**, 387–3891. (Provides a discussion of the term "Public Good.")

Questions for Discussion

12.1 Compare the methods and datasets used to monitor intraseasonal-to-interannual climate variability to those used to assess climate change.

12.2 Explain the rationale for using probabilistic formats for climate forecasts. Discuss methods of verifying such forecasts and the connection between climate monitoring and forecast verification.

12.3 What role do climate models play in forecasting intraseasonal-to-interannual climate variability? What about decadal variability? Climate change? How might these roles change over the next decade?

12.4 What risks imperil the current climate monitoring and prediction efforts? Do you expect our ability to understand and predict climate variability and change to improve or degrade over the coming 10 years? How about 50 years? Describe the various risk factors in the United States and elsewhere.

Appendix A A Short Guide to Some Statistics Used in Climate Analysis

A.1 Introduction

Climate analysis requires the use and interpretation of standard statistical tools. In this book, we assume that the reader is familiar with these tools, and cite standard references for those wishing to learn more details about them or to brush up on statistical methods. For those less familiar with statistics, we present here brief descriptions of common statistical terms and tools, but without the mathematical details. This short Appendix is something more than a glossary, but less than a tutorial. Our aim is to jog the memories of those who have had a formal course on statistics, provide the uninitiated reader with the concepts behind various statistical tools, and discuss some of the caveats in their interpretation. Among the standard references in atmospheric studies are *Statistical Methods in the Atmospheric Sciences* (Wilks, 2011) and *Statistical Analysis in Climate Research* (von Storch and Zwiers, 2002).

A.2 Basic Statistics

Earth's climate is often described in terms of the simple statistics briefly defined here.

Mean The arithmetical average of data over an area or over some time span, or both. An example is the average monthly temperature at a location. Sometimes referred to as the "expected" value in the statistical literature.

Anomaly The difference of a data value, or an observation, from the mean of the historical data.

Median The midpoint value in a collection of data that are ranked from the smallest to the largest value. Half the data are less than the median and half greater.

Mode If data are binned into intervals of equal size, the mode is the interval with the largest number of occurrences. In some cases, called bimodal, two different nonadjacent bins will have a larger number of occurrences than adjacent bins.

Histogram Graph of the number of occurrences of data of a certain value, or within a certain interval, on the vertical

	axis versus values of the data on the horizontal axis. The vertical axis may be displayed as a percentage or proportion of the number of occurrences. Histograms of observed data are often analyzed under the assumption that they are sampled from a background distribution that is one of the theoretical PDFs discussed in this appendix.
Stationary	A dataset is stationary when the statistics of subsets of the data are the same as those of the full dataset. For instance, the mean and standard deviation of one 30-year sample in 100 years of data will be essentially the same as any another 30-year sample.
Probability Density Function (PDF)	This terminology is most often used in discussing theoretical distributions of data and in conjunction with statistical tests of significance. Climate data are often assumed to have mathematically well-defined distributions, such as the Gaussian and Gamma.

The **Gaussian** or **Normal Distribution** is likely the most familiar. In a normal distribution, the mean, median, and mode are equal. The data are distributed symmetrically about the mean, with about 68% (95%) of the data falling within one (two) standard deviation(s) of the mean. This PDF shape is the familiar "bell curve."

In **skewed** distributions, for example the **Gamma** distribution shown in Chapter 2 (Figure 2.1), the data are not symmetrical about the mean but spread out more strongly toward one side. For example, in daily precipitation data the peak in the histogram very often occurs toward the left-had side of the graph because the most frequent rainfall occurrences tend to be relatively small, but a few days, those with thunderstorms for example, have much larger precipitation amounts.

A.3 Correlations

Simple (Linear) Correlations	A statistical measure of the strength of the linear relationship between two quantities. Correlation is sometimes misinterpreted as establishing cause and effect.
Autocorrelations (Lag Correlations)	Correlations between a quantity and the same quantity at a different time. Historically used to identify potential empirical climate prediction tools.
Pattern Correlations	Correlations between pairs of maps, treating the data values at map locations, for instance, as if they were points in two different time series. An important nuance for the practical user is that data values on a map can represent areas of different sizes, depending on the map projection used. Pattern correlations in such cases must be adjusted for this to avoid incorrectly emphasizing certain areas.

Variance — The variance is a measure of the "noisiness" of the data. Historically, it has often been referred to as the "spread" in meteorological studies. Mathematically, the variance of a quantity is estimated as the mean of the squared anomalies (differences from the mean) in the data. The **covariance** is the degree to which variance in one data set is related to that in another and is estimated as the mean of the products of the anomalies.

Variance Accounted For — The fraction, or percentage, of the total variance associated with a phenomenon. Often referred to as variance "explained." For instance, the annual cycle "accounts for" (i.e. is responsible for producing) the bulk of monthly temperature variance in midlatitudes. In correlation analysis, the variance accounted for goes as the square of the correlation. For example, a correlation of 0.5 (0.3) accounts for 25% (9%) of the variance.

Standard Deviation — The positive square root of the variance.

A.4 Spectral (Fourier) Analysis

Our purpose in introducing spectral analysis in this book is to give those not familiar with it some idea of what it does and to identify some of the issues that must be considered in interpreting the analysis. Fourier (1768–1830) showed that any time series can be decomposed as a series of sines and cosines, and that the original data can be recovered exactly as the sum of those series. Spectral analysis, also discussed in Chapter 2, identifies the relative contributions of variance at various temporal scales to the total variance. Often, the analysis is presented as either variance, or logarithm of the variance, on the vertical axis and either frequency, or period, on the horizontal axis.

Spectral analysis and its interpretation are somewhat of an art. The completeness and length of the input data, how gaps in data were filled, and several other factors will influence the analysis. The uncertainty in any spectral estimate depends on how many frequencies are averaged to obtain the estimate. Greater frequency resolution implies less averaging and so is coupled to greater uncertainty in the magnitude of the spectral value. Once the period considered is a substantial fraction of the total length of the time series the spectral estimates become very uncertain.

A.5 EOF and Related Analysis

EOFs, PCA, and the like – Empirical Orthogonal Functions (also known as Principal Component Analysis) and others are the names given to a collection of statistical techniques that are often used in the analysis of a set of historical charts. The subjects of the analysis could be digitized versions of hand-analyzed charts, maps generated by numerical models, satellite observations, or virtually any dataset that can be put into matrix form. Interpretation of the analyses is simplified if the data fields are complete in space and time. If there are spatial or temporal gaps in the datasets they are usually filled with empirical estimates. In practice, the data, say for an EOF analysis, are arranged into a matrix where, in most cases, each row represents the points on a map, at a single time, and the columns represent different observation times. The matrix elements are transformed into a covariance or correlation matrix for which the complete set of eigenvalues and eigenvectors are computed. The eigenvectors (the "empirical orthogonal functions") can then be interpreted as characteristic spatial and temporal patterns, and the corresponding eigenvalues can be used to calculate the percentage of the total variance of the dataset explained by each such pair.

Each of these statistical tools produces a set of characteristic patterns accompanied by a time series, where the amplitudes of the time series represent the strength of a pattern at any time in the data record. These techniques order the patterns from the most important to the least important, where importance is defined as the percentage of the total variance represented by the pattern. The techniques generally differ from each other in technical details, such as how quantities are scaled. These powerful analysis tools are generally available in standard software packages and many of the software packages provide documentation that is helpful in interpreting the output. An introduction to the mathematical details and some examples can be found in von-Storch and Zwiers (2002), Wilks (2011), or online at the website listed under "EOF and SVD" in the References.

Example: An EOF analysis of the sea surface temperature in the Tropical Pacific with the annual cycle removed (Figure A.1) shows that the leading pattern, the one having the largest amount of variance, is associated with ENSO. The mapped quantities are elements of the eigenvector corresponding to the largest eigenvalue. The time series associated with this pattern (Figure A.2), computed as projections of the original map data onto the leading eigenvector, indicates how the strength of this pattern varies in the historical record.

Likewise, the next leading pattern (Figure A.3) corresponds to the second-largest eigenvalue, and accounts for the largest share of the remaining variance. Its time series (Figure A.4) has been ascribed to multi-decadal variability. EOF and similar analyses will always produce as many pattern/time series pairs as the lesser of the number of points in space or in time, but the higher order pairs (those representing the least variance) are often of no physical significance. In a set of random fields of data, one of the EOF patterns will account for more variance than all the others strictly by chance, another pattern will account for slightly less variance than the first but more than the remaining patterns, and so on.

A.6 Statistical Significance

All statistical tools employed in climate analysis should be accompanied by guidance on how to estimate the uncertainties associated with them. Related to the uncertainty are

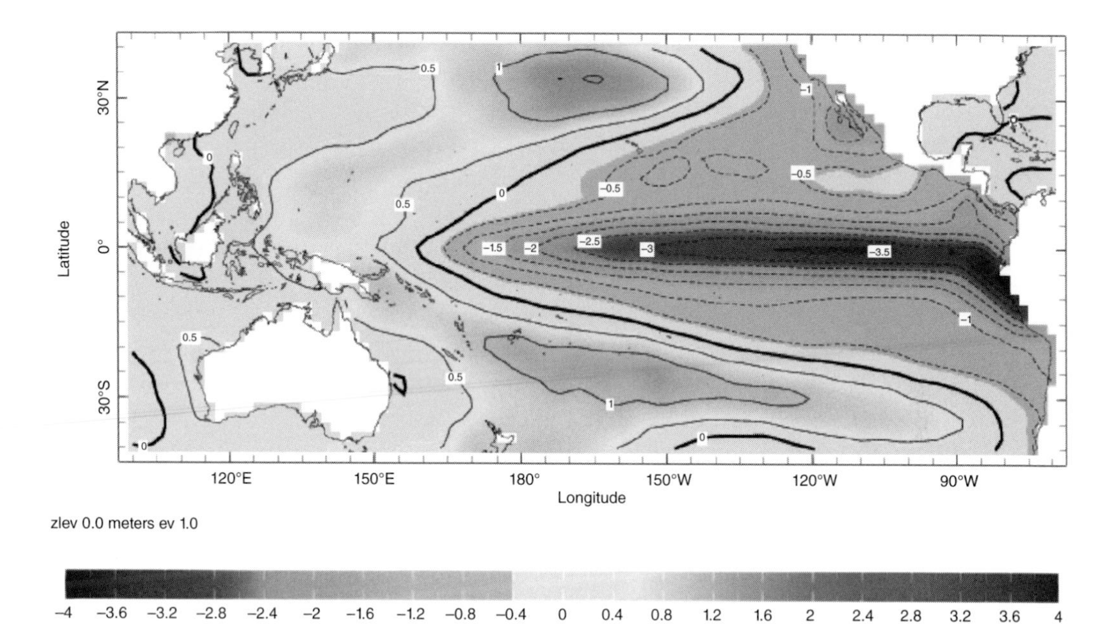

Figure A.1 Leading EOF pattern for the Pacific Ocean SST using ERSSTv3b monthly SST anomalies for 1981 to 2010 (Smith et al., 2008). This pattern accounts for 35% of the variance remaining after the annual cycle was removed. (Courtesy of M. Bell, IRI Data Library). A black-and-white version of this figure appears in some formats. For the color version, please refer to the plate section.

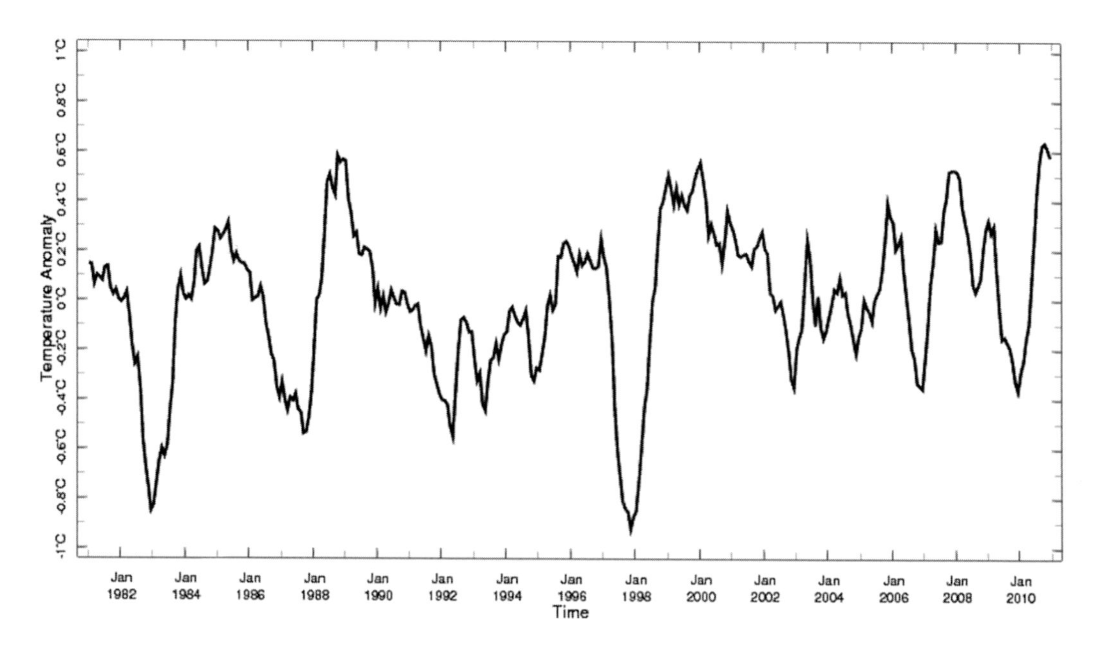

Figure A.2 The time series corresponding to the leading EOF pattern. (Courtesy of M. Bell, IRI Data Library).

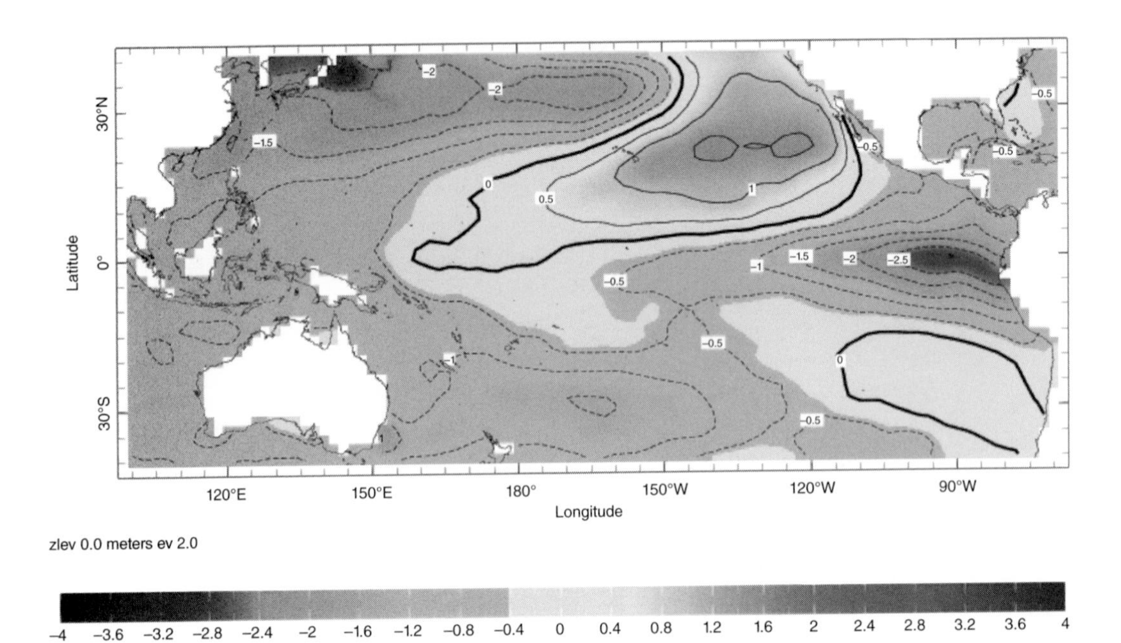

Figure A.3 The second EOF pattern for the Pacific SST accounting for 12% of the variance. (Courtesy of M. Bell, IRI Data Library). A black-and-white version of this figure appears in some formats. For the color version, please refer to the plate section.

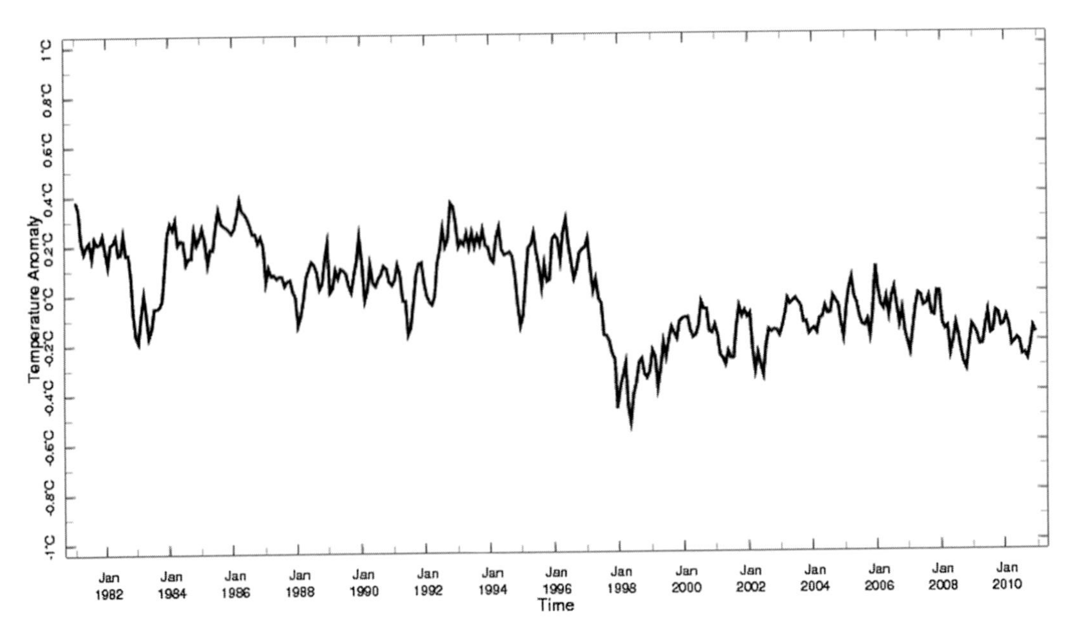

Figure A.4 The time series associated with the second EOF pattern of Pacific SST. (Courtesy of M. Bell, IRI Data Library).

estimates of the fraction, or percentage, of total variance accounted for by the analysis and the statistical significance. Ways to estimate statistical significance are very often included in the documentation for the statistical tool. The statistical significance relates to the probability that the result being discussed could have arisen strictly by chance provided that the data meet the assumptions inherent in the analysis. Most climate data can be pre-processed to closely approximate the assumptions inherent in the statistical analysis but no dataset is perfect. Many climate studies assume that the data are distributed according to the Gaussian, or Normal, distribution discussed earlier, even though, in many cases, the actual data distributions only very roughly fit the ideal. Thus, the actual statistical significance may not conform to the theoretical estimates.

Monte Carlo Testing – Since the dawn of the computer age, some statistical analyses have moved away from the assumption that the data fit some theoretical distribution and have moved toward testing the data based on their empirical distribution. These Monte Carlo[1] schemes rely on re-sampling the data to arrive at empirical estimates of statistical significance (Livezey and Chen, 1983). The availability of high-speed computers has made Monte Carlo testing less onerous and easier to design.

An advantage of Monte Carlo tests is that they do not require that the data be distributed according to one of the analytically defined frequency distributions. Monte Carlo testing requires that the data be randomly re-sampled for a very large number of realizations (all possible combinations if that is practical) to determine the likelihood that a given result may have occurred by chance.

False Discovery Rate (FDR) – Recently, Wilks (2017) has reviewed a technique widely used in other disciplines outside of the geophysical sciences to test significance when there are several hundreds, or thousands, of tests to be evaluated simultaneously. For example, such a test might be based on correlations plotted on a grid. In climate studies this could be, say, mapped correlations of teleconnection patterns, with gridpoint correlation, each tested for the plausibility that the underlying ("population") relationship is zero. The use of FDR is an alternative to Monte Carlo significance testing of mapped correlations that is not as computationally demanding.

[1] In Monte Carlo testing the data are randomly sampled thousands of times to estimate the percentage of the time that a statistical result would occur by chance. For example, if a phenomenon occurs in association with 7 out of 10 ENSOs the Monte Carlo test might sample the data a thousand times to see what percentage of the time the phenomenon occurs 7 or more times in a randomly selected set of 10 years.

Appendix B Vorticity/Divergence, Stream Function/Velocity Potential

The discussion that follows gives a bare bones mathematical description of these quantities. More detailed discussion can be found in many standard meteorological texts, e.g. Wallace and Hobbs 2006, or in mathematical textbooks that include a section on vector calculus. In meteorology, the vorticity at a given pressure level is formally defined as the curl of the wind vector.

$$\nabla \times \mathrm{V} = \zeta \qquad \text{B.1}$$

Typically, the horizontal components of the vorticity are taken to be small so that Eq B.1 reduces to

$$\zeta = \mathbf{k}\left(\frac{\partial v}{\partial x} - \frac{\partial u}{\partial y}\right) \qquad \text{B.2}$$

In the early numerical models the vorticity was computed using a finite difference version of Eq B.2 at grid points. Contemporary models have become considerably more sophisticated but the computation of interest is still for the local vorticity.

Likewise, in meteorology the divergence at a given pressure level is formally defined by

$$\mathrm{Div} = -\nabla \cdot \mathrm{V} \qquad \text{B.3}$$

Sign notation is such that a negative value of Div denotes convergence.

The stream function represents the rotational part of the flow. In this mathematical decomposition (shown for simple two-dimensional flow for simplicity) can be written as

$$\nabla \cdot \mathrm{V} = \frac{\partial u}{\partial x} + \frac{\partial v}{\partial y} = 0 \qquad \text{B.4}$$

where the stream function (ψ) is found by solving $u = -\partial\psi/\partial y$ and $v = \partial\psi/\partial x$.

The stream function, Eq B.4, is associated with the rotational component of the wind, and accounts for a much larger fraction of the total large-scale wind field than the velocity potential discussed below.

The velocity potential is defined by the divergent part of the flow, mathematically the component of the wind for which the curl of the wind is zero. One of the fundamental theorems of vector calculus states that if the curl of a vector is zero than the vector can be written as the gradient of a scalar field. Or in mathematical notation

$$\text{If } V = \nabla\varphi \quad \text{where } \varphi \text{ is a scalar field then } \nabla \times V = 0 \qquad \text{B.5}$$

The velocity potential is the 'φ' in Eq B.5. Very often maps of the velocity potential also display the "divergent wind", the gradient of φ as in Eq B.5.

Appendix C Preliminary Examination of the Data

A number of excellent datasets can be found in various archives referenced in this book; however, none is perfect. Data centers very often have more than one version of a particular dataset. The various versions may include the raw, or unadjusted, data as well as data adjusted for changes in location, instrumentation, or observing practices. Some versions of the data may include a complete record for all of the locations in the network, irrespective of the quality of the data or the amount of data missing from the original observations. These complete data records are generally constructed by filling in missing or suspect data with estimates obtained through interpolation or statistical methods. A preliminary examination of the data is necessary to determine the extent to which estimates and adjustments have been made. The discussion in this appendix provides a broad outline of the practical considerations that arise when constructing climate datasets.

C.1 Data Plotting Tools and Software

As discussed in Chapter 2, visual examination of the data is one of the best ways to determine whether they are appropriate for their intended use. Simple graphics, such as time series and scatter plots, will show if the data behave reasonably. The analyst should ask questions such as: Are the data in the expected units? What are the ranges of extremes? Are there large temporal gaps in the data? Unexpected or inappropriate answers to these questions will highlight the need for additional examination of the supporting observations and documentation. Time series plots permit immediate identification of data gaps, sudden changes in the character of the data record, and unrealistic/unphysical extreme values, as illustrated in Figures 2.2 and 2.3. Even if the data being examined have already been adjusted for these problems we recommend that the analyst plot both the raw data and adjusted data time series on the same graph to gain insight on the characteristics of the data adjustments.

There are several online resources that provide plotting software, for example GrADS (Doty and Kinter, 1995) and other tools listed in the References. In many cases, the data center websites listed in the references offer the option of plotting data available on these sites but do not provide plotting software. Some investment in time and effort will be required to learn how to identify and find the appropriate data, and to access and download the data in formats compatible with commercial or other available plotting software packages. A few sites, such as the IRI/LDEO Data Library, do provide access to data, flexible plotting, and analysis capabilities, as well as guidance and tutorials

C.2 Addressing Data Gaps

Almost all historical time series of observations have periods with missing data. Time series with no gaps may have had gaps in the original records that were filled using one, or some combination, of the methods discussed in this appendix. It is a common practice at data centers to indicate missing observations in the raw time series by (an unrealistic) missing value indicator such as −999. Simple plots of the data will identify these missing value indicators in a dataset and enable the analyst to avoid including them in computations of statistics. Caution is required, particularly in the use of older records and those from non-meteorological sources, which may not have explicitly indicated missing values and, instead, have left “blanks” in the record. Early automated attempts to digitize data from such sources sometimes interpreted the “blank” as a “zero” value. If the data documentation for older data does not make it clear that this possibility was addressed the data should to be plotted or otherwise examined to identify an unrealistic frequency of occurrence of zero values, for example in summer temperatures near the Gulf of Mexico.

Most statistical analyses require complete datasets to facilitate interpretation of the analysis. So, some realistic estimate for the missing values in data gaps should be made in order for the analysis to proceed. There are several standard methods used to fill gaps, and in most cases the method used is described in the metadata that accompany the dataset. If the missing observations occur at isolated single points, they can be replaced through interpolation from adjacent values, by replacing the missing value with the mean over the whole, or some part, of the time series, or by replacing the missing value with a value chosen randomly from the same time series. Each of these methods will introduce a small difference in statistics computed from the gap-filled time series compared to the statistics from a truly complete data set, but in most cases these differences will be small as long as the number of gaps is small compared to the total amount of data. Some empirical analysis is recommended to determine how the methods of filling gaps and percentage of gaps influence the analysis.

The method used to replace missing values must take account of any factors that systematically modulate the dataset, such as the diurnal or seasonal cycles. In constructing data plots, it is often useful to examine the deviations in observed data values from some reference, or base, state. When examining several years of data, for example, it is useful to remove the mean annual cycle, as was discussed in Chapter 2. In many cases construction of base-period means will need to be an iterative process, since initial computation of the mean state or mean annual or diurnal cycles may include data values that are in error or biased in some way and will have to be removed in order to get more accurate estimates of the true mean.

In cases where there are longer data gaps there are fewer, and less satisfactory, options for filling gaps. Replacement of missing data by random selection of values from other parts of the record or replacement with the mean value is unlikely to be appropriate. In such cases, estimates to fill gaps at one location might be obtained from statistical relationships between data at the location with missing data and the data at nearby observing sites. Linear regression or multiple linear regression have been used to fill in the gaps in some datasets. It is not uncommon to find that no single observing site in a dataset has complete data. That is, each of N nearby sites has gaps somewhere in their records. In such cases, multiple regression might be used to form a virtual data record to represent the temperature for some location or region (Vose et al., 2014). These methods are more applicable to data that vary smoothly over space, for example temperature anomalies, rather than temperature.

Some analyses may require the examination of hundreds or thousands of observing sites, making it impractical to examine each data record visually. A useful approach in such cases is to examine the aggregate statistics of the data. For example, construction of frequency distribution functions can identify outliers and thus reduce the number of individual observing sites that require closer examination. Careful application of statistics can be used to identify data gaps, large random errors, abrupt changes in overall values (change in bias), and other artifacts in the data.

Appendix D Components of the Mean Water Budget

The components of the water budget can be estimated in a number of different ways, often by investigation of the atmospheric and terrestrial branches of the hydrologic cycle obtained from observations (Rasmusson 1967, 1968; Ropelewski and Yarosh 1998).

The terrestrial branch of the water budget can be written as

$$\{\Delta S\} = \{P - E\} - D \qquad \text{D.1}$$

where ΔS represents the change in storage, i.e. the water that goes into the water table

P is the precipitation rate

E is the sum of the evaporation from the surface and transpiration rates of moisture from plants

and D is the discharge, often referred to as "runoff," into streams, lakes, and ponds. The curly brackets denote an average over an area.

In practice, the precipitation is measured at a location, averages of several locations in an area, or by satellite estimates. The discharge is estimated from stream gage data. The change in storage, S, is not well measured and may be estimated as a residual difference between the precipitation and sum of the evapotranspiration and discharge or assumed to be zero. In estimating the residual it is assumed that terms in the water budget are in balance, i.e. increases in one budget quantity are balanced by decreases in another. The water budget also includes an atmospheric branch that can be written as:

$$\frac{\partial\{W\}}{\partial t} = \frac{1}{A}\int F_n dC = -\{P - E\} \qquad \text{D.2}$$

where: W is average atmospheric water content over the area of interest and the partial derivative refers to the time rate of change of water held by the atmosphere, A is the surface area under consideration. The integral represents the amount of moisture converging into or diverging, from the area, calculated over the depth of the troposphere. The integral is evaluated as the sum over all pressure levels from the surface to the top of the troposphere. The term F_n is the vapor transport normal to the contour of the area and C designates that the integral is a contour integral. The contour integral is mathematically equivalent to the flux of moisture either diverging from or converging into the analysis area.

P is precipitation, generally available as a measurement, and E represents the sum of evaporation and evapotranspiration, for which direct measurements are generally not available. This quantity is often estimated using either empirical or numerical models, or as a residual.

The moisture budget described here is sensitive to uncertainties in the measurements and is best suited for computation of average monthly, seasonal, and longer averages over large areas, i.e. those exceeding 10^6 sq km. The observational studies are limited by the lack of uniform network of surface evaporation and subsurface water content, including soil moisture and ground water levels, i.e. the storage.

Glossary

Advection: Transported by the wind or water.
Albedo: The fraction of incoming solar radiation that is reflected from a surface.
Analysis: The decomposition of a complex substance or topic into its component parts in order to improve understanding.
Anemometers: Instruments that measure wind direction and speed.
Annual Cycle: The component of climate variability that depends on the time of the year but is independent of the year.
Anomaly: The difference of a data value, or an observation, from the mean over some period of the data, or observation.
Atmosphere: The gaseous envelope extending from the surface to several hundred kilometers. Its characteristics, including temperature, density, composition, and movement, exhibit variations on a wide range of time and space scales.
Austral: Referring to the Southern Hemisphere, as in Austral, or Southern Hemisphere, summer.
Barometers: Instruments that measure atmospheric pressure.
Biosphere: The label applied to all living things as a group.
Blackbody: A theoretical construct describing a surface that absorbs all, and reflects none, of the radiation falling upon it.
Boreal: Referring to Northern Hemisphere, as in Boreal, or Northern Hemisphere, summer.
Canonical Correlation Analysis: A statistical procedure for identifying relationships between two sets of data. Sometimes referred to as Canonical Variate Analysis. Used in climate analysis to create predictive relationships among parameters, for example between sea surface temperature and precipitation.
CESM: The NCAR Community Earth System Model (CESM) consisting of five component models, Atmosphere, Ocean, Land Surface, Land-Ice, Sea-Ice, and a coupler. These models may be run either as standalone or coupled to any number of the other four components.
Climate: The mean state and variations of the physical environment of Earth's surface and nearby atmosphere and ocean.
Climate Analysis: The science and practice of creating and examining complete and comprehensive datasets to describe, understand, and predict the state and evolution of the climate system.

Climate System: Includes the surface of the globe, the oceans, and most of the atmosphere as well as internal and external forcing, together with the interactions among these across a range of time scales. The components of any system are coupled to one another.

Correlations: A statistical measure of the strength of relationship between two quantities. Correlation is often misinterpreted as establishing cause and effect.

Coupling: The linkage between two entities, such as components of the climate system, whose behaviors are influenced relatively strongly by each other.

Covariance: The variance (see Appendix A) shared by two datasets and is estimated as the mean of the products of the anomalies.

Cryosphere: Consists of the persistent ice and snow on Earth's surface. It includes both sea ice and ice and snow on land. Highly transient ice and snow, such as drifting icebergs or short-lived snow cover on land, are not included but seasonal changes in them are.

Cyclonic: Rotation about the vertical having the same sense as the Earth's rotation. Counterclockwise in the Northern Hemisphere, clockwise in the Southern Hemisphere as viewed from above. (After the AMS Glossary of Meteorology, 2012.)

Data Assimilation: The combining of diverse data, possibly sampled at different times and intervals and different locations, into a unified and consistent description of a physical system, such as the state of the atmosphere. (AMS Glossary of Meteorology, 2012.)

Dew Point: The temperature at which a parcel of air at constant pressure reaches a relative humidity of 100% and moisture begins to condense.

Diurnal Cycle: The variations in weather over 24 hours associated with the rotation of the earth. The mean diurnal cycle and its annual cycle through the year characterize the local climate.

Downscaling: The process of estimating the value of a parameter at a specific location from the values contained in a reanalysis or forecast model grid.

Emergent Behaviors: Complex phenomena that arise from the interaction of components of systems, but which are not observed in the individual components in isolation.

Evaporation: The process through which surface liquid water, from water bodies or land, enters the atmosphere as vapor. For convenience, it is generally used to include sublimation, in which ice transforms to vapor.

Evapotranspiration: The term used for evaporation facilitated by plants in which water is transported from the roots to the leaves and evaporated there, thus enabling nutrient transport.

Free Atmopshere: That portion of Earth's atmosphere where the effect of surface friction on the air motion is negligible, and in which the air is usually treated as an ideal fluid. (After the AMS Glossary of Meteorology.)

Freeboard: The difference between water level and top of the floating ice.

Gaussian or Normal Distribution: is likely the most familiar. In a normal distribution, the mean, median, and mode are equal. The data distributed symmetrically about the mean, with about 68% (95%) of the data falling within one (two) standard deviation(s) of the mean.

Geopotential Height: Defined on the National Weather Service website as " ... roughly the height above sea level of a pressure level. For example, if a station reports that the 500 mb [500 hPa] height at its location is 5600 m, it means that the level of the atmosphere over that station at which the atmospheric pressure is 500 mb is 5600 meters above sea level. This is an estimated height based on temperature and pressure data."

GRACE: Gravity Recovery and Climate Experiment.

Hindcasts: Forecasts made from past initial data to predict past events.

Humidity: The amount of water vapor in the air.

In situ: Measurements taken at observation sites as opposed to remotely sensed.

Land Surface: In the context of climate analysis includes everything that contacts the atmosphere and is not included in the ocean and the crysosphere.

Leads: Linear cracks in the ice, with widths ranging from a few meters or less to a kilometer or more. Leads form when sea ice fields diverge as they move. The water in leads begins to freeze almost immediately during the cold seasons. (After the definition in the NSIDC website.)

Mass Balance: For ice sheets and glaciers this is the difference between the mass added by snowfall and the mass loss due to melting.

Noise: The bias and random variations in observations introduced by the observing system, the human interaction (if any), and the recording and storage mechanisms. In addition, weather variations themselves are sometimes referred to as noise in the context of climate, since they can be thought of as high frequency variations about the more slowly varying climate signal.

Objective Analysis: An analysis that is free from any direct subjective influences resulting from human experience, interpretation, or bias. (From AMS Glossary of Meteorology.)

Ocean/Sea: For purposes of this book, ocean will refer to any permanent body of water that is large enough to interact with the atmosphere so as to influence the large-scale climate.

Operational: In the context of weather forecasting refers to regular, generally periodic, activities that are associated with forecasts issued to the public. Since such forecasts are time-sensitive and must be issued as quickly as possible, operational activities are often referred to as real-time, meaning as quickly as possible. Several weather observations gathered to support operational activities have been archived for use in climate analysis.

Outgoing Longwave Radiation (OLR): The radiative character of energy radiated from the warmer earth to cooler space, often derived from window channel measurements from satellites in polar orbit (from the AMS Glossary of Meteorology). In climate analysis, OLR from operational polar orbiting meteorological satellites is used as a semi-quantitative index of deep convection.

Phase: The change from any of the solid, liquid, or gaseous states to any of the others. Most commonly referred to in reference to ice, liquid water, and water vapor.

Pixel: The smallest resolved area in an image. The term originated in the early days of computerized image processing.

Polynya: A large area of open water surrounded by pack ice, resulting from interactions with oceanic and atmospheric circulations and heat exchanges. They may persist for several months into years.

Precipitation: The processes that transfer water in either solid or liquid form from the atmosphere to the surface.

Quality control: When referring to the use of observations in creating an analysis, the process by which erroneous observations are eliminated or, in some cases, corrected.

Radiances: The measurement of radiation as a function of wavelength.

Radiometers: Instruments that measure the amount of radiation coming from the atmosphere and surface in different parts of the electromagnetic spectrum. Radiometers may be part of a surface observational site, or used as remote sensors aboard aircraft or satellites.

Relative Humidity: The ratio of the observed specific humidity to the saturation-specific humidity at the observed temperature and pressure.

Representativeness Error: Correct observations that are capturing phenomena that are not representative of the surrounding environment on the scales appropriate to a numerical model.

Roughness: Refers to the extent to which the land surface influences the winds above it.

Sea Ice Concentration: A measure of the percentage of an area of the sea covered by ice.

Snow Course: A permanent site to monitor snow pack conditions, including liquid water equivalent, at representative locations in mountainous regions. Four or more measurements are typically made along 300 m (1000 ft) long courses. (After the USDA website and AMS Glossary of Meteorology.)

Solar Constant: The measure of the intensity of solar radiation, integrated over all wavelengths, at the average distance of Earth's orbit.

Specific Humidity: The ratio of the mass of water vapor to the total mass in a parcel of air.

Spectral Windows: Radiation at these wavelengths passes through without much change in the same manner that visible light passes through a glass window.

Statistical Significance: Relates to the probability that the result being discussed could have arisen strictly by chance provided that the data meet the assumptions inherent in the analysis.

System: A collection of coupled entities.

Teleconnection: A statistical linkage between climate variations occurring between widely separated regions of the globe and most often identified by a significant correlation. The term was first used by Sir Gilbert Walker in his seminal studies of global pressure patterns.

Thermocline: A layer of water where the vertical temperature gradient is greater than the gradients above and below it. Thermocline also refers to the layer in which such a gradient occurs. (After the AMS Glossary of Meteorology.)

Total Solar Irradiance (TSI): The total amount of radiation emitted from the sun across all wavelengths.

Weather: As distinguished from climate, weather consists of the short-term (minutes to days) variations in the atmosphere. Popularly, weather is thought of in terms of temperature, humidity, precipitation, cloudiness, visibility, and wind. (Source: the AMS Glossary of Meteorology.)

References

Adler, R. F., and A. J. Negri, 1988: A Satellite Infrared Technique to Estimate Tropical Convective and Stratiform Rainfall. *J. Appl. Meteor.*, **27**, 30–51.

Adler, R. F., C. Kidd, G. Petty, M. Morissey, and H. M. Goodman, 2001: Intercomparison of Global Precipitation Products: The Third Precipitation Intercomparison Project (PIP–3). *Bull. Amer. Meteor. Soc.*, **82**, 1377–1396.

Adler, R. F. et al., 2003: The Version-2 Global Precipitation Climatology Project (GPCP) Monthly Precipitation Analysis (1979–Present). *J. Hydrometeor.*, **4**, 1147–1167.

Alley, W. M., 1984: The Palmer Drought Severity Index: Limitations and Assumptions. *J. Clim. And Appl. Meteor.*, **23**, 1100–1109.

AMS, 2013: Drought: An Information Statement of the American Meteorological Society. *Bull. Amer. Meteor. Soc.*, **94**, 1932–1936.

Angell, J. K., 1988: Variations and Trends in Tropospheric and Stratospheric Global Temperatures, 1958–87. *J. Clim.* **1**, 1296–1313.

Angell, J. K., and J. Korshover, 1964: Quasi-Biennial Variations in Temperature, Total Ozone, and Tropospheric Height. *J. Atmos. Sci.*, **21**, 479–492.

Angell, J. K., and J. Korshover. 1983: Global Temperature Variations in the Troposphere and Stratosphere, 1958–82. *Mon. Wea. Rev.* **111**, 901–21.

Arguez, Anthony, and Scott Applequist, 2013: A Harmonic Approach for Calculating Daily Temperature Normals Constrained by Homogenized Monthly Temperature Normals. *J. Atmos. Oceanic Technol.*, **30**, 1259–1265.

Arkin, P. .A., 1979: The Relationship between Fractional Coverage of High Cloud and Rainfall Accumulations during Gate over the B-Scale Array. *Mon. Wea. Rev.*, **107**, 1382–1387.

Arkin, P. A., 1982: The Relationship Between Interannual Variability in the 200 mb Tropical Wind Field and the Southern Oscillation. *Monthly Weather Review*, **110**, 1393–1404.

Arkin, P. A., 1989: The Global Precipitation Climatology Project. *Adv. Space Res.*, **9**, 311–316.

Arkin, P. A., and B. Meisner, 1987: The Relationship between Large-Scale Convective Rainfall and Cold Cloud over the Western Hemisphere during 1982–84. *Mon. Wea. Rev.*, **115**, 51–74.

Arkin, P. A., and P. Xie, 1994: The Global Precipitation Climatology Project: First Algorithm Intercomparison Project. *Bull. Amer. Meteor. Soc.*, **75**, 401–419.

Arkin, P. A., R. Joyce, and J. E. Janowiak, 1994: IR Techniques: GOES Precipitation Index. *Remote Sens. Rev.*, **11**, 107–124.

Arkin, P. A., Kopman, J. D., and Reynolds, R. W. 1983. 1982-1983 El Niiio/Southern Oscillation Event quick look atlas (Available through the NOAA Central Library, SSMC3, 1315 East West Highway, Silver Spring, MD 20910).

Arkin, P. A., T. M. Smith, M. R. P. Sapiano, and J. Janowiak, 2010: The Observed Sensitivity of the Global Hydrological Cycle to Changes in Surface Temperature. *Environ. Res. Lett.*, **5**, 035201.

Arrhenius, S., 1896: On the Influence of Carbonic Acid in the Air upon the Temperature of the Ground. *Philosophical Magazine and Journal of Science (Fifth Series)*, **41**, 237–275. See www.rsc.org/images/Arrhenius1896_tcm18-173546/pdf.

Assel, R. A., 2005: Classification of Annual Great Lakes Ice Cycles: Winters 1973–2002. *J. Clim.*, **18**, 4895–4905.

Baldwin, M. P., and T. J. Dunkerton, 1999: Propagation of the Arctic Oscillation from the Stratosphere to the Troposphere. *J. Geophys. Res.*, **104**, 30937–30946.

Baldwin, M. P. and co-authors, 2001: The Quasi-Biennial Oscillation. *Revi. of Geophys.*, **39**, 179–229.

Barnston, A. G., and R. E. Livezey, 1987: Classification, Seasonality and Persistence of Low-Frequency Atmospheric Circulation Patterns. *Mon. Wea. Rev.*, **115**, 1083–1126.

Barnston, A. G., and R. E. Livezey, 1989: A Closer Look at the Effect of the 11-Year Solar Cycle and the Quasi-Biennial Oscillation on Northern Hemisphere 700 mb Height and Extratropical North American Surface Temperature. *J. Clim.*, **2**, 1295–1313.

Barnston, A. G., and R. E. Livezey, 1991: Statistical Prediction of January-February Mean Northern Hemisphere Lower Tropospheric Climate from the 11-Year Solar Cycle and the Southern Oscillation for West and East QBO Phases. *J. Clim.*, **4**, 249–262.

Barnston, A. G., and C. F. Ropelewski, 1992: Prediction of ENSO Episodes Using Canonical Correlation Analysis. *J. Clim.*, **5**, 1316–1345.

Barnston, A. G., M. K. Tippett, M. L. L'Heureux, S. Li, and D. G. DeWitt, 2012: Skill of Real-Time Seasonal ENSO Model Predictions during 2002–11: Is Our Capability Increasing? *Bull. Amer. Meteor. Soc.*, **93**, 631–651.

Barrett, E. C., 1970: The Estimation of Monthly Rainfall from Satellite Data. *Mon. Wea. Rev.*, **98**, 322–327.

Barrett, B. S., and L. M. Leslie, 2009: 2009: Links between Tropical Cyclone Activity and Madden-Julian Oscillation Phase in the North Atlantic and Northeast Pacific Basins. *Mon. Wea. Rev.*, **137**, 727–744.

Barry, R., and T. Y. Gan, 2011: *The Global Cryosphere: Past, Present, Future*, Cambridge University Press, 472 pp., ISBN 978-0-521-15685-1.

Becker, E. J., E. H. Berbery, and R. W. Higgins, 2011: Modulation of Cold-Season U.S. Daily Precipitation by the Madden-Julian Oscillation. *Jour. of Clim.*, 24. 5157–5166.

Becker, F. and Z. L. Li, 1990: Temperature-Independent Spectral Indices in Thermal Infrared Bands. *Remote Sensing of the Environment*, 32, 17–33.

Behrangi, A., and G. Stephens, 2014: An Update on the Oceanic Precipitation Rate and Its Zonal Distribution in Light of Advanced Observations from Space. *J. Clim.*, 27, 3957–3965.

Bengtsson, L., and J. Shukla, 1988: Integration of Space and In Situ Observations to Study Global Climate Change. *Bull. Amer. Meteor. Soc.*, **69**, 1130–1143.

Berlage, H. P., 1966: The Southern Oscillation and World Weather. Meded. Verh. K. Ned. Meteor. Inst. 152pp.

Bieniek, P. A., U. S. Bhat, L. A. Rundquist, S. D. Lindsey, X. Zhang, and R. L. Thoman, 2011: Large-Scale Climate Controls of Interior Alaska River Ice Breakup. *J. Clim.*, **24**, 286–297.

Birkett, C. M., C. Reynolds, B. Beckley, and B. Doorn, 2010: From Research to Operations: The USDA Global Reservoir and Lake Monitor, Chapter 2 in *Coastal Altimetry*, ed. S. Vignudelli,

A. G. Kostianoy, P. Cipollini and J. Benveniste, Springer Publications, ISBN 978-3-642-12795-3.

Bjerknes, J., 1966: A Possible Response of the Atmospheric Hadley Circulation to Equatorial Anomalies of Ocean Temperature. *Tellus*, 18, 820–829.

Bjerknes, J., 1969: Atmospheric Teleconnections from the Equatorial Pacific. *Mon. Wea. Rev.*, **98**, 820–829.

Blackmon, M. L., J. M. Wallace, N.-C. Lau, and S. L. Mullen, 1977: An Observational Study of the Northern Hemisphere Wintertime Circulation. *J. Atmosph. Sci.*, **34**, 1040–1053.

Blunden, J. and D. S. Arndt (Eds.), 2013: The State of the Climate in 2012. *Bull. Amer. Meteor. Soc.*, **94**, 240, Appendix 2. Relevant Datasets and Sources, pp. S205–S211 (Lists the urls for a large number of datasets used in climate monitoring).

Bourlès, B. et al., 2008: THE PIRATA PROGRAM: History, Accomplishments, and Future Directions. *Bull. Amer. Meteor. Soc.*, 89, 1111–1125.

Boyer, T. P. et al., 2013: World Ocean Database 2013. Silver Spring, MD, NOAA Printing Officce, 208pp. (NOAA Atlas NESDIS, 72).

Bradley, R. S. and P. D. Jones, 1993: 'Little Ice Age' Summer Temperature Variations: Their Nature and Relevance to Recent Global Warming Trends. *Holocene*, **3**, 376.

Broecker, W., 2010: *The Great Ocean Conveyor.* Princeton University Press. 154 pp.

Broecker, W., 1998: The Paleoocean Circulation during the Last Deglaciation. A Bi-Polar Seesaw? *Paleocenography*, **13**, 119–121.

Brohan P., J. J. Kennedy, I. Harris, S. F. B. Tett and P. D. Jones, 2006: Uncertainty Estimates in Regional and Global Observed Temperature Changes: A New Data Set from 1850. *J. Geophys. Res.*, **111**, D12106, DOI:10.1029/2005JD006548.

Budyko, M. I., 1969: The Effect of Solar Radiation Variations on the Climate of the Earth. *Tellus*, **21**, 611–619.

Büntgen, U., D. C. Frank, D. Nievergelt, and J Esper, 2006: Summer Temperature Variations in the European Alps, a.d. 755–2004. *J. Clim.*, **19**, 5606–5623.

Bushugyev, A.V., 2012: WMO Sea Ice Nomenclature, 12 pp. Available from the *World Meteor-oGreenland Survey,* 38 pp., ISBN87–91144-00–0logical Organization. (Check date and details.)

Byrd, G. P., 1985: An Adjustment for the Effects of Observation Time on Mean Temperature and Degree-Day Computations. *J. Clim. Appl. Meteor.*, **24**, 869–874.

Cahill, T. (Ed.) 1998: *South*, Lyons Press Edition, ISBN 1-55821-783-5.

Callendar, G. S., 1938: The Artificial Production of Carbon Dioxide and Its Influence on Temperature. *Quart, J. Roy. Meteor. Soc.* **64**, 223–237.

Callendar, G. S., 1961: Temperature Fluctuations and Trends over the Earth. *Quart. J. Roy. Meteor. Soc.*, **87**, 1–12.

Camp, C. D. and K.-K. Tung, 2007: The Influence of the Solar Cycle and QBO on the Late-Winter Stratospheric Polar Vortex. *J. Atmos. Sci.*, **64**, 1267–1283.

Cane, M. A., S. E. Zebiak, and S. C. Dolan, 1986: Experimental Forecasts of El Niño. *Nature*, **321**, 827–832.

Chang, C-P. (Ed.), 2011: The Global Monsoon System: Research and Forecast, World Meteorological Organiztion, WMO/TD **1266**, (TMRP Report 70).

Chapman, W. and National Center for Atmospheric Research Staff (Eds.) Last modified 20 Aug 2013. "The Climate Data Guide: Walsh and Chapman Northern Hemisphere Sea Ice." Retrieved from https://climatedataguide.ucar.edu/climate-data/walsh-and-chapman-northern-hemisphere-sea-ice.

Charney, J. W., J. Quirk, S–h Chow, and J. Kornfield, 1977: A Comparative Study of the Effects of Albedo Change on Drought in Semi–Arid Regions. *J. Atmos. Sci.*, **34**, 1366–1385.

Charney, J. et al., 1979: Carbon Dioxide and Climate: A Scientific Assessment. *National Acad. Sci.*, Washington, DC 18 pp. Available from the Climate Research Board and online Chaney_report.pdf.

Chelliah, M. and C. F. Ropelewski, 2000: Reanalyses-Based Tropospheric Temperature Estimates: Uncertainties in the Context of Global Climate Change Detection. *J. Clim.*, **13**, 3187–3205.

Ciais, P. et al., 2014:. Current Systematic Carbon-Cycle Observations and the Need for Implementing a Policy-Relevant Carbon Observing System. *Biogeosciences*, **11**, 3547–3602.

Cinquini, L. et al., 2014: The Earth System Grid Federation: An Open Infrastructure for Access to Distributed Geospatial Data. *Futur. Gener. Comput. Syst.*, 36, 400–417, DOI:10.1016/j.future.2013.07.002). http://linkinghub.elsevier.com/retrieve/pii/S0167739X13001477.

Cohen, J., M. Barlow, P.J. Kushner, and K. Saito, 2007: Stratospheric-Tropospheric Coupling and Links with Eurasian Land Surface Variability. *J. Clim.*, **20**, 5335–5343.

Collins, M, R. Knutti et al., 2014: Long-term climate change: Projections, Commitments, and Irreversibility. *IPCC Assessment Report 5, Working Group 1, Chapter 12.*, 108 pp. Cambridge University Press.

Comiso, J. C., D. J. Cavalieri, and T. Markus 2003. Sea Ice Concentration, Ice Temperature, and Snow Depth Using AMSR-E Data. *IEEE Transactions on Geoscience and Remote Sensing*, **41**, 243–252.

Compo, G. P. et al., 2011: The Twentieth Century Reanalysis Project. *Quart. J. Roy. Meteorol. Soc.*, **137**, 1–28. DOI: 10.1002/qj.776.

Conkright, M. E. et al., 2002: World Ocean Atlas 2001: Objective Analyses, Data Statistics, and Figures, CD-ROM, Documentation. National Oceanographic Data Center, Silver Spring, MD, 17 pp.

Conover, J. H., 1985: Highlights of the History of the Blue Hill Observatory and the Early Days of the American Meteorological Society. *Bulletin of the American Meteorological Society*, **66**, 30–37.

Dai, A., K. E. Trenberth, and T. Qian, 2004: A Global Dataset of Palmer Drought Severity Index for 1870–2002: Relationship with Soil Moisture and Effects of Surface Warming. *J. Hydrometeor.*, **5**, 1117–1130.

Dai, A., T. R. Karl, B. Sun, and K. E. Trenberth, 2006: Recent Trends in Cloudiness over the United States: A Tale of Monitoring Inadequacies. *Bull. Amer. Meteor. Soc.*, **87**, 597–606.

Dalrymple, G.B., 1991: *The Age of the Earth*, Stanford University Press, 492 pp., ISBN 978-0804-71569-0.

Daly, C., R. P. Neilson, and D. L. Phillips, 1994: A Statistical–Topographic Model for Mapping Climatological Precipitation over Mountainous Terrain. *J. Appl. Meteor.*, **33**, 140–158.

De Fries, R. S., M. Hansen, J. R. G. Townshend, and R. Sohlberg, 1998: Global Land Cover Classications at 8 km Spatial Resolution: The Use of Training Data Derived from Landsat Imagery in Decision Tree Classifiers. *Int.J. Remote Sensing*, **19**, 3141–3168.

DeGaetano A. T., 2000: A Serially Complete Simulated Observation Time Metadata File for U.S. Daily Historical Climatology Network Stations. *Bull. Amer. Meteor. Soc.*, **81**, 49–67.

Dewey, K. F. and R. Heim, 1980: A Digital Archive of Northern Hemisphere Snow Cover from November 1966 to December 1980. *Bull. Amer. Metsoc.*, **63**, 1132–1141.

Dole, R. et al., 2011: Was There a Basis for Anticipating the 2010 Russian Heat Wave? *Geophys. Res. Lett.*, **38**, DOI: 10.1029/2010GL046582.

Dorigo, W. A., and Coauthors, 2011: The International Soil Moisture Network: A Data Hosting Facility for Global In Situ Soil Moisture Measurements. *Hydrol. Earth Syst. Sci.*, **15**, 1675–1698.

Dorman, C. E. and R. H. Bourke, 1979: Precipitation over the Pacific Ocean, 30°S to 60°N. *Mon. Wea. Rev.*, **107**, 896–910.

Dorman, C. E. and R. H. Bourke, 1981: Precipitation over the Atlantic Ocean, 30°S to 70°N. *Mon. Wea. Rev.*, **109**, 554–563.

Doty, B. and J. L. Kinter III, 1995: *Geophysical Data Analysis and Visualization Using GrADS. Visualization Techniques in Space and Atmospheric Sciences*, eds. E. P. Szuszczewicz and J. H. Bredekamp. (NASA, Washington, DC), 209–219.

Durack, P. J. and S. E. Wijffels 2010: Fifty-Year Trends in Global Ocean Salinities and Their Relationships to Broad-Scale Warming. *J. Clim.*, **23**, 4342–4362.

Durre, I., T. Reale, D. Carlson, J. Christy, M. Uddstrom, M. Gelman, and P. Thorne, 2005: Improving the Usefulness of Operational Radiosonde Data. *Bull. Amer. Meteor. Soc.*, **86**, 411–418.

Ebert, E. E., M. J. Manton, P. A. Arkin, R. J. Allam, G. E. Holpin, and A. Gruber, 1996: Results from the GPCP Algorithm Intercomparison Programme. *Bull. Amer. Meteor. Soc.*, **82**, 2773–2785.

Ek, M. and Holtslag, 2004: Influence of Soil Moisture on Boundary Layer Cloud Development. *J. Hydrometeorol.*, **5**, 86–99.

Ek, M. B., K. E. Mitchell, Y. Lin, E. Rogers, P. Grunmann, V. Koren, G. Gayno, and J. D. Tarplay, 2003: Implementation of Noah Land-Surface Model Advances in the NCEP Operational Mesoscale Eta Model. *J. Geophys. Res.*, **108**, 8851, doi:10.1029/2002JD003296.

Ekman, V. W. 1905. On the Influence of the Earth's Rotation on Ocean Currents. *Arch. Math. Astron. Phys.*, **2**, 1–52.

Emery, W. J., J. S. Castro, G. A. Wick, P. Schuessel, and C. Donlon, 2001: Estimating Sea Surface Temperature from Infrared Satellite and In Situ Temperature Data. *Bull. Ametsoc.*, **82**, 2773–2785.

Epstein, E. S., 1985: Statistical Inference and Prediction in Climatology: A Baysian Approach, *Meteor. Monogr.*, **20**, 204.

Ewen, T., A. Grant, and S. Bronnimann, 2008: Monthly Upper-Air Dataset for North America Back to 1922 from the Monthly Weather Review. *Mon. Wea. Rev.*, **136**, 1792–1805.

Fan, Y., and H. van den Dool, 2004: Climate Prediction Center Global Monthly Soil Moisture Data Set at 0.5° Resolution for 1948 to Present. *J. Geophys. Res.*, 109, D10102, doi 10.1029/2003JD004345.

Fan, Y. and H. van den Dool, 2008: A Global Monthly Land Surface Air Temperature Analysis for 1948-Present. *J. Geophys. Res.*, **113**, D01103, doi:10.1029/2007JD008470.

Fan,Y., H. M. van den Dool, and W. Wu, 2011: Verification and Intercomparison of Multi-Model Simulated Land Surface Hydrological Data Sets over the United States. *J. Hydrometeor.*, **12**, 531–555.

Fasullo, J. T. and K. E. Trenberth, 2008a: The Annual Cycle of the Energy Budget, Part I: Global Mean and Land-Ocean Exchanges. *J. Clim.*, **21**, 2297–2312.

Fasullo, J. T. and K. E. Trenberth, 2008b: The Annual Cycle of the Energy Budget, Part II: Meridional Structures and Poleward Transports. *J. Clim.*, **21**, 2313–2325.

Ferrel, W., 1889: *A Popular Treatise on the Winds*. MacMillian and Co., London.

Ferraro, R. R., 1997: SSM/I Derived Global Rainfall Estimates for Climatological Applications, *J. Geophys. Res.*, **102**, 16715–16735.

Fetterer, F., K. Knowles, W. Meier, M. Savoie, and A. K. Windnagel, 2017: updated daily. *Sea Ice Index, Version 3*. Boulder, Colorado USA. NSIDC: National Snow and Ice Data Center. doi: http://dx.doi.org/10.7265/N5K072F8.

Fleagle, R. G. and J. A. Businger, 1980: *An Introduction to Atmospheric Physics*, 2nd Edition, Academic Press, New York. 432 pp.

Folland, C. K., 1988: Numerical Models of the Raingauge Exposure Problem, Field Experiments and an Improved Collector Design. *Quart. J. Roy. Meteor. Soc.*, **114**, 1485–1516.

Folland, C. K., and D. E. Parker, 1995: Correction of Instrumental Biases in Historical Sea Surface Temperature Data. *Quarterly Journal of the Royal Meteorological Society*, 121, 319–367.

Francis, J., 2017: Why Are Arctic Linkages to Extreme Weather Still Up in the Air? *Bull. Amer. Meteor. Soc.*, **98**, 2551–2557.

Free, M., D. J. Seidel, J. K. Angell, J. Lanzante, I. Durre, and T. C. Peterson, (2005): Radiosonde Atmospheric Temperature Products for Assessing Climate (RATPAC): A New Data Set of Large-Area Anomaly Time Series. *J. Geophys. Res.*, **110**, D22101.

Garratt, J. R., A. J. Prata, L. D. Rotstayn, B. J. McAvaney, and S. Cusack, 1998: The Surface Radiation Budget over Oceans and Continents. *J. Clim.*, **11**, 1951–1968.

Gates, L., 1992: AMIP: The Atmospheric Model Intercomparison Project. *Bull. Amer. Meteor. Soc.*, **73**, 1962–1970.

Ge, Q-S, J-Y Zheng, Z-X Hao, P-Y Zhang, and W-C Wang, 2005: Reconstruction of Historical Climate in China: High-resolution Precipitation Data from Qing Dynasty Archives. *Bull.Amer. Meteor. Soc.*, **86**, 671–679.

Global Soil Data Task Group, 2000: Global Gridded Surfaces of Selected Soil Charactersitics International Geosphere-Biosphere Program – Data and Information System (IGBP-DIS).

Gifford, H. Miller, John R. Southon, Chance Anderson, Helgi Björnsson, Thorvaldur Thordarson, Aslaug Geirsdottir, Yafang Zhong, Darren J Larsen, Bette L Otto-Bliesner, Marika M Holland, David Anthony Bailey, Kurt A. Refsnider, and Scott J. Lehman, 2012: Abrupt Onset of the Little Ice Age Triggered by Volcanism and Sustained by Sea-Ice/Ocean Feedbacks. *Geophysical Research Letters*, **2012**; DOI: 10.1029/2011GL050168.

Greene, A. M, L. Goddard, and R. Cousins, 2011: Web Tool Deconstructs Variability in Twentieth-Century Climate. *Eos Trans. AGU*, **92**, 397, doi:10.1029/2011EO450001.

Griffith, C. G., W. L. Woodley, P. G. Grube, D. W. Martin, Stout, D. N. Sikdar, 1978: Rain Estimation from Geosynchronous Satellite Imagery: Visible and Infrared Studies. *Mon. Wea. Rev.*, **106**, 1153–1171.

Grody, N. C., 1991: Classification of Snow Cover and Precipitation using the Special Sensor Microwave/Imager (SSM/I). *J. Geophys. Res.*, **96**, 7423–7435.

Gutman, G., D. Tarpley, A. Ignatov, and S. Olson, 1995: The Enhanced NOAA Global Land Data Set from the Advanced Very High Resolution Radiometer. *Bull. Amer. Meteor. Soc.*, **76**, 1141–1156.

Guttman N. B. and G. Quayle, 1996: A Historical Perspective of U.S. Climate Divisions. *Bull. Amer. Meteor. Soc.*, **77**, 293–303.

Hadley, D., 1735: Concerning the Cause of the General Trad-Winds. *Phil. Trans.*, **29**, 58–62.

Hahn, C. J., W. B. Rossow, and S. G. Warren, 2001: ISCCP Cloud Properties Associated with Standard Cloud Types Identified in Individual Surface Observations. *J. Clim.*, **14**, 11–28.

Halide, H. and P. V. Ridd, 2008: Complicated Models Do not Significantly Outperform Very Simple ENSO Models. *Int. Jour.of Clim.*, **28**, 219–233.

Halley, E., 1686: An Historical Account of the Trade-Winds and Monsoons Observable in the Seas between and Near the Tropicks with an Attempt to Assign the Physical Cause of Said Winds. *Phil. Trans.*, **26**, 153–168.

Halpert, M. S. and C. F. Ropelewski, 1992: Surface Temperature Patterns Associated with the Southern Oscillation. *J. Clim.*, **5**, 577–593.

Halpert, M. S., G. D. Bell, V. E. Kousky, and C. F. Ropelewski, 1996: Climate Assessment for 1995. *Bull. Amer. Meteor. Soc.*, **77**, S1–S44.

Hamming, R. W., 1986: *Numerical Methods for Scientists and Engineers*, Second Edition, 721 pp. Dover Reprint of original McGraw-Hill (1973).

Hansen, J. and S. Lebedeff, 1987: Global Trends of Measured Surface Air Temperature. *J. Geophys. Res.*, **92**, 13345–13372.

Hansen, M. C., R. S. Defries, J. R. G. Townshend, and R. Sohlberg, 2000: Global Land Cover Classification at 1 km Spatial Resolution Using A Classification Tree Approach. *Int. J. Remote Sensing*, **21**,1331–1364.

Harris, I., P. D. Jones, T. J. Osborn, and D. H. Lister, 2014: Updated High-Resolution Grids of Monthly Climatic Observations: The CRU TS3.10 Dataset. *International Journal of Climatology*, **34**, 623–642.

Hastenrath, S., 1991: *Climte Dynamics of the Tropics*, Kluwer Academic Publishers, 488 pp., ISBN 0-7923-1213-9.

Hastings, David A. and Paula K. Dunbar, 1999. Global Land One-kilometer Base Elevation (GLOBE) Digital Elevation Model, Documentation, Volume 1.0. Key to Geophysical Records Documentation (KGRD) 34. National Oceanic and Atmospheric Administration, National Geophysical Data Center, 325 Broadway, Boulder, Colorado 80303, U.S.A. Web address below.

Hayes, Michael J., Mark D. Svoboda, Donald A. Wilhite, and Olga V. Vanyarkho, 1999: Monitoring the 1996 Drought Using the Standardized Precipitation Index. *Bull. Amer. Meteor. Soc.*, **80**, 429–438.

Hayes, S. P., L. J. Mangum, J. Picaut, A. Sumi, and K. Takeuchi, 1991: TOGA-TAO: A Moored Array for Real-Time Measurements in the Tropical Pacific Ocean. *Bull. Amer. Meteor. Soc.*, **72**, 338–347.

Heim, R. R., 2002: A Review of Twentieth-Century Drought Indices Used in the United States. *Bull. Amer. Meteor. Soc.*, **83**, 1149–1165.

Hersbach, H. and D. Dee, 2016: ERA5 reanalysis is in production. ECMWF Newsletter 147, April 2016, can be found at www.ecmwf.int/en/newsletter/147/news/era5-reanalysis-production.

Hildenbrandsson, H. H., 1897: Quelques recherches sur le centres d'action de l'atmosphere. *Kon. Svenska Vetens.-Acad. Handl.*, **29**, 36.

Hoffman, R. N., N. Prive, and M. Bourassa, 2017: Comments on "Reanalyses and Observations: What's the Difference?" *Bull. Amer. Meteor. Soc.*, **98**, 2455–2459.

Hollinger, S. E. and S. A. Isard, 1994: A Soil Moisture Climatology of Illinois. *J. Clim.*, **7**, 822–833.

Holton, J. R. and R.S. Lindzen, 1972: An Updated Theory of the Quasi-Biennial Oscillation in the Tropical Stratosphere. *J. Atmos. Sci.*, **29**, 1076–1080.

Horel, J. D. and J. M. Wallace, 1981: Planetary-Scale Atmospheric Phenomena Associated with the Southern Oscillation. *Mon. Wea. Rev.*, **109**, 813–829.

Huang, J., H. M. van den Dool, and A. G. Barnston, 1996: Long-Lead Seasonal Temperature Prediction Using Optimal Climate Normals. *J. Clim.*, **9**, 809–817.

Huang, J., H. M. van den Dool, and K. P. Georgarakos, 1996: Analysis of Model-Calculated Soil Moisture over the United States (1931–1993) and Long-Range Temperature Forecasts. *J. Clim.*, **9**, 1350–1362.

Huffman, G. J., R. F. Adler, P. A., A. Chang, R. Ferraro, A. Gruber, J. Janowiak, A. McNab, B. Rudolf, and U. Schneider, 1997: The Global Precipitation Climatology Project (GPCP) Combined Precipitation Data Set. *Bull. Amer. Meteor. Soc.*, **78**, 5–20.

Huffman, G. J., R. F. Adler, D. T. Bolvin, G. Gu, E. J. Nelkin, K. P. Bowman, E. F. Stocker, and D. B. Wolff, 2007: The TRMM Multi-Satellite Precipitation Analysis: Quasi-Global, Multi-Year, Combined-Sensor Precipitation Estimates at Fine Scale. *J. Hydrometeor.*, **8**, 38–55.

Huffman, G.J., D. T. Bolvin, E. J. Nelkin, 2015: Day 1 IMERG Final Run Release Notes. http://pmm.nasa.gov/sites/default/files/document_files/IMERG_FinalRun_Day1_release_notes.pdf.

Huler, S., 2004: *Defining the Wind: The Beaufort Scale and How a Nineteenth-Century Admiral Turned Science into Poetry*. Crown Publishers, Random House.

Hurrell, J. W. and K. E. Trenberth, 1999: Global Sea Surface Temperature Analyses: Multiple Problems and Their Implications for Climate Analysis, Modeling, and Reanalysis. *Bull. Amer. Meteor. Soc.*, 80, 2661–2678.

Hurrell J. W., Y. Kushnir, G. Ottersen, and M. Visbeck, eds., 2003: *The North Atlantic Oscillation: Climate Significance and Environmental Impact*, American Geophysical Union, Geophysical Monograph Series, **134**, 279 pp.

Hurrell J. W., K. E. Trenberth, S. J. Brown, and J. R. Christy, 2000: Comparison of Tropospheric Temperatures from Radiosondes and Satellites: 1979–98. *Bull. Amer. Meteor. Soc.*, **81**, 2165–2177.

Hurrell, J.W. et al., 2006: Atlantic Climate Predictability and Variability: A CLIVAR Perspective. *J. Clim.*, **19**, 5100–5121.

Hurst, H. E., 1951: Long Term Storage Capacity of Reservoirs. *Trans. Am. Soc. Civ. Eng.*, **116**, 770.

Huybers, P., 2006: Early Pleistocene Glacial Cycles and the Integrated Summer Insolation Forcing. *Science*, **313**, 508–511.

Janis, M. J., 2002: Observation-Time-Dependent Biases and Departures for Daily Minimum and Maximum Air Temperatures. *J. Appl. Meteor.*, **41**, 588–603.

Janowiak, J. E., C. F. Ropelewski, and M. S. Halpert, 1986: The Precipitation Anomaly Classification: A Method for Monitoring Regional Precipitation Deficiency and Excess on a Global Scale. *J. Clim. Appl. Meteor.*, **25**, 565–574.

Jones P. D., T. M. L. Wigley, and P. M. Kelly, 1982: Variations in Surface Air Temperatures: Part 1. Northern Hemisphere, 1881–1980. *Mon. Wea. Rev.*, **110**, 59–70.

Jones, P. D., S. C. B. Raper, and T. M. L. Wigley, 1986a: Southern Hemisphere Surface Air Temperature Variations: 1851–1984. *J. Clim. Appl. Meteorol.*, **25**, 1213–1230.

Jones, P.D., S. C. B. Raper, R. S. Bradley, H. F. Diaz, P. M. Kelly, and T. M. L. Wigley, 1986b: Northern Hemisphere Surface Air Temperature Variations: 1851–1984. *J. Clim. Appl. Meteorol.*, **25**, 161–179.

Joyce, R. J., J. E. Janowiak, P. A. Arkin, and P. Xie, 2004: CMORPH: A Method That Produces Global Precipitation Estimates from Passive Microwave and Infrared Data at High Spatial and Temporal Resolution. *J. Hydrometeor.*, **3**, 487–503.

Jones, P.D., D.H. Lister, T.J. Osborn, C. Harpham, M. Salmon, and C. P. Morice, 2012: Hemispheric and Large-Scale Land-Surface Air Temperature Variations: An Extensive Revision and an Update To 2010. *Journal of Geophysical Research*, 117, D05127 (doi:10.1029/2011JD017139).

Kalnay, E. et al., 1996: The NCEP/NCAR 40-year Reanalysis Project. *Bull. Amer. Met. Soc.*, **77**, 437–471.

Kalnay, E., 2003: *Atmospheric modelling, data assimilation and predictability*. Cambridge University Press, pp. xxii + 341. ISBNs 0 521 79179 0, 0 521 79629 6.

Kanamitsu, M., W. Ebisuzaki, J. Woollen, S.-K. Yang, J. J. Hnilo, M. Fiorino, and G. L. Potter, 2002: NCEP–DOE AMIP-II Reanalysis (R-2). *Bull. Amer. Meteorol. Soc.*, **83**, 1631–1643.

Kaplan A., Y. Kushnir, and M. A. Cane, 2000: Reduced Space Optimal Interpolation of Historical Marine Sea Level Pressure: 1854-1992. *J. Clim.*, 13, 2987–3002.

Kaplan, A., Y. Kushnir, M. A. Cane, and M. B. Blumenthal, 1997: Reduced Space Optimal Analysis for Historical Datasets: 136 Years of Atlantic Sea Surface Temperatures. *J. Geophys. Res.*, **102**, 27,835–27,860.

Karl, T. R., P. D. Jones, R. W. Knight, G. Kukla, N. Plummet, V. Razuvayev, K. P. Gallo, J. Lindesay, R. J. Charlson, and T. C. Peterson, 1993: A New Perspective on Recent Global Warming – Asymmetric Trends of Daily Maximum and Minimum Temperatures. *Bull. Am. Meteorol. Soc.*, **74**, 1007–1023.

Karl, T. R., Claude N. Williams Jr., Pamela J. Young, and Wayne M. Wendland, 1986: A Model to Estimate the Time of Observation Bias Associated with Monthly Mean Maximum, Minimum and Mean Temperatures for the United States. *J. Clim. Appl. Meteor.*, **25**, 145–160.

Karlsen, H. G., J. Bille-Hansen, K. Q. Hansen, H. S. Anderson, and H. Skourup, 2001: Distribution and variability of icebergs in the eastern Davis Strait 63°N to 68°N., *Greenland Survey*, 38 pp., ISBN 87-91144-00-0.

Kasahara, A. and W. M. Washington, 1971: General Circulation Experiments with a Six-Layer NCAR Model, Including Orography, Cloudiness and Surface Temperature Calculations. *J. Atmos. Sci.*, **28**, 657–701. doi: http://dx.doi.org/10.1175/1520–0469(1971)028<0657:GCEWAS>2.0.CO;2.

Kidder, S. Q. and T. H. von der Haar, 1995: *Satellite Meteorology – An Introduction*. Academic Press, New York. 466 pp.

Kiehl, J. T. and K. E. Trenberth, 1997: Earth's Annual Global Mean Energy Budget. *Bull. Amer. Meteor. Soc.*, **78**, 197–208.

Kiladis, G. N. and H. van Loon, 1988: The Southern Oscillation. Part VII: Meteorological Anomalies over the Indian and Pacific Sectors Associated With the Extremes of the Oscillation. *Mon. Wea. Rev.*, 116, 120–136.

Klotzbach, P. J., 2014: The Madden–Julian Oscillation's Impacts on Worldwide Tropical Cyclone Activity. *J. Clim.*, **27**, 2317–2330, https://doi.org/10.1175/JCLI-D-13–00483.1.

Kohl, E. and J. A. Knox, 2016: My Drought is Different from Your Drought: A Case Study of Policy Implications of Multiple Ways of Knowing Drought. *Weather, Climate, and Society*, **4**, 373–388.

Kolb, R., 1997: *Blind Watchers of the Sky*, Oxford University Press, 338 pp., ISBN 0-19-286203-0.

Können, G. P., P. D. Jones, M. H. Kaltofen, and R. J. Allan, 1998: Pre-1866 Extensions of the Southern Oscillation Index Using Early Indonesian and Tahitian Meteorological Readings. *J. Clim.*, **11**, 2325–2339.

Köppen, W., 1881: Über mehrjährige Perioden der Witterung – III. Mehrjährige Änderungen der Temperatur 1841 bis 1875 in den Tropen der nördlichen und südlichen gemässigten Zone, an den Jahresmitteln. untersucht. *Zeitschrift der Österreichischen Gesellschaft für Meteorologie*, **XVI**, 141–150.

Kummerow, C., Y. Hong, W. S. Olson, S. Yang, R. F. Adler, J. McCollum, R. Ferraro, G. Petty, D.-B. Shin, and T. T. Wilheit, 2001: The Evolution of the Goddard Profiling Algorithm

(GPROF) for Rainfall Estimation from Passive Microwave Sensors. *J. Appl. Meteor.*, **40**, 1801–1820.

Kutzbach, J. E., Ruddiman, W. F., Vavrus, S. J., and J. Philipoon, 2010: Climate Model Simulation of Anthropogenic Influence on Greenhouse-Induced Climate Change (Early Agriculture to Modern): The Role of Ocean Feedbacks. *Climatic Change*, **99**, 351–381.

Laloyaux, P., E. de Boisseson, and P. Dahlgren, 2017: CERA-20C: An Earth system approach to climate reanalysis. ECMWF Newsletter 150, January 2017. This reference may be found at www.ecmwf.int/en/newsletter/150/meteorology/cera-20c-earth-system-approach-climate-reanalysis.

Landsberg, H. E., J. M. Mitchell, Jr., H. L. Crutcher, and F. T. Quinlan, 1963: Surface Signs of the Biennial Atmospheric Pulse. *Mon. Wea. Rev.*, 91, 549–556.

Lawrimore, J. H., M. J. Menne, B. E. Gleason, C. N. Williams, D. B. Wuertz, R. S. Vose, and J. Rennie, 2011: An Overview of the Global Historical Climatology Network Monthly Mean Temperature Data Set, Version 3. *J. Geophys. Res.*, **116**, D19121, doi:10.1029/2011JD016187.

Lawford, R. G. et al., 2004: Advancing Global and Continental-Scale Hydrometerology: Contributions of GEWEX Hydrometeorology Panel. *Bull. Amer. Meteor. Soc.*, **85**, 1917–1930.

Lee, D. M., 1980: On monitoring rainfall deficiencies in semidesert regions. *The Threatened Drylands – Regional and Systematic Studies of Desertification*, Mabbutt and Berkowicz, eds., Fujinomiya. [Available from the Australian Bureau of Meteorology, Melbourne, Australia.]

Legler, D., H. J. Freeland, R. Lumpkin, G. Ball, M. J. McPhaden, S. North, R. Cowley, G. Goni, U. Send, and M. Merrifield, 2015: The Current Status of the Real-Time In Situ Global Ocean Observing System for Operational Oceanography. *J. Operational Oceanogr.*, **8**(S2), 189–200. (Also available online at the address given in the Website list below) This paper can also be viewed at www.tandfonline.com/doi/full/10.1080/1755876X.2015.1049883.

Lethbridge, M., 1967: Precipitation Probability and Satellite Radiation Data. *Mon. Wea. Rev.*, **95**, 487–490.

Le Treut, H, R. Somerville, U. Cubasch, Y. Ding, C. Maritzen, A. Mokssit, T. Peterson, and M. Prather, 2007: In the Historical Overview section of *Climate Change 2007: The Physical Science Basis, Contribution of Working Group I to the Fourth Assessment Report of the Intergovernmental Panel on Climate Change*, Cambridge University Press, Cambridge, United Kingdom and New York, NY.

Levitus, S. and T. Boyer, 1994: *World Ocean Atlas 1994*, Vol. 2: Oxygen. NOAA Atlas NESDIS 2, U.S. Gov. Printing Office, Wash., D.C., 186 pp.

Levitus, S., R. Burgett, and T. Boyer, 1994: *World Ocean Atlas 1994*, Vol. 3: Salinity. NOAA Atlas NESDIS 3, U.S. Gov. Printing Office, Wash., D.C., 99 pp.

Liang, X., D. P. Lettenmaier, E. F. Wood, and S. J. Burges, 1994: A Simple Hydrologically Based Model of Land Surface Water and Energy Fluxes for GSMs. *J. Geophys. Res.*, **99**(D7), 14,415–14,428.

Lindzen, R. S. and J. R. Holton, 1968: A Theory of the Quasi-Biennial Oscillation. *J. Atmos. Sci.*, **25**, 1095–1107.

List, R. J., 1951: *Smithsonian Meteorological Tables*, Sixth Edition, Published by the Smithsonian Institution, 527 pp. Also available online.

Livezey, R. E. and W. Y. Chen, 1983: Statistical Field Significance and its Determination by Monte Carlo Techniques. *Mon. Wea. Rev.*, **111**, 46–59.

Lorenc, A. and F. Rawlins, 2005: Why Does 4D-Var Beat 3D-Var? *Q. J. R. Meteorol. Soc.*, **131**, 3247–3257.

Lorenz, E. N., 1962: Deterministic Nonperiodic Flow, *J. Atmos. Sci.*, 20, 130–141.

Lorenz, E. N., 1967: *The Nature and Theory of the General Circulation of the Atmosphere*. World Meteorological Organization, Geneva, Switzerland, 161 pp.

Lorenz, E. N., 1970: Climatic Change as a Mathematical Problem. *J. Appl. Meteor.*, 9, 325–329.

Lyon, B., 2004: The Strength of El Niño and the Spatial Extent of Tropical Drought. *Geophys. Res. Lett.*, **31**, L21204. doi:10.1029/2004GL020901.

Lyon, B. and A. G. Barnston, 2005: ENSO and the Spatial Extent of Interannual Precipitation Extremes in Tropical Land Areas. *J. Clim.*, **18**, 5095–5109.

MacDonald, S., 2005: A Global Profiling System for Improved Weather and Climate Prediction. *Bull. Amer. Meteor. Soc.*, **86**, 1747–1764.

Madden, R. A., and P. Julian, 1971: Detection of a 40–50 Day Oscillation in The Zonal Wind. *J. Atmos. Sci.*, **28**, 702–708.

Madden, R. A., and P. Julian, 1972: Description of Global-Scale Circulation Cells in the Tropics with a 40–50 Day Period. *J. Atmos. Sci.*, **29**, 109–123.

Madden, R. A. and P. R. Julian, 1994: Observations of the 40–50 Day Tropical Oscillation: A Review. *Mon. Wea. Rev.*, **122**, 814–837.

Mahrt, L. and H. Pan, 1984: A Two-Layer Model of Soil Hydrology. *Bound.-Layer Meteor.*, **29**, 1–20.

Maiden, M. E. and S. Greco, 1994: NASA's Pathfinder Data Set Programme: Land Surface Parameters. *Int. J. Remote Sensing*, **15**, 3333–3345.

Malone, T. F. (Ed.), 1951: *Compendium of Meteorology*, Amer. Meteor. Soc., 967–975 (Climatology – A Synthesis of Weather, C.S. Durst Author).

Manabe, S., J. Smagorinsky, and R. F. Strickler, 1965: Simulated Climatology of a General Circulation Model with a Hydrologic Cycle. *Mon. Wea. Rev.*, **93**, 769–798. doi: http://dx.doi.org/10.1175/1520–0493(1965)093<0769:SCOAGC>2.3.CO;2.

Mandelbrot, B. B. and J. R. Wallis, 1968: Noah, Joseph and Operational Hydrology. *Water Resour. Res.*, **4**, 909–918.

Manley, G., 1974: Central England Temperatures: Monthly Means 1659 to 1973. *Quart. J. Roy. Meteor. Soc.*, **100**, 389–405.

Mann, M. E., 2002: Little Ice Age. *Encyclopedia of Global Environmental Change*, **1**, 504–509, ISBN 0-471-97796-9.

Mantua, N. J., S. R. Hare, Y. Zhang, J. M. Wallace, and R. C. Francis, 1997: A Pacific Interdecadal Climate Oscillation with Impacts on Salmon Production. *Bulletin of the American Meteorological Society*, 78, 1069–1079.

Mantua, N. J. and S. R. Hare, 2002: The Pacific Decadal Oscillation. *J. of Oceanography*, **58**, 35–44.

Marko, J. R., D. B. Fissel, P. Wadhams, P. M. Kelly, and R. D. Brown, 1994: Iceberg Severity off Eastern North America: Its Relationship to Sea Ice Variability and Climate Change. *J. Clim.*, **7**, 1335–1351.

Mason, S. J. and L. Goddard, 2001: Probabilistic Precipitation Anomalies Associated with ENSO. *Bull. Amer. Meorol. Soc.*, **82**, 619–638.

Mather, J. H. and J. W. Voyles, 2013: The ARM Climate Research Facility: A Review of Structure and Capabilities. *Bull. Amer. Meteor. Soc.*, **94**, 377–392.

Mathews, J. B. R. and J. B. Mathews, 2012: Comparing Historical and Modern Methods of Sea Surface Temperature Measurement – Part 2: Field Comparison in the Central Tropical Pacific. *Ocean Sci.*, **9**, 695–711.

Maury, M. F., 1855: *The Physical Geography of the Sea*. Harper and Brothers, New York.

McGehee, R. and C. Lehman, 2012: A Paleoclimate Model of Ice-Albedo Feedback Forced by Variations in Earth's Orbit. *SIAM J. Applied Dynamical Systems*, **11**, 684–707.

McKee, T. B., N. J. Doesken, and J. Kleist, 1993: The relationship of drought frequency and duration to time scales. Preprints, *Eighth Conf. on Applied Climatology*, Anaheim, CA, Amer. Meteor. Soc., 179–184.

McPhaden, M. J. et al., 1998: The Tropical Ocean Global Atmosphere (TOGA) Observing System: A Decade of Progress. *J. Geophys. Res.*, 103, 14,169–14,240.

Meehl, G. A., C. Covey, B. McAvaney, M. Latif, and R. J. Stouffer, 2005: Overview of the Coupled Model Intercomparison. *Project. Bull. Amer. Meteor. Soc.*, 86, 89–93.

Meehl, J. et al., 2009: Decadal Predictability. *Bull. Amer. Met. Soc.*, **90**, 1467–1485.

Meehl, G. et al., 2014: Decadal Climate Prediction – An Update from the Trenches. *Bull. Amer. Meteor. Soc.*, **95**, 243–267.

Meinshausen, M. et al., 2011: The RCP Greenhouse Gas Concentrations and Their Extensions from 1792 to 2300. *Climatic Change*, **109**, 213–241. Also available online at Springerlink.com.

Meisinger, C. L., 1922: The Preparation and Significance of Free-Air Pressure Maps for the Central and Eastern United States. *Monthly Weather Review*, **50**, 453–468.

Meisner, B. and P. A. Arkin, 1987: Spatial and Annual Variations in the Diurnal Cycle of Large-Scale Tropical Convective Cloudiness and Precipitation. *Mon. Wea. Rev.*, **115**, 2009–2032.

Meko, D. M., et al. 2007. Medieval Drought in the Upper Colorado River Basin. *Geophysical Research Letters*, 34, L10705, doi:10.1029/2007GL029988.

Meng, H., J. Dong, R. Ferraro, B. Yan, L. Zhao, C. Kongoli, N.-Y. Wang, and B. Zavodsky, 2017: A 1DVAR-based snowfall rate retrieval algorithm for passive microwave radiometers, *J. Geophys. Res. Atmos.*, 122, 6520–6540, https://doi.org/10.1002/2016JD026325.

Mesa, O. J. and G. Poveda, 1993: The Hurst Effect: The Scale of Fluctuation Approach. *Water Resour. Res.*, **29**, 3395–4002.

Milankovitch, M. 1941: Canon of insolation and the ice age problem (in Serbian). K. Serb. Acad. Beorg. Spec. Publ. 132 (English translation by the Israel Program for Scientific Translations, Jerusalem, 1969).

Miller, G. H. et al., 2012: Abrupt Onset of the Little Ice Age Triggered by Volcanism and Sustained by Sea-Ice/Ocean Feedbacks. *Geophys. Res. Lett.*, 39, L02708, doi: 10.1029/2011GL050168.

Mitchell, J. M., 1961: Recent Secular Changes of Global Temperature. *Ann. New York, Acad. Sci.*, **95**, 235–250.

Mitchell, J. M., 1963: On the World-Wide Patterns of Global Temperature. *Arid Zone Res.*, **2**, 161–181.

Moninger, W. R., R. D. Mamrosh, and P. M. Pauley, 2003: Automated Meteorological Reports from Commercial Aircraft. *Bull. Amer. Metero. Soc.*, **84**, 203–216.

Namias, J., 1968: Long Range Weather Forecasting – History, Current Status and Outlook. *Bull. Amer. Meteor. Soc.*, **49**, 438–470.

National Academy Press, 1982: Climate in Earth History: Studies in Geophysics. A PDF may be downloaded at www.nap.edu/catalog/11798/climate-in-earth-history-studies-in-geophysics.

Negri A. J. and R. F. Adler, 1993: An Intercomparlson of Three Satellite Infrared Rainfall Techniques over Japan and Surrounding Waters. *J. Appl.Meterol.*, **32**, 357–373.

Nespor, V. and B. Sevruk, 1999: Estimation of Wind-Induced Error of Rainfall Gauge Measurements Using a Numerical Simulation. *J. Atm. and Oceanic Tech.*, **16**, 450–464.

Newman, M. et al., 2016: The Pacific Decadal Oscillation Revisited. *J. Clim.*, **29**, 4399–4427.

Nishida, M., A. Shimizu, T. Tsuda, C. Rocken, and R. H. Ware, 2000: Seasonal and Longitudinal Variations in the Tropical Tropopause Observed with the GPS Occultation Technique (GPS/MET). *J. Meteorol. Soc. Jpn.*, **78**, 691–700.

Nobre, C. A., P. J. Sellers, and J. Shulka, 1991: Amazonian Deforestation and Regional Climate Change. *J. Clim.*, **4**, 957–988.

Ohmura, A. eta l., 1998: Baseline Surface Radiation Network (BSRN)/WCRP): New Precision Radiometry for Climate Research. *Bull. Amer. Meteor. Soc.*, **79**, 2115–2136.

Oort, A. H., and E. M. Rasmusson, 1971: Atmospheric Circulation Statistics. **NOAA Professional Paper No. 5**, U.S. Dept. of Commerce, 323 pp.

Oreskes, N., K. Shrader-Frechette, and K. Belitz, 1994: Verification, Validation, and Confirmation of Numerical Models in the Earth Sciences. *Science*, **263**, 641–646.

Palmer, W. C., 1965: Meteorological drought. *Office of Climatology Res.Paper,* **45**. U.S. Weather Bureau (now the US National Climate Service). 58 pp.

Palmer, W. C., 1968: Keeping Track of Crop Moisture Nationwide: The New Crop Moisture Index. *Weatherwise*, **21**, 156–161.

Panofsky, H., 2014: Analyzing Atmospheric Behavior. *Physics Today*, **67**, 38–41.

Parker, D. E., C. K. Folland, and M. Jackson, 1995: Marine Surface Temperature: Observed Variations and Data Requirements. *Clim. Change*, **31**, 559–600.

Pearson, P. N. and P. R. Palmer, 2000: Atmospheric Carbon Dioxide Concentrations over the Past 60 Million Years. *Nature*, **406**, 695–699.

Peixoto, J. P. and A. H. Oort, 1992: *The Physics of Climate*, American Institute of Physics, 520 pp. ISBN 978-0883187128.

Penland, C. and T. Magorian, 1993: Prediction of Niño 3 Sea Surface Temperatures Using Linear Inverse Modeling. *J. Clim.*, **6**, 106.

Persson, A. 2017: The Story of the Hovmöller Diagram: An (Almost) Eyewitness Account. *Bull. Amer. Meteor. Soc.*, **98**, 948–957.

Petit J. R. et al., 1999: The Climate and Atmospheric History of the Past 420,000 Years from the Vostok Ice Core, Antarctica. *Nature*, **399**, 429–436.

Petterssen, S. and P. A. Calabrese, 1959: On Some Weather Influences Due to Warming of the Air by the Great Lakes in Winter. *J.Meteor.*, **16**, 646–652.

Philander, S. G., 1990: *El Niño, La Niña, and the Southern Oscillation.* ix + 293 pp. Academic Press (Harcourt Brace Jovanovich), San Diego, New York, Berkeley, Boston, London, Sydney, Tokyo, Toronto. ISBN 0 12 553235 0. International Geophysics Series Vol. 46.

Phillips, N. A., 1956: The General Circulation of the Atmosphere: A Numerical Experiment. *Quart. J. Roy. Meteor. Soc.*, **82**(352), 123–154.

Pielke Sr., R. A., R. Mahmood, and C. McAlpine, 2016: Land's Complex Role In Climate Change. *Physics Today*, **69**(11), 40.

Pinker, R. T. and L. A. Corio, 1984: Surface Radiation Budget from Satellites. *Mon. Wea. Rev.*, **112**, 209–215.

Plummer, C. C., D. McGaery, and D. H. Carlson, 2003: *Physical Geology*, 574pp, ISBN 0-07-240-246-6.

Pope, A., W. G. Rees, A. J. Fox, and A. Fleming, 2014: Open Access Data in Polar and Cryospheric Remote Sensing. *Remote Sens.*, **6**, 6183–6220.

Pope, A., P. Wagner, R. Johnson, J. Shutler, J. Baeseman, and L. Newman, 2017: Community Review of Southern Ocean Satellite Data Needs. *Antarctic Sci.*, **29**(2), 97–138.

Prager, E. J. and S. A. Earle, 2000: *The Oceans*. 314pp. McGraw-Hill, ISBN 0-07-135253-8.

Pruppacher, H. R., and J. D. Klett, 2010: *Microphysics of Clouds and Precipitation*, Springer Science and Business Media, 954pp., ISBN 0306481006.9780306481000.

Quayle, R. G., D. R. Easterling, T. R. Karl, and P. Y. Hughes, 1991: Effects of Recent Thermometer Changes in the Cooperative Station Network. *Bull. Amer. Meteor. Soc.*, **72**, 1718–1723.

Quiring, S. M., T. W. Ford, J. K. Wang, A. Khong, E. Harns, T. Lindgren, D. W. Goldberg, and Z. Li, 2016: The North American Soil Moisture Database. *Bull. Amer. Meteor. Soc.*, **97**, 1441–1459.

Rasmusson, E. M., 1967: Atmospheric Water Vapor Transport and the Water Balance of North America: Part I. Characteristics of the Water Vapor Field. *Mon. Wea. Rev.*, **95**, 403–426.

Rasmusson, E. M., 1968: Atmospheric Water Vapor Transport and the Water Balance of North America: Part II. Large-scale Water Balance Investigations. *Mon. Wea. Rev.*, **96**, 720–734.

Rasmusson, E. M., 1998: Tribute to Jerome Namias: The Pioneering Years. *Bull. Amer. Meteorol. Soc.*, **79**, 1083–1087.

Rasmusson, E. M. and T. H. Carpenter, 1982: Variations in Tropical Seas Surface Temperature and Surface Wind Fields Associated with the El Nino/Southern Oscillation. *Mon. Wea. Rev.*, **110**, 354–384.

Rasmusson, E. M., P. Arkin, J. Jalickee, and W. Chen, 1981: Biennial Variations in Surface Temperatures over the United States as Revealed by Singular Decomposition. *Mon. Wea. Rev.*, **109**, 587–598.

Rasmussen, R. et al., 2012: How Well Are We Measuring Snow: The NOAA/FAA/NCAR Winter Precipitation Test Bed: Solid Precipitation Test Bed. *Bull. Amer. Met. Soc.*, **93**, 811–829.

Reed, R. J., W. J. Campbell, L. A. Rasmusssen, and R. G. Rogers, 1961: Evidence of a Downward Propagating Annual Wind Reversal in the Equatorial Stratosphere. *J.Geophys. Sci.*, **66**, 813–818.

Reed, R. J., and E. E. Recker, 1971: Structure and Properties of Synoptic-Scale Wave Disturbances in the Equatorial Western Pacific. *J. Atmos. Sci.*, **28**, 1117–1133.

Reichle, R. H., R. D. Koster, J. Dong, and A. A. Berg, 2004: Global Soil Moisture from Satellite Observations, Land Surface Models and Ground Data: Implications for Data Assimilation. *J. Hydrometeor.*, **5**, 430–442.

Reinking, R. F. et al., 1993: The Lake Ontario Winter Storms (LOWS) Project. *Bull. Amer. Meteor. Soc.*, **74**, 1828 –1949.

Rennie, J. J. et al., 2014: The International Surface Temperature Initiative Global Land Surface Databank: Monthly Temperature Data Release Description and Methods. *Geosci. Data Journal*, DOI: 10.1002/gdj3.8.

Reynolds, R. W., 1988: A Real-Time Global Sea Surface Temperature Analysis. *J. Clim.*, **1**, 75–87.

Reynolds, R. W. and D. C. Marsico, 1993: An Improved Real-Time Global Sea Surface Temperature Analysis. *J. Clim.*, **6**, 114–119.

Reynolds, R. W. and T. M. Smith, 1994: Improved Global Sea Surface Temperature Analyses Using Optimum Interpolation. *J. Clim.*, **7**, 929–948.

Reynolds, R. W. et al., 2007: Daily High-Resolution-Blended Analyses for Sea Surface Temperature. *J. Clim.*, 20, 5473–5496.

Richards, F. and P. A. Arkin, 1981: On the Relationship between Satellite Observed Cloud Cover and Precipitation. *Mon. Wea. Rev.*, **109**, 1081–1093.

Rigor, I. G., J. M. Wallace, and R. L. Colony, 2002: Response of Sea Ice to the Arctic Oscillation. *J. Clim.*, **15**, 2648–2663.

Roads, John O., Shyh-C. Chen, Alexander K. Guetter, and Konstantine P. Georgakakos, 1994: Large-Scale Aspects of the United States Hydrologic Cycle. *Bull. Amer. Meteor. Soc.*, **75**, 1589–1610.

Robinson, D., 1993: Hemispheric Snow Cover from Satellites. *Ann. Glaciol.*, 17, 367–371.

Robock, Alan and Jianping Mao, 1995: The Volcanic Signal in Surface Temperature Observations. *J. Climate*, **8**, 1086–1103.

Robock, A. 2000: Volcanic Eruptions and Climate. *Rev. Geophys.*, **38**, 191–219.

Robock, A., Y. Konstantin, Y. Vinnikov, G. Srinivasan, J. K. Entin, S. E. Hollinger, N. A. Speranskay, S. Liu, and A. Namkhai, 2000: The Global Soil Moisture Data Bank, *Bull. Amer. Metsoc.*, **81**, 1281–1299.

Rodell, M. et al., 2004: The Global Land Data Assimilation System. *Bull. Amer. Meteor. Soc.*, **85**, 381–394. DOI: 10.1175/BAMS-85-3-381.

Rodell, M., J. Chen, H. Kato, J. S. Famiglietti, J. Nigro, and C. R. Wilson, 2007: Estimating Ground Water Storage Changes in the Mississippi River Basin (USA) Using GRACE. *Hydrogeol. J.* **15**, 159–166.

Romanovsky, V., M. Burgess, S. Smith, K. Yoshikawa, and J. Brown, 2002: Permafrost Temperature Records: Indicators of Climate Change. *EOS, AGU Transactions*, **83**, 589–594.

Ropelewski, C.F., 1983: Spatial and Temporal Variations in Antarctic Sea Ice (1973–1982). *J. Clim. Appl. Met.*, **22**, 470–473.

Ropelewski, C. F., J. E. Janowiak, and M. S. Halpert, 1985: The Analysis and Display of Real Time Surface Climate Data. *Mon. Wea. Rev.*, **113**, 1101–1106.

Ropelewski, C. F. and M. S. Halpert, 1986: North American precipitation and temperature patterns associated with the El Nino/Southern Oscillation (ENSO). *Mon. Wea. Rev.*, 114, 2352–2362.

Ropelewski C. F., and M. S. Halpert, 1987: Global and Regional Precipitation Patterns Associated with the El Nino/Southern Oscillation. *Mon. Wea. Rev.*, **115**, 1606–1626.

Ropelewski C. F., and P. D. Jones, 1987: An Extension of the Tahiti-Darwin Southern Oscillation Index. *Mon. Wea. Rev.*, **115**, 2161–2165.

Ropelewski C. F., and M. S. Halpert, 1989: Precipitation Patterns Associated with the High Index Phase of the Southern Oscillation. *J. Climate*, **2**, 268–284.

Ropelewski C. F. and M. S. Halpert, 1991: Greenhouse-Gas-Induced Climatic Change: A Critical Appraisal of Simulations and Observation. In *The Southern Oscillation and Northern Hemisphere Temperature Variability*. (M. E. Schlesinger, ed.) Elsevier Press, 369–376.

Ropelewski, C. F., M. S. Halpert, and J. Wang:, 1992: Observed Tropospheric Biennial Variability and Its Relationship to the Southern Oscillation. *J. Clim.*, **5**, 594–614.

Ropelewski, C. F., 1995: Long-Term Observations of Land Surface Characteristics. In *Long-term climate monitoring by the global climate observing system*. (T. R. Karl Ed.) *Climatic Change*, **31**, 415–425.

Ropelewski, C. F. and E. S., Yarosh, 1998: The Observed Mean Annual Cycle of Moisture Budgets over the Central United States (1973–92). *J. Clim.*, **11**, 2180–2190.

Ropelewski, C. F. and C. F. Folland, 2000: Prospects for the prediction of meteorological drought. In *Drought: A Global Assessment*, Vol. 1, (D. A. Wilhite, ed.) Routledge Hazards and Disasters Series, Routledge Press pp. 21–41.

Ropelewski, C. F, D. S Gutzler, R. W. Higgins, and C. R. Mechoso, 2005: The North American Monsoon System. In The Global Monsoon System: Research and Forecast. World Meteorological Organization Technical Document 1266, 207–218, Hangzhou, China, Nov. 2–6, 2004.

Ropelewski, C. F. and M. A. Bell, 2008: Shifts in the Statistics of Daily Rainfall in South America Conditional on ENSO Phase. *J. Clim.*, **21**, 849–865.

Ropelewski, C. F. and P. A. Arkin, 2017: Advances in Climate Analysis and Monitoring: Reflections on 40 years of Climate Diagnostics and Prediction Workshops. *Bull. Amer. Meteor. Soc.*, **98**, 461–471.

Rossow, W. B. and R. A. Schiffer, 1999: Advances in Understanding Clouds from ISCCP. *Bull. Amer. Meteor. Soc.*, **80**, 2261–2287.

Royer, D. L., 2006: CO_2 – Forced Climate Thresholds during the Phanerzoic. *Geochim. Cosmochim. Acta*, **70**, 5665–5675.

Ruddiman, W. F., 2005: *Plows, Plagues, and Petroleum: How Humans Took Control of Climate*, Princeton University Press. ISBN 978-0-691-14634-8, 225 pp.

Sagan, C. and Mullen, G., 1972: Earth and Mars: Evolution of Atmospheres and Surface Temperatures. *Science*, **177**(4043), 52–56.

Saha S. et al., 2010: The NCEP Climate System Forecast System Reanalysis. *Bull. Amer. Meteor. Soc.*, **91**, 1015–1057.

Sanchez, P. A. et al., 2009: Digital Soil Map of the World. *Science*, **324**, 680–681.

Santanello, J., P. Dirmeyer, C. Ferguson, K. Findell, A. Tawfik, A. Berg, M. Ek, P. Gentine, B. Guillod, C. van Heerwaarden, J. Roundy, and V. Wulfmeyer, 2017: Land-Atmosphere Interactions: The LoCo Perspective. *Bull. Amer. Meteor. Soc.* **99**, DOI:10.1175/BAMS-D-17-0001.1, in press.

Sarachik, E. S. and M. A. Cane, 2010: *The El Niño-Southern Oscillation Phenomenon*, Cambridge University Press, London, 384pp.

Sargent, R. G. 2013: Verification and Validation of Simulation Models. *Jour. of Simulation*, **7**, 12–24.

Savarin, A., 2016: Pathways to better prediction of the Madden-Julian Oscillation over the Indian Ocean., MS Thesis, U. of Miami, 108pp.

Schaefer, G. L. and R. F. Paetzold, 2000: SNOTEL (SNOpack TELemetry) and SCAN (Soil Climate Analysis Network). Presented at the Conference on Automated Weather Stations for Applications in Agriculture and Water Resources Management: Current Use and Future Perspectives, Lincoln, NB, NRCS/USDA, March 2002.

Schaal, L. A. and R. F. Dale, 1977: Time of Observation Temperature Bias and "Climatic Change". *J. Appl. Meteor.*, **16**, 215–222.

Schlesinger et al., 1979: Terminology for Model Credibility. *J. Simul.*, 32(3), 103–104.

Schmid, C., R. L. Molinari, R. Sabina, Y-H Daneshzadeh, X. Xia, E. Forteza, and H Yang, 2007: The Real-Time Data Management System for Argo Profiling Float Observations. *J. Atmos. Oceanic Tech.*, **24**, 1608–1628.

Schubert, S. D., R. B. Rood, and J. Pfaendtner, 1993: An Assimilated Dataset for Earth-Science Applications. *Bull. Amer. Meteor. Soc.*, 74, 2331–2342.

Scott, R. W., E. C. Krug, and S. L. Burch, 2010: Illinois Soil Moisture under Sod Experiment. *J. Hydrometeor.*, **11**, 683–704.

Seager, Richard, 2007: The Turn of the Century North American Drought: Global Context, Dynamics, and Past Analogs. *J. Clim.*, **20**, 5527–5552.

Segal, M., J. P. Garratt, R. A. Pielke, and Z. Ye, 1991: Scaling and Numerical Model Evaluations of Snow-Coveer Effects on the Generation of Mesoscale Circulations. *J. Atmos. Sci.*, **48**, 1024–1042.

Segre, E. 2007: *From falling bodies to radio waves: Classical Physicists and their discoveries*, Dover publications (paperback). Originally published 1984.

Shakelton, E. 1918: *South:The Story of Shakelton's Last Expedition, 1914–1917.*, html version from Project Gutenberg www.gutenberg.net.

Shaw, T. A. and J. Perlwitz, 2013: The life cycle of Northern Hemisphere downward wave coupling between the stratosphere and troposphere., **26**, 1745–1763.

Shenx, W. E and E. R. Kreins, 1970: A Comparison between Observed Winds and Cloud Motions Derived from Satellite Infrared Measurements. *J. Appl. Meteorol.*, **9**, 702–710.

Shepherd, J. M., T. Anderson, L. Bounoua, A. Horst, C. Mitra, and C. Strother, 2013: Urban Climate Archipelagos: A new framework for urban-climate interactions. *IEEE Earthzine*, published online at www.earthzine.org/2013/11/29/urban-climate-archipelagos-a-new-frame work-for-urban-impacts-onclimate/.

Siemann, A. L., G. Coccia, M. Pan, and E. F. Wood, 2016: Development and Analysis of Long-Term, Global, Terrestrial Land Surface Dataset Based on HIRS Satellite Retrievals. *J. Clim.*, 29, 3589–3606.

Silver, N. 2012: *The signal and the noise: why so many predictions fail-but some don't*, The Penguin Press, 534pp. ISBN 978-1-59420-1.

Skinner, J. H., L. Lye, and S. E. Bruneau, 2010: Climatic influences on the annual iceberg flux off the coast of Newfoundland., *CSCE 2010 General Conference.*, Winnipeg, Manitoba, June 9–12, 2010. Canadian Society for Civil Engineering, www.proceedings.com/ 0199.html.

Smith, G. L., A. C. Wilber, S. K. Gupta, and P. W. Stackhouse, 2002: Surface Radiation Budget and Climate Classification. *J. Clim.*, **15**, 1175–1188.

Smith, T. M. and R. W. Reynolds, 2002: Bias Corrections for Historical Sea Surface Temperatures Based on Marine Air Temperatures. *J. Clim.*, **15**, 73–87.

Smith, T. M. and R. W. Reynolds, 2004a: Improved Extended Reconstruction of SST 1854–1997. *J. Climate*, **17**, 2466–2477.

Smith, T. M., and R. W. Reynolds, 2004b: Reconstruction of Monthly Mean Oceanic Sea Level Pressure Based on COADS and Station Data (1854–1997). *J. Oceanic Atmos. Tech.*, **21**, 1272–1282.

Smith, T. M.and R. W. Reynolds, 2005: A Global Merged Land-Air-Sea-Surface Temperature Reconstruction Based on Historical Observations (1880–1997). *J. Clim.*, **18**, 2021–2036.

Smith, T. M., R. W. Reynolds, T. C. Peterson, and J. Lawrimore, 2008: Improvements to NOAA's Historical Merged Land-Ocean Surface Temperature Analysis (1880–2006). *J. Clim.*, **21**, 2283–2296.

Smith, T. M., M. R. P. Sapiano, and P. A. Arkin, 2009a, Modes of Multi-Decadal Oceanic Precipitation Variations from a Reconstruction and AR4 Model Output for the 20th Century. *Geophys. Res. Lett.*, **36**, L14708, DOI:10.1029/2009GL039234.

Smith, T. M., P. A. Arkin, and M. R. P. Sapiano, 2009b, Reconstruction of Near-Global Annual Precipitation Using Correlations With Sea Surface Temperature and Sea Level Pressure. *J. Geophys. Res.*, **114**, D12107, DOI:10.1029/2008JD011580.

Smith, T. M., P. A. Arkin, M. R. P. Sapiano, and C.-Y. Chang, 2010: Merged Statistical Analyses of Historical Monthly Precipitation Anomalies Beginning 1900. *J. Clim.*, **23**, 5755–5770.

Smith, T. M., R. W. Reynolds, R. E. Livezey, and D. C. Stokes, 1996: Reconstruction of Historical Sea Surface Temperatures Using Empirical Orthogonal Functions. *J. Clim.*, **9**, 1403–1420.

Soil Survey Staff, 1999: Soil Taxonomy: A basic system of soil classification for making and interpreting of soil surveys. *Agriculture Handbook*, 436, 871 pp. (U.S. Dept of Agri. National

Resources Conservation Service) Available from the Superintendent of Documents, U.S. Government Printing Office, Washington, DC, 20402.

Solomon, S. et al. (eds.), 2007: *Contributions of Working Group I to the Fourth Assessment of the Intergovernmental Panel on Climate Change*, Cambridge University Press, 337–383.

Solomon, S., D. Qin, M. Manning, Z. Chen, M. Marquis, K. B. Averyt, M. Tignor, and H. L. Miller (eds.), 2007: *Climate Change 2007: The Physical Science Basis, Contribution of Working Group I to the Fourth Assessment Report of the Intergovernmental Panel on Climate Change*, Cambridge University Press, Cambridge, United Kingdom and New York, NY, 996 pp. ISBN 978 0 521 705967.

Sorooshian S., K-L Hsu, X. Gao, H.V. Gupta, B. Imam, and D. Braithwaite, 2000: Evaluation of PERSIANN System Satellite–Based Estimates of Tropical Rainfall. *Bull. Amer. Meteor. Soc.*, **81**, 2035–2046.

Spencer, R. W., 1986: A Satellite Passive 37-GHz Scattering-Based Method for Measuring Oceanic Rain Rates. *J. Clim. Appl.*, **25**, 754–766.

Stanhill, Gerald and Shabtai Cohen, 2005: Solar Radiation Changes in the United States during the Twentieth Century: Evidence from Sunshine Duration Measurements. *J. Clim.*, **18**, 1503–1512.

Stickler A., A. N. Grant, T. Ewen, T. F. Ross, R. S. Vose, J. Comeaux, P. Bessemoulin, K. Jylhä, W. K. Adam, P. Jeannet, A. Nagurny, A. M. Sterin, R. Allan, G. P. Compo, T. Griesser, and S. Brönnimann, 2010: The Comprehensive Historical Upper-Air Network. *Bull. Amer. Meteor. Soc.*, **91**, 741–751.

Stocker, T. F., D. Qin, G.-K Plattner, M. Tignor, S. K. Allen, J. Boshung, A. Nauls, Y. Xia, V. Bex, and P. M. Midgley (eds.), 2013: (In the Summary for Policy Makers), *Climate Change 2013: The Physical Science Basis, Contribution of Working Group I to the Fifth Assessment Report of the Intergovernmental Panel on Climate Change*, Cambridge University Press, Cambridge, United Kingdom and New York, NY.

Stommel, H. and E. Stommel, 1983: *Volcano Weather: The Story of 1816, the Year without a Summer*, South Seas Press, 177 pp., ISBN 0-015160-71-4.

von Storch, H. and F. W. Zwiers, 2002: *Statistical Analysis in Climate Research*, Cambridge University Press, 496 pp., ISBN 0521012309.

Strangeways, I., 2006: *Precipitation: Theory, Measurement and Distribution*. Cambridge University Press, 302 pp., ISBN 1139460013.9781139460019.

Sweet, W. J, Park, J. Marra, C. Zervas, and S. Gill, 2014: Sea Level Rise and Nuisance Flood Frequency Changes around the United States. *NOAA Technical Report NOS CD-OPS 073*, 66 pp. Available online at http://tidesandcurrents.noaa.gov/publications/NOAA_Techincal_Report_NOS_COOPS_073.pdf.

Tapley, B. D., S. Bettadpur, M. Watkins, and Ch. Reigber, 2004: The Gravity Recovery and Climate Experiment; Mission Overview and Early Results. *Geophys. Res. Lett*, **31**, L09607.

Taylor, R. C., 1973: An Atlas of Pacific Islands Rainfall, Hawaii Institute of Geophysics Data Report no. 25, University of Hawaii, Honolulu, Hawaii.

Tarpley, J. D., S. R. Schnider, and R. L. Money, 1982: Global Vegetation Indices from the NOAA-7 M.

Teixeira, J., D. Waliser, R. Ferraro, P. Gleckler, T. Lee, and G. Potter, 2014: Satellite Observations for CMIP5: The Genesis of Obs4MIPs. *Bull. Amer. Meteor. Soc.*, **95**, 1329–1334.

Thompson, D. W. J. and J. M. Wallace, 2000: Annular Modes in the Extratropical Circulation. Part I: Month-to-Month Variability. *J. Clim.*, **13**, 1000–1016.

Thompson, J., 1892: On the Grand Currents of the General Circulation. *Phil. Trans. Roy. Soc., A*, 183, 653–684.

Thompson, L., E. Mosley-Thompson, M. E. Davis, P.-N. Lin, K. Henderson, and T. A. Mashiotta, 2003: Tropical Glacier and Ice Core Evidence of Climate Change on Annual to Millennial Time Scale. *Clim. Change*, **59**, 137–155.

Thompson, L. E., 2004: High Altitude and Mid- and Low-Latitude Ice Core Records: Implications for Our Future. In *Earth Paleoenvironments: Record Preserved in Mid and High Latitude Glaciers*, L. D. Cecil et al., (eds). Kluwer Academic Publishers.

Thorne et al., 2017: Towards an Integrated Set of Surface Meteorological Observations for Climate Science and Applications. *Bull. Amer. Meteor. Soc.*, DOI:10.1175/BAMS-D-16-0165.1, in press.

Tian, Y, C. D. Peters-Lidard, R. F. Adler, T. Kubota, and T. Ushio, 2010: Evaluation of GSMaP Precipitation Estimates over the Contiguous United States. *J. Hydrometeor*, **11**, 566–573.

Tolle, M. E., S. Engler, and H-J. Panitz, 2017: Impact of Abrupt Land Cover Changes by Tropical Deforestation on Southeast Asian Climate and Agriculture. *J. Clim.*, **30**, 2587–2600.

Tower, W. S., 1903: Mountain and Valley Breezes. *Mon. Wea. Rev.*, **31**, 528–529.

Trenberth, K. E., 1990: Recent Observed Interdecadal Climate Changes in the Northern Hemisphere. *Bull. Amer. Meteor. Soc.*, **71**, 988–993.

Trenberth, K. E. and C. J. Guillemot, 1995: Evaluation of the Global Atmospheric Moisture Budget as Seen from Analyses. *J. Clim.*, **8**, 2255–2272.

Trenberth, K. E., D. P. Stepaniak, J. W. Hurrell, and M. Fiorino, 2001: Quality of Reanalyses in the Tropics. *J. Clim.*, **14**, 1499–1510.

Trenberth, K. E. and D. J. Shea, 2006: Atlantic Hurricanes and Natural Variability in 2005. *Geoph.l Res. Lett.*, **33**, L12704.

Trenberth, K., Fasullo, J., and Kiehl, J., 2009: Earth's Global Energy Budget. *Bull. Amer. Meteor. Soc.*, **90**, 311–323.

Tucker, C. J., 1979: Red and Photographic Infrared Linear Combinations for Monitoring Vegetation. *Remote Sens. Environ.*, **8**, 127–150.

Tucker, G. B., 1961: Precipitation over the North Atlantic Ocean. *Quart. J. Roy. Meteor. Soc.*, 87, 147–158.

Tucker, G. B., 1962: Reply. *Quart. J. Roy. Meteor. Soc.*, 88, 188.

USDA and USDoC (US Department of Agriculture and US Department of Commerce), 2014: *Weekly Weather and Crop Bulletin*, **101**, Jan 2014, 26 pp. Also available at www.usda.gov/oce/weather/.

van Loon, H., 1965: A climatological study of the atmospheric circulation in the Southern Hemisphere during the IGY, Part I: 1 July 1957-31 March 1958. *J. Appl. Meteor. Cli.*, 4, 479–491.

van Loon, H., 1979: The Association between Latitudinal Temperature Gradient and Eddy Transport. Part I: Transport of Sensible Heat in Winter. *Mon. Wea. Rev.*, 107, 525–534.

van Loon, H. and K. Labitzke, 1988: Association between the 11-Year Solar Cycle, the QBO, and the Atmosphere. Part II: Surface and 700 mb in the Northern Hemisphere in Winter. *J. Clim.*, **1**, 905–920.

Vasvalingam. M. and J. D. Tandy, 1972: The Neutron Method for Measuring Soil Moisture Content – A Review. *J. Soil Sci.*, **23**, 499–511.

Vinnikov, K. Y. 1977: Procedures for Acquisition of Data on the Variations of Northern Hemisphere Surface Air Temperature during 1881–1975 (In Russian). *Meteor. Gidrol.*, **9**, 110–114.

Vose, S. R., C. N. Williams Jr., T. C. Peterson, T. R. Karl, and D. R. Easterling, 2003: An Evaluation of the Time of Observations Bias Adjustment in the U.S. Historical Climatology Network. *Geoph. Res. Ltrs.*, **30**, 2026, CLM 3–1-CLM3–4.

Vose R. S., Scott Applequist, Mike Squires, Imke Durre, Matthew J. Menne, Claude N. Williams Jr., Chris Fenimore, Karin Gleason, and Derek Arndt, 2014: Improved Historical Temperature and Precipitation Time Series for U.S. Climate Divisions. *J. Appl. Meteor. Climatol.*, **53**, 1232–1251.

Wahr, J., D. Wingham, and C. Bentley. 2000. A Method of Combining ICESat and GRACE Satellite Data to Constrain Antarctic Mass Balance. *J. Geophys. Res.*, **105**(B7), 16,279–16,294.

Walker, G. T., 1924: Correlation of the Seasonal Variations in Weather IX: A Further Study of World Weather. *Mem. Indian Meteor. Dep.*, **24**, 275–332.

Walker, G. T., 1928: World Weather. *Quart. J. Roy. Meteor. Soc.*, 29–87.

Walker, G. T. and E. W. Bliss, 1932: World Weather V. *Mem. Roy. Metero. Soc.*, 53–84.

Walker, G. T. and E. W. Bliss, 1937: World Weather VI. *Mem. Roy. Metero. Soc.*, 119–139.

Wallace, J. M. and D. S. Gutzler, 1981: Teleconnections in the Geopotential Height Field during the Northern Hemisphere Winter. *Mon. Wea. Rev.*, **109**, 784–812.

Wallace, J. M. and P. V. Hobbs, 2006: *Atmospheric Science*, 2nd Edition, Academic Press, 483 pp.

Wallace, J. M. I. M. Held, D. W., Thompson, K. E. Trenberth, and J. E. Walsh, 2014: Global Warming and Winter Weather. *Science*, **343**, 729–730.

Walsh, J. E. and C. M. Johnson, 1979: An Analysis of Arctic Sea Ice Fluctuation 1953–1977. *J. Phys. Ocean.*, **9**, 580–591.

Walsh J. E., D. R. Tucek, and M. R. Peterson, 1982: Seasonal Snow Cover and Short Term Climate Fluctuations over the United States. *Mon. Wea. Rev.*, **110**, 1474–1486.

Wang, J., X. Bai, H. Hu, A. Clites, M. Colton, and B. Lofgren 2012: Temporal and Spatial Variability of Great Lakes Ice Cover, 1973–2010. *J. Clim.*, **25**, 1318–1329.

Wang, P. K., 2013: *Physics and Dynamics of Clouds and Precipitation*, Cambridge University Press, New York. 452 pp.

Warren, S. G., R. Eastman, and C. J. Hahn, 2015: Cloud Climatology. In *Encyclopedia of Atmospheric Sciences*, Oxford University Press.

Watson, C. S., N. J. White, J. A. Church, M. A. King, R. J. Burgette, and B. Legresy, 2015: Unabated Global Mean Sea-Level Rise over the Satellite Altimeter Era. *Nat. Clim. Change*, **5**, 565–568.

Wheelan, C., 2014*: Naked Statistics: Stripping the Dread from Data*, 282 pp., W.W. Norton and Company.

Wick, G. A., J. J. Bates, and D. J. Scott, 2002: Satellite and Skin-Layer Effects on the Accuracy of Sea Surface Temperature Measurements from the GOES Satellites. *J. Atmos. Oceanic Technol.*, 19, 1834–1848.

Wijffels, S. E. et al., 2008: Changing Expendable Bathythermograph Fall Rates and Their Impact on Estimates of Thermosteric Sea Level Rise. *J. Clim.*, 21, 5657–5672.

Wilheit T. T., A. T. C. Chang, and L. S. Chiu, 1991: Retrieval of Monthly Rainfall Indices from Microwave Radiometric Measurements Using Probability Distribution Functions. *Jour. Atmos. Ocean. Tech.*, **8**, 118–136.

Wilks, D. S., 2011: *Statistical Methods in Meteorology*, 3rd Edition, Academic Press, Elsevier, 704 pp. ISBN-13: 978–0–12–385022–5.

Wilks, D. S., 2017: The Stippling Shows Statistically Significant Grid Points. *Bull. Amer. Meteor. Soc.*, **97**, 2263–2273.

Willett, H. C., 1950: Temperature trends of the past century. *Cent. Proc. Roy. Meteor. Soc.*, 195–206.

Williams, C. N., R. S. Vose, D. R. Easterling, and M. J. Menne, cited 2006: United States historical climatology network daily temperature, precipitation, and snow data. ORNL/CDIAC-118, NDP-070, Carbon Dioxide Information Analysis Center, Oak Ridge National Laboratory. [Available online at cdiac.ornl. gov/ftp/ndp070/ndp070.txt.]

WMO, 1970: The Beaufort Scale of Wind Force (Technical and operational aspects). Commission for Marine Meteorology, Report. on Marine Science Affairs 3, 22pp. (Available from the World Meteorological Organization, Case Postale 5, Geneva, Switzerland).

WMO, 2008: Guide to Meteorological Instruments and Methods of Observation WMO –No. 8, Section I, Chapter 7 and Chapter 8. (Available through the WMO or online) www.wmo.int/pages/prog/gcos/documents/gruanmanuals/CIMO/CIMO_Guide_7th_Edition-2008.pdf.

WMO Resolution 40 www.wmo.int/pages/prog/www/ois/Operational_Information/Publications/Congress/Cg_XII/res40_en.html.

WMO Resolution 60 www.wmo.int/pages/prog/sat/meetings/documents/CM-13_Doc_03-01_WMO-Data-Policies.pdf.

Woodruff, S. D., R. J. Slutz, R. L. Jenne, and P. M. Steurer, 1987: A Comprehensive Ocean-Atmosphere Data Set. *Bull. Amer. Meteor. Soc.*, 68, 1239–1250.

Woodruff, S. D., S. J. Worley, S. J. Lubker, Z. Ji, J. E. Freeman, D. I. Berry, P. Brohan, E. C. Kent, R. W. Reynolds, S. R. Smith, and C. Wilkinson, 2011: ICOADS Release 2.5: Extensions and enhancements to the surface marine meteorological archive. *Int. J. Climatol.* (CLIMAR-III Special Issue), **31**, 951–967 (DOI:10.1002/joc.2103).

Wyrtki, K., 1975: El Niño: The Dynamic Response of the Equatorial Pacific Oceanto Atmospheric Forcing. *J. Phys. Oceanogr.*, **5**, 572–584.

Xie, P. and P.A. Arkin 1996: Gauge-based monthly analysis of global land precipitation from 1971 to 1994. *J. Climate*, **9**, 840–858.

Xie, P., and P. A. Arkin, 1996: Analyses of global monthly precipitation using gauge observations, satellite estimates and numerical model predictions. *J. Clim.*, **9**, 840–858.

Xie, P. and P. A. Arkin, 1997: Global precipitation: A 17-year monthly analysis based on gauge observations, satellite estimates, and numerical model outputs. *Bull. Amer. Meteor. Soc.*, **78**, 2539–2558.

Xie, P. and P. A. Arkin 1998: Global Monthly Precipitation Estimates from Satellite-Observed Outgoing Longwave Radiation. *J. Climate*, **11**, 137–164.

Xie, P. and A.-Y. Xiong, 2011: A Conceptual Model for Constructing High-Resolution Gauge-Satellite Merged Precipitation Analyses. *J. Geophys. Res.*, 116, D21106, DOI:10.1029/2011JD016118.

Xie, P., R. Joyce, S. Wu, S.-H. Yoo, Y. Yarosh, F. Sun, and R. Lin, 2017: Reprocessed, Bias-Corrected CMORPH Global High-Resolution Precipitation Estimates from 1998. *J. Hydrometeor.*, **18**, 1617–1651.

Yang et al., 2013: A Multiscale Soil Moisture and Freeze-Thaw Monitoring Network on the Third Pole. *Bull. Amer. Meteor. Soc.*, **94**, 1907–1914.

Zebiak, Stephen E. and Mark A. Cane, 1987: A Model El Niño–Southern Oscillation. *Mon. Wea. Rev.*, **115**, 2262–2278.

Zhang, Y, J. M. Wallace and D. S. Battisti, 1997: ENSO-like Interdecadal Variability: 1900–93. *J. Clim.*, **10**, 1004–1020.

Zhang, T, R. G. Barry, K. Knowles, J. A. Heginbottom, and J. Brown, 1999: Statistics and Characteristics of Permafrost and Ground-Ice Distribution in the Northern Hemisphere. *J. Polar Geogr.*, **23**, 132–154.

Zielinski, A., P. A. Mayewski, L. D. Meeker, S. Whitlow, M. S. Twickler, and K. Taylor, 1996: Potential Atmospheric Impact of the Toba Mega-Eruption ~71'000 Years Ago. *Geoph. Res. Ltrs.* **23**, 837–840.

Zweng, M. M, J.R. Reagan, J.I. Antonov, R. A. Locarnini, A. V. Mishonov, T. P. Boyer, H. E. Garcia, O. K. Baranova, D. R. Johnson, D. Seidov, and M. M. Biddle, 2013. *World Ocean Atlas 2013*, Volume 2: Salinity. S. Levitus, Ed., A. Mishonov Technical Ed.; NOAA Atlas NESDIS 74, 39 pp.

Websites

We have tried to insure that the URLs listed below are correct and current as of the publication date. However, the very nature of the Internet dictates that web addresses will be modified or change over time.

Alaska Satellite Facility www.asf.alaska.edu/asf-tutorials/sar-faq/

AMIP – Atmospheric Model Inter-comparison Project https://pcmdi.llnl.gov/mips/amip/home/overview.html

ARM – Atmospheric Radiation Measurement Program – Climate Research Facility www.archive.arm.gov/

ASOS – Automated Surface Observation System description www.weather.gov/media/lmk/pdf/educational_pages/ASOSandClimateObservations__What_Is_ASOS.pdf

ASOS – Automated Surface Observation System User's guide (1998) – pdf can be downloaded from www.nws.noaa.gov/asos/pdfs/aum-toc.pdf

Blue Hill Observatory – http://bluehill.org/observatory/

BoM – Australian Bureau of Meteorology www.bom.gov.au/climate/

Buoy Data

Moored Buoys Data
www.pmel.noaa.gov/gtmba/
Drifting Buoy Data
www.aoml.noaa.gov/phod/dac/index.php

CDB – Climate Diagnostics Bulletin, Published by the CPC Monthly since 1983 and available electronically from 1999 at www.cpc.ncep.noaa.gov/products/CDB

CERA20C – ECMWF 20th Century Reanalysis version C http://apps.ecmwf.int/datasets/data/cera20c-enda/levtype=sfc/type=an/

Copernicus Climate Change Service https://climate.copernicus.eu/

Climate Data Guide Comparison of Reanalysis Products (NCAR/UCAR)

https://climatedataguide.ucar.edu/climate-data/atmospheric-reanalysis-overview-comparison-tables

Climate Data Records from Environmental Satellites – Interim Report www.nap.edu/catalog/10944/climate-data-records-from-environmental-satellites-interim-report

Climate Forecast and Monitoring www.climate.gov/

Climate Prediction Center (CPC/NOAA) www.cpc.ncep.noaa.gov/

Climatic Research Unit, University of East Anglia, www.cru.uea.ac.uk/

CRU Climatic Research Unit data – The Climate Data Guide: CRU TS3.21 Gridded precipitation and other meteorological variables since 1901. Retrieved from https://climatedataguide.ucar.edu/climate-data/cru-ts321-gridded-precipitation-and-other-meteorological-variables-1901.

Cold Regions Research and Engineering Laboratory (CRREL) www.crrel.usace.army.mil

Compendium of Meteorology https://archive.org/details/compendiumofmete00amer

COOP – Cooperative Observer Network Data for the United States
www.ncdc.noaa.gov/data-access/land-based-station-data/land-based-datasets/cooperative-observer-network-coop

Digital Soils Data Base (IGBP-DIS) www.daac.ornl.gov/

ENSO and Seasonal Climate Forecasts
Climate Prediction Center/NCEP/NOAA www.cpc.noaa.gov/
European Centre for Medium-Range Weather Forecasts www.ecmwf.int/en/forecasts
International Research Center for climate and society (IRI) https://iri.columbia.edu/our-expertise/climate/forecasts/seasonal-climate-forecasts/

ENSO Monitoring websites (NOAA)
www.cpc.ncep.noaa.gov/products/analysis_monitoring/enso_advisory/index.shtml
www.climate.gov/enso

ENSO Blog (NOAA)
www.climate.gov/news-features/department/enso-blog

EOF and SVD – A Manual for EOF and SVD Analyses of Climatic Data (H.Bjornsson and S.A. Venega) C^2GCR Report 97–1, Feb 1997 www.jsg.utexas.edu/fu/siles/EOFSVD.pdf

EPA 2014, Environmental Protection Agency Ozone website. http://cfpub.epa.gov/airnow/index.cfm?action=gooduphigh.index

ERA20c – ECMWF 20th Century Reanalysis Data http://apps.ecmwf.int/datasets/data/era20c-ofa/

ESRL/PSD Earth System Research Laboratory/Physical Science Division/NOAA. Climate and Weather Data www.esrl.noaa.gov/psd/data/

European Space Agency (ESA) Climate Change Initiative (CCI) Land Cover project www.esa-landcover-cci.org/

European Space Agency (ESA) Earth Observation Data https://earth.esa.int/web/guest/home

GADR – Global Argo Data Repository
www.nodc.noaa.gov/argo/

GHCN – Global Historical Climate Network NCEI/NOAA www.ncdc.noaa.gov/data-access/land-based-station-data/land-based-datasets/global-historical-climatology-network-ghcn

Global Land Surface Facility (GLCF) http://glcf.umd.edu/data/landcover/description.shtml Data are available at 1°, 8 km and 1 km resolution. Land surface is characterized into 13 classes.

Global Multi-resolution Terrain Elevation Data (GMTED2010) http://eros.usgs.gov/elevation-products Vertical resolution of roughly 25–40m depending on horizontal resolution either on ½, ¼ to 1/8 degree Data poleward of 60^oN may be incomplete. https://lta.cr.usgs.gov/GMTED2010. The U. S. Geological Survey (USGS) provides a National Elevation Data set for the United States at horizontal resolutions 10 m to 30 m. The USGS also provides global DEM data (ref GMTED2010) at comparable horizontal resolutions and vertical resolutions of 25 m to 30 m. Data are incomplete north of 60° N.

Global Reservoir and Lake Monitor
https://ipad.fas.usda.gov/cropexplorer/global_reservoir/

Glossary of Meteorology, American Meteorological Society
http://glossary.ametsoc.org

GRACE (Gravity Recovery and Climate experiment. www.nasa.gov/mission_pages/Grace/

Great Lakes Environmental Research Laboratory (GLERL) www.glerl.noaa.gov/data

Harvard School of Design for general information about using digital elevation models. www.gsd.harvard.edu/gis/manual/dem/

Hurrell, James & National Center for Atmospheric Research Staff (Eds). Last modified 28 Aug 2013. The Climate Data Guide: Hurrell North Atlantic Oscillation (NAO) Index (station-based). Retrieved from https://climatedataguide.ucar.edu/climate-data/hurrell-north-atlantic-oscillation-nao-index-station-based.

IMD – India Meteorological Department – www.imd.gov.in/pages/monsoon_main.php

Integrated Climate Data Center (ICDC) – Land Type Data are distributed by the Land Processes Distributed Active Archive Center (LP DAAC), located at the U.S. Geological Survey (USGS) Earth Resources Observation and Science (EROS) Center http://lpdaac.usgs.gov and also distributed in netCDF format by the Integrated Climate Data Center (ICDC, http://icdc.zmaw.de) University of Hamburg, Hamburg, Germany.

International Geosphere Biosphere Programme (IGBP) www.igbp.net/

International Soil Moisture Network (ISMN) www.ipf.tuwein.ac.at/insitu

International Surface Irradiance Study (ISIS) datasets www.esrl.noaa.gov/gmd/grad/isis/index.html

International Surface Pressure Databank (ISPD) http://reanalyses.org/observations/international-surface-pressure-databank

IPCC History www.ipcc.ch/organization/organization_history.shtml

IPCC Third Assessment Report (TAR) www.ipcc.ch/ipccreports/tar/wg1/index.php?idp=38

IPWG – International Precipitation Working Group http://ipwg.isac.cnr.it/data/datasets3.html

IRI/LDEO Data Library, http://iridl.ldeo.columbia.edu/

IRI Maproom – http://iridl.ldeo.columbia.edu/maproom/

JRA-55C reanalysis https://climatedataguide.ucar.edu/climate-data/jra-55c-reanalysis-using-conventional-observations

JRA-55AMIP http://dias-dmg.tkl.iis.u-tokyo.ac.jp/dmm/doc/JRA55_AMIP-DIAS-en.html

Long-term Data Record (NDVI and other products) (https://ltdr.modaps.eosdis.nasa.gov/cgi-bin/ltdr/ltdrPage.cgi)

MERRA-2 Reanalysis https://disc.sci.gsfc.nasa.gov/uui/datasets?keywords=%22MERRA-2%22

MJO Monitoring www.cpc.ncep.noaa.gov/products/precip/CWlink/MJO/mjoupdate.pdf

NASA – Earth Data Centers https://earthdata.nasa.gov/

NASA – Land cover types http://earthobservatory.nasa.gov

NASA Global Change Master Directory http://gcmd.nasa.gov

NASA – USGS Land Products DAAC https://lpdaac.usgs.gov/products/

National Climatic Data Center (NCDC) Paleoclimatology web site www.ncdc.noaa.gov/paleo/abrupt/data4.html

The National Map – https://nationalmap.gov/landcover.html

National Climatic Data Center (NCDC) drought indices www.ncdc.gov/oa/climate/research/prelim/drought/palmer.html

NCDC maps of seasonal climate change www.ncdc.noaa.gov/oa/climate/research/trends.html#top

National Drought Mitigation Center, University of Nebraska, Lincoln. http://drought.unl.edu/Planning/Monitoring/ComparisonofIndicesIntro/PDSI.aspx

National Geophysical Data Center (NGDC) www.ngdc.noaa.gov/ and www.ngdc.noaa.gov/mgg/topo/globe.html

National Ice Center – www.natice.noaa.gov Ref for the Interactive Multisensor Snow and Ice Mapping System (IMS)

Natural Resources Conservation Service (NRCS) www.wcc.nrcs.usda.gov

National Snow and Ice Data Center – NSIDC, 2011: Global lake and rice ice phenology. Internal development version accessed by NSIDC staff, December 2011. http://nsidc.org/data/lake_river_ice

NLCD, National Land Cover Database, 2006: Produced by the U.S. Geological Survey in conjunction with the Multi-Resolution Land Characteristics Consortium (MRLC). Available for no charge at www.mrlc.gov

NOAA Central Library Photos www.photolib.noaa.gov

NOAA Central Library Weather Maps www.lib.noaa.gov/collections/imgdocmaps/daily_weather_maps.html

NOAA Earth System Research Laboratory (ESRL) www.esrl.noaa.gov/psd/index.html)

Normalized Difference Vegetation Index (NDVI) data and more information at http://gcmd.nasa.gov/records/GCMD_EOSWEBSTER_NOAANASA_Path_NDVI.html or http://iridl.ldeo.columbia.edu/SOURCES/NASA/GES-DAAC/PAL/.vegetation/.pal_ndvi.html or https://lta.cr.usgs.gov/NDVI

North American Soil Moisture Database (NASMD) http://soilmoisture.tamu.edu/

NSIDC – National Snow and Ice Data Center http://nsidc.org/

NSIDC – Nomenclature http://nsidc.org/cryosphere/seaice/data/terminology.html

PMEL – Pacific Marine Environmental Laboratory www.pmel.noaa.gov/

PRISM – Parameter-elevation Regressions on Independent Slopes Model http://prism.oregonstate.edu

Radiosondes – NCAR historical summary www.eol.ucar.edu/homes/junhong/Ency-radiosonde.pdf

Radiosonde Atmospheric Temperature Products for Assessing Climate – NCDC www.ncdc.noaa.gov/oa/climate/ratpac/

Rutgers Global Snow laboratory http://climate.rutgers.edu/snowcover/

SCAN (Soil Climate Analysis Network) www.wcc.nrcs.usda.gov/scan

Snow Measurement Guidelines www.nws.noaa.gov/os/coop/reference/Snow_Measurement_Guidelines.pdf

Sea Surface Temperature (SST) data

Blended Satellite and *in situ* data www.esrl.noaa.gov/psd/data/gridded/data.noaa.oisst.v2.html

or

www.ncdc.noaa.gov/oisst

I*n situ* SST Data

https://data.nodc.noaa.gov/cgi-bin/iso?id=gov.noaa.ncdc:C00884

or

www.ncdc.noaa.gov/data-access/marineocean-data/extended-reconstructed-sea-surface-temperature-ersst-v4

Soil Moisture Operational Products System (SMOPS) www.ospo.noaa.gov/Products/land/smops/

Standard Rain Gage – www.weather.gov/iwx/coop_8inch

STAR – NOAA Center for Satellite Applications and Research

www.star.nesdis.noaa.gov/

Sunshine hours tables http://aa.usno.navy.mil/data/docs/Dur_OneYear.php

Surface-based cloud observations https://atmos.washington.edu/CloudMap/WebO/index.html

Tokyo Climate Center ds.data.jma.go.jp/tcc/tcc/

United Kingdom Meteorological Office (UKMO) Unified Model –

www.metoffice.gov.uk/research/modelling-systems/unified-model

U.S. Climate Reference Network
www.ncdc.noaa.gov/crn/
U.S Coast Guard Navigation Center
https://navcen.uscg.gov/?pageName=
iipHowDoTheLabradorAndGulfStreamCurrentsAffectIcebergsInTheNorthAtlanticOcean
U.S. Drought Monitor
http://droughtmonitor.unl.edu/
The U.S. Drought Monitor Facts Sheet, NOAA, 2012 www.nws.noaa.gov/om/csd/graphics/content/outreach/brochures/FactSheet_Drought.pdf
USGS National Elevation Dataset 3–10m resolutions available at no cost on http://nationalmap.gov/viewer.html. This data set also includes Land Cover.
USGS Fact Sheet Colorado River, 2004: 3062 Version 2 http://pubs.usgs.gov/fs/2004/3062
U.S. Department of Agriculture – National Resources Conservation Service www.nrcs.usda.gov/wps/portal/nrcs/site/national/home/
Weekly Weather and Crop Bulletin www.usda.gov/oce/weather/
World Radiation Monitoring Center www.bsrn.awi.de/
WMO (2003) www.wmo.int/pages/prog/wcp/documents/Guidefulltext.pdf
WMO Guide to Climatological Practices (2011) Can be downloaded from www.wmo.int/pages/prog/wcp/ccl/guide/guide_climat_practices.php

Index